Aboul-Ella Hassanien, Ajith Abraham,
Athanasios V. Vasilakos, and Witold Pedrycz (Eds.)

Foundations of Computational Intelligence Volume 1

T0191509

Studies in Computational Intelligence, Volume 201

Editor-in-Chief

Prof. Janusz Kacprzyk
Systems Research Institute
Polish Academy of Sciences
ul. Newelska 6
01-447 Warsaw
Poland
E-mail: kacprzyk@ibspan.waw.pl

Further volumes of this series can be found on our homepage: springer.com

Vol. 178. Swagatam Das, Ajith Abraham and Amit Konar
Metaheuristic Clustering, 2009
ISBN 978-3-540-92172-1

Vol. 179. Mircea Gh. Negoita and Sorin Hintea
Bio-Inspired Technologies for the Hardware of Adaptive Systems, 2009
ISBN 978-3-540-76994-1

Vol. 180. Wojciech Mitkowski and Janusz Kacprzyk (Eds.)
Modelling Dynamics in Processes and Systems, 2009
ISBN 978-3-540-92202-5

Vol. 181. Georgios Miaoulis and Dimitri Plemenos (Eds.)
Intelligent Scene Modelling Information Systems, 2009
ISBN 978-3-540-92901-7

Vol. 182. Andrzej Bargiela and Witold Pedrycz (Eds.)
Human-Centric Information Processing Through Granular Modelling, 2009
ISBN 978-3-540-92915-4

Vol. 183. Marco A.C. Pacheco and Marley M.B.R. Vellasco (Eds.)
Intelligent Systems in Oil Field Development under Uncertainty, 2009
ISBN 978-3-540-92999-4

Vol. 184. Ljupco Kocarev, Zbigniew Galias and Shiguo Lian (Eds.)
Intelligent Computing Based on Chaos, 2009
ISBN 978-3-540-95971-7

Vol. 185. Anthony Brabazon and Michael O'Neill (Eds.)
Natural Computing in Computational Finance, 2009
ISBN 978-3-540-95973-1

Vol. 186. Chi-Keong Goh and Kay Chen Tan
Evolutionary Multi-objective Optimization in Uncertain Environments, 2009
ISBN 978-3-540-95975-5

Vol. 187. Mitsuo Gen, David Green, Osamu Katai, Bob McKay, Akira Namatame, Ruhul A. Sarker and Byoung-Tak Zhang (Eds.)
Intelligent and Evolutionary Systems, 2009
ISBN 978-3-540-95977-9

Vol. 188. Agustín Gutiérrez and Santiago Marco (Eds.)
Biologically Inspired Signal Processing for Chemical Sensing, 2009
ISBN 978-3-642-00175-8

Vol. 189. Sally McClean, Peter Millard, Elia El-Darzi and Chris Nugent (Eds.)
Intelligent Patient Management, 2009
ISBN 978-3-642-00178-9

Vol. 190. K.R. Venugopal, K.G. Srinivasa and L.M. Patnaik
Soft Computing for Data Mining Applications, 2009
ISBN 978-3-642-00192-5

Vol. 191. Zong Woo Geem (Ed.)
Music-Inspired Harmony Search Algorithm, 2009
ISBN 978-3-642-00184-0

Vol. 192. Agus Budiyono, Bambang Riyanto and Endra Joelianto (Eds.)
Intelligent Unmanned Systems: Theory and Applications, 2009
ISBN 978-3-642-00263-2

Vol. 193. Raymond Chiong (Ed.)
Nature-Inspired Algorithms for Optimisation, 2009
ISBN 978-3-642-00266-3

Vol. 194. Ian Dempsey, Michael O'Neill and Anthony Brabazon (Eds.)
Foundations in Grammatical Evolution for Dynamic Environments, 2009
ISBN 978-3-642-00313-4

Vol. 195. Vivek Bannore and Leszek Swierkowski
Iterative-Interpolation Super-Resolution Image Reconstruction:
A Computationally Efficient Technique, 2009
ISBN 978-3-642-00384-4

Vol. 196. Valentina Emilia Balas, János Fodor and Annamária R. Várkonyi-Kóczy (Eds.)
Soft Computing Based Modeling in Intelligent Systems, 2009
ISBN 978-3-642-00447-6

Vol. 197. Mauro Birattari
Tuning Metaheuristics, 2009
ISBN 978-3-642-00482-7

Vol. 198. Efrén Mezura-Montes (Ed.)
Constraint-Handling in Evolutionary Optimization, 2009
ISBN 978-3-642-00618-0

Vol. 199. Kazumi Nakamatsu, Gloria Phillips-Wren, Lakhmi C. Jain, and Robert J. Howlett (Eds.)
New Advances in Intelligent Decision Technologies, 2009
ISBN 978-3-642-00908-2

Vol. 200. Dimitri Plemenos and Georgios Miaoulis
Visual Complexity and Intelligent Computer Graphics Techniques Enhancements, 2009
ISBN 978-3-642-01258-7

Vol. 201. Aboul-Ella Hassanien, Ajith Abraham, Athanasios V. Vasilakos, and Witold Pedrycz (Eds.)
Foundations of Computational Intelligence Volume 1, 2009
ISBN 978-3-642-01081-1

Aboul-Ella Hassanien, Ajith Abraham,
Athanasios V. Vasilakos, and Witold Pedrycz (Eds.)

Foundations of Computational Intelligence Volume 1

Learning and Approximation

 Springer

Prof. Aboul-Ella Hassanien
College of Business Administration
Quantitative and Information System
Department
Kuwait University
P.O. Box 5486
Safat, 13055
Kuwait
E-mail: abo@cba.edu.kw

Dr. Ajith Abraham
Norwegian University of Science &
Technology
Center of Excellence for Quantifiable
Quality of Service
O.S. Bragstads plass 2E
7491 Trondheim
Norway
E-mail: ajith.abraham@ieee.org

Athanasios V. Vasilakos
Department of Computer and
Telecommunications Engineering
University of Western Macedonia
Agios Dimitrios Park
50 100 Kozani
Greece
E-mail: vasilako@ath.forthnet.gr

Prof. Dr. Witold Pedrycz
University of Alberta
Dept. Electrical and Computer Engineering
Edmonton, Alberta T6J 2V4
Canada
E-mail: pedrycz@ee.ualberta.ca

ISBN 978-3-642-10164-9   e-ISBN 978-3-642-01082-8

DOI 10.1007/978-3-642-01082-8

Studies in Computational Intelligence   ISSN 1860949X

Typeset & *Cover Design:* Scientific Publishing Services Pvt. Ltd., Chennai, India.

Printed in acid-free paper

9 8 7 6 5 4 3 2 1

springer.com

Preface

Foundations of Computational Intelligence

Volume 1: Learning and Approximation: Theoretical Foundations and Applications

Learning methods and approximation algorithms are fundamental tools that deal with computationally hard problems and problems in which the input is gradually disclosed over time. Both kinds of problems have a large number of applications arising from a variety of fields, such as algorithmic game theory, approximation classes, coloring and partitioning, competitive analysis, computational finance, cuts and connectivity, inapproximability results, mechanism design, network design, packing and covering, paradigms for design and analysis of approximation and online algorithms, randomization techniques, real-world applications, scheduling problems and so on. The past years have witnessed a large number of interesting applications using various techniques of Computational Intelligence such as rough sets, connectionist learning; fuzzy logic; evolutionary computing; artificial immune systems; swarm intelligence; reinforcement learning, intelligent multimedia processing etc.. In spite of numerous successful applications of Computational Intelligence in business and industry, it is sometimes difficult to explain the performance of these techniques and algorithms from a theoretical perspective. Therefore, we encouraged authors to present original ideas dealing with the incorporation of different mechanisms of Computational Intelligent dealing with Learning and Approximation algorithms and underlying processes.

This edited volume comprises 15 chapters, including an overview chapter, which provides an up-to-date and state-of-the art research on the application of Computational Intelligence for learning and approximation.

The book is divided into 4 main parts:

Part-I: Function Approximation
Part-II: Connectionist Learning
Part-III: Knowledge Representation and Acquisition
Part-IV: Learning and Visualization

Part I entitled Function Approximation contains four chapters that deal with several techniques of Computational Intelligence in application to function approximation and machine learning strategies.

In Chapter 1, "Machine Learning and Genetic Regulatory Networks: A Review and a Roadmap", Fogelberg and Palade, review the Genetic Regulatory Networks (GRNs) field and offer an excellent roadmap for new researchers to the field. Authors describe the relevant theoretical and empirical biochemistry and the different types of GRN inference. They also discuss the data that can be used to carry out GRN inference. With this background biologically-centered material, the chapter surveys previous applications of machine learning techniques and computational intelligence to the domain of the GRNs. It describes clustering, logical and mathematical formalisms, Bayesian approaches and some combinations of those. All techniques are shortly explained theoretically, and important examples of previous research using each are highlighted.

Chapter 2, "Automatic Approximation of Expensive Functions with Active Learning", by Gorissen et al. presents a fully automated and integrated global surrogate modeling methodology for regression modeling and active learning. The work brings together important insights coming from distributed systems, artificial intelligence, and modeling and simulation, and shows a variety of applications. The merits of this approach are illustrated with several examples and several surrogate model types.

In Chapter 3, "New Multi-Objective Algorithms for Neural Network Training applied to Genomic Classification Data" Costa et.al., present the LASSO multiobjective algorithm. The algorithm is able to develop solutions of high generalization capabilities for regression and classification problems. The method was first compared with the original Multi-Objective and Early-Stopping algorithms and it achieved better results with high generalization responses and reduced topology. To measure the effective number of weights for each algorithm a random selection procedure were applied to both Multiobjective algorithm (MOBJ) and Early-Stopping solutions. Experimental results illustrate the potential of the algorithm to train and to select compact models.

In Chapter 4, "An Evolutionary Approximation for the Coefficients of Decision Functions within a Support Vector Machine Learning Strategy", authored by Ruxandra et al. bring support vector machines together with evolutionary computation, with the aim to offer a simplified solving version for the central optimization problem of determining the equation of the hyperplane deriving from support vector learning. The evolutionary approach suggested in this chapter resolves the complexity of the optimizer, opens the 'black-box' of support vector training and breaks the limits of the canonical solving component.

Part II entitled Connectionist Learning contains Six chapters that describe several computational intelligence techniques in Connectionist Learning including Meta-Learning, Entropy Guided Transformation, Artificial Development and Multi-Agent Architecture.

In Chapter 5, "Meta-learning and Neurocomputing – A New Perspective for Computational Intelligence", Castiello deals with the analysis of computational mechanisms of induction in order to assess the potentiality of meta-learning methods versus the common base-learning practices. Firstly a formal investigation of inductive mechanisms is presented, sketching a distinction between fixed and dynamical bias learning. Then a survey is presented with suggestions and examples, which have been proposed in the literature to increase the efficiency of common learning algorithms. To explore the meta-learning possibilities of neural network systems, knowledge-based neurocomputing techniques are also considered.

Chapter 6, by Mashinchi et al. "Three-term Fuzzy Back-propagation", a fuzzy proportional factor is added to the fuzzy BP's iteration scheme to enhance the convergence speed. The added factor makes the proposed method more dependant on the distance of actual outputs and desired ones. Thus in contrast with the conventional fuzzy BP, when the slop of error function is very close to zero, the algorithm does not necessarily return almost the same weights for the next iteration. According to the simulation's results, the proposed method is superior to the fuzzy BP in terms of performance error.

Dos Santos and Milidiu in Chapter 7, "Entropy Guided Transformation Learning", presents Entropy Guided Transformation Learning (ETL)", propose a new machine learning algorithm for classification tasks. ETL generalizes Transformation Based Learning (TBL) by automatically solving the TBL bottleneck: the construction of good template sets. ETL uses the information gain in order to select the feature combinations that provide good template sets. Authors describe the application of ETL to two language independent Text Mining pre-processing tasks: part-of-speech tagging and phrase chunking.

Artificial Development is a field of Evolutionary Computation inspired by the developmental processes and cellular growth seen in nature. Multiple models of artificial development have been proposed in the past, which can be broadly divided into those based on biochemical processes and those based on a high level grammar. Chapter 8, "Artificial Development", by Chavoya covers the main research areas pertaining to artificial development with an emphasis toward the systems based on artificial regulatory networks. To this end, a short introduction to biological gene regulatory networks and their relationship with development is first presented. The sections on the various artificial development models are followed by a section on the canonical problem in artificial and natural development known as the French flag problem.

In Chapter 9, "Robust Training of Artificial Feed-forward Neural Networks" by Moumen et. al. present several methods to enhance the robustness of neural network training algorithms. First, employing a family of robust statistics estimators, commonly known as M-estimators, the back-propagation algorithm is reviewed and evaluated for the task of function approximation and dynamical model identification. As these M-estimators sometimes do not have sufficient insensitivity to data outliers, the chapter next resorts to the statistically more robust estimator of the least median of squares, and develops a stochastic algorithm to minimize a related cost function.

Chapter 10, "Workload Assignment In Production Networks By Multi-Agent Architecture", authored by Renna and Argoneto deals with low level production planning with the of allocating the orders to the distributed plant. The orders assigned by the medium level have to be allocated to the plant to manufacture the products. Coordination among the plants is a crucial activity to pursue high level performance. In this context, three approaches for coordinating production planning activities within production networks are proposed. Moreover, the proposed approaches are modeled and designed by a Multi Agent negotiation model. In order to test the functionality of the proposed agent based distributed architecture for distributed production planning; a proper simulation environment has been developed.

Knowledge Representation and Acquisition is the Third Part of this Volume. It contains two chapters discussing some computational approaches in knowledge representation.

Chapter 11, "Extensions to Knowledge Acquisition and Effect of Multimodal Representation in Unsupervised Learning" by Daswin et al. present the advances made in the unsupervised learning paradigm (self organizing methods) and its potential in realizing artificial cognitive machines. The initial Sections delineate intricacies of the process of learning in humans with an articulate discussion of the function of thought and the function of memory. The self organizing method and the biological rationalizations that led to its development are explored in the second section. Further, the focus is shifted to the effect of structure restrictions on unsupervised learning and the enhancements resulting from a structure adapting learning algorithm. New means of knowledge acquisition through adaptive unsupervised learning algorithm and the contribution of multimodal representation of inputs to unsupervised learning are also illustrated.

In Chapter 12, "A New Implementation for Neural Networks in Fourier-Space" by El-Bakry and Hamada, presents fast neural networks for pattern detection. Such processors are designed based on cross correlation in the frequency domain between the input image and the input weights of neural networks. This approach is developed to reduce the computational steps required by these fast neural networks during the search process. The principle of divide and conquer strategy is applied through image decomposition. Each image is divided into small in size sub-images and then each one is tested separately by using a single fast neural processor. Furthermore, faster pattern detection is obtained by using parallel processing techniques to test the resulting sub-images at the same time using the same number of fast neural networks.

The Final Part of the book deals with the Learning and Visualization. It contains three chapters, which discusses the dissimilarity analysis, dynamic Self-Organizing Maps and Hybrid Learning approach.

In Chapter 13, "Dissimilarity Analysis and Application to Visual Comparisons", by   Aupetit et al.   discuss the embedding of a set of data into a vector space when an unconditional pairwise dissimilarity w between data. The vector space is endowed with a suitable pseudo-Euclidean structure and the data embedding is built by extending the classical kernel principal component analysis. This embedding is unique, up to an isomorphism, and injective if and only if w separates the data. This construction takes advantage of axis corresponding to negative Eigen values to develop pseudo-Euclidean scatter plot matrix representations. This new visual tool is applied to compare various dissimilarities between hidden Markov models built from person's faces.

Chapter 14, "Dynamic Self-Organising Maps: Theory, Methods and Applications" authored by Arthur et al. provide a comprehensive description and theory of Growing Self-Organising Maps (GSOM), which also includes recent theoretical developments. Methods of clustering and identifying clusters using GSOM are also introduced here together with their related applications and results. In addition, an encompassing domain for GSOM, applications that have been explored in this book chapter range from bioinformatics to sensor networks to manufacturing processes.

In Chapter 15, "Hybrid Learning Enhancement of RBF Network with Particle Swarm Optimization", by Noman et.al. propose RBF Network hybrid learning with Particle Swarm Optimization (PSO) for better convergence, error rates and classification results. RBF Network hybrid learning involves two phases. The first phase is a structure identification, in which unsupervised learning is exploited to determine the RBF centers and widths. This is done by executing different algorithms such as k-mean clustering and standard derivation respectively. The second phase is parameters estimation, in which supervised learning is implemented to establish the connections weights between the hidden layer and the output layer. This is done by performing different algorithms such as Least Mean Squares (LMS) and gradient based methods. The incorporation of PSO in RBF Network hybrid learning is accomplished by optimizing the centers, the widths and the weights of RBF Network.

We are very much grateful to the authors of this volume and to the reviewers for their great effort by reviewing and providing useful feedback to the authors. The editors would like to express thanks to Dr. Thomas Ditzinger (Springer Engineering Inhouse Editor, Studies in Computational Intelligence Series), Professor Janusz Kacprzyk (Editor-in-Chief, Springer Studies in Computational Intelligence Series) and Ms. Heather King (Editorial Assistant, Springer Verlag, Heidelberg) for the editorial assistance and excellent collaboration to produce this important scientific work. We hope that the reader will share our joy and will find the volume useful

December 2008

Aboul Ella Hassanien, Egypt
Ajith Abraham, Trondheim, Norway
Athanasios V. Vasilakos, Greece
Witold Pedrycz, Canada

Contents

Part III: Knowledge Representation and Acquisition

Part IV: Learning and Visualization

Part I
Function Approximation

Machine Learning and Genetic Regulatory Networks: A Review and a Roadmap

Christopher Fogelberg and Vasile Palade

Abstract. Genetic regulatory networks (GRNs) are causal structures which can be represented as large directed graphs. Their inference is a central problem in bioinformatics. Because of the paucity of available data and high levels of associated noise, machine learning is essential to performing good and tractable inference of the underlying causal structure.

This chapter serves as a review of the GRN field as a whole, as well as a roadmap for researchers new to the field. It describes the relevant theoretical and empirical biochemistry and the different types of GRN inference. It also describes the data that can be used to perform GRN inference. With this biologically-centred material as background, the chapter surveys previous applications of machine learning techniques and computational intelligence to GRN inference. It describes clustering, logical and mathematical formalisms, Bayesian approaches and some combinations. Each of these is shortly explained theoretically, and important examples of previous research using each are highlighted. Finally, the chapter analyses wider statistical problems in the field, and concludes with a summary of the main achievements of previous research as well as some open research questions in the field.

1 Introduction

Genetic regulatory networks (GRN) are large directed graph models of the regulatory interactions amongst genes which cause the phenotypic states of biological organisms. Inference of their structure and parameters is a central

Christopher Fogelberg and Vasile Palade
Oxford University Computing Laboratory, Wolfson Building, OX1-3QD, UK

Christopher Fogelberg
Oxford-Man Institute, OX1-4EH, UK
e-mail: `christopher.fogelberg@comlab.ox.ac.uk`

A.-E. Hassanien et al. (Eds.): Foundations of Comput. Intel. Vol. 1, SCI 201, pp. 3–34.
springerlink.com   © Springer-Verlag Berlin Heidelberg 2009

problem in bioinformatics. However, because of the paucity of the training data and its noisiness, machine learning is essential to good and tractable inference. How machine learning techniques can be developed and applied to this problem is the focus of this review.

Section 2 summarises the relevant biology, and section 3 describes the machine learning and statistical problems in GRN inference.

Sections 4 discusses biological data types that can be used, and section 5 describes existing approaches to network inference. Section 6 describes important and more general statistical concerns associated with the problem of inference, and section 7 provides a brief visual categorisation of the research in the field. Section 8 concludes the survey by describing several open research questions.

Other reviews of GRN include [25], [19] and [26]. However, many of these are dated. Those that are more current focus on presenting new research findings and do not summarise the field as a whole.

2   The Underlying Biology

A GRN is one kind of regulatory (causal) network. Others include protein networks and metabolic processes[77]. This section briefly summarises the cellular biology that is relevant to GRN inference.

2.1   Network Structure and Macro-characteristics

GRN have a *messily robust* structure as a consequence of evolution[107]. This subsection discusses the known and hypothesised network-level characteristics of GRNs. Subsection 2.2 describes the micro-characteristics of GRNs.

A GRN is a directed graph, the vertices of this graph are genes and the edges describe the regulatory relationships between genes. GRN may be modeled as either directed[101] or undirected[112] graphs, however the true underlying regulatory network is a directed graph. Recent[6] and historical[57] research shows that GRN are not just random directed graphs. Barabasi and Oltvai [6] also discusses the statistical macro-characteristics of GRN.

The Out-degree (k_{out}), and In-degree (k_{in})

GRN network structure appears to be neither random nor rigidly hierarchical, but *scale free*. This means that the probability distribution for the out-degree follows a power law[6; 57]. I.e., the probability that i regulates k other genes is $p(k) \approx k^{-\lambda}$, where usually $\lambda \in [2, 3]$. Kauffman's [57] analysis of scale free Boolean networks shows that they behave as if they are on the cusp of being highly ordered and totally chaotic. Barabasi and Oltvai [6] claims that "being on the cusp" contributes to a GRN's evolvability and adaptability.

These distributions over k_{in} and k_{out} means that a number of assumptions have been made in previous research to simplify the problem and make it more tractable.

For example, the exponential distribution over k_{in} means that most genes are regulated by only a few others. Crucially, this average is not a maximum. This means that techniques which strictly limit k_{in} to some arbitrary constant (e.g. [101; 109]) may not be able to infer all networks. This compromises their explanatory power.

Modules

Genes are organised into modules. A module is a group of genes which are functionally linked by their phenotypic effects. Examples of phenotypic effects include protein folding, the *cell development cycle*[97], glycolysis metabolism[96], and amino acid metabolism[5].

Evolution means that genes in the same module are often physically proximate and *co-regulated*, even *equi-regulated*. However, a gene may be in multiple modules, such genes are often regulated by different genes for each module[6; 55].

One or two genes may be the main regulators of all or most of the other genes in the module. It is crucial to take these "hub" genes into account, else the model may be fragile and lack biological meaning[97].

Genetic networks are enormously redundant. For example, the Per1, Per2 and Per3 genes help regulate circadian oscillations in many species. Knocking out one or even two of them produces no detectable changes in the organism[59]. This redundancy is an expected consequence of evolution[6].

Known modules range in size from 10 to several hundred genes, and have no characteristic size[6].

Motifs

This subsubsection discusses motifs. A motif is a sub-graph which is repeated more times in a GRN than would be expected if a graph with its edge distributions were randomly connected[59]. For example, the feed-forward triangle shown in figure 1(b) frequently occurs with module-regulatory genes, where one module-regulatory gene binds to another and then both contribute to the module's regulation[5].

Auto-regulation (usually self-inhibition[24]) is also over-represented, as are the *cascade* and *convergence* motifs. Each of these three is illustrated in figure 1. The biases in network structures that motifs represent can be used to guide, describe and evaluate network inference.

Like modules, motifs often overlap. In addition, they are strongly conserved during evolution[17; 50].

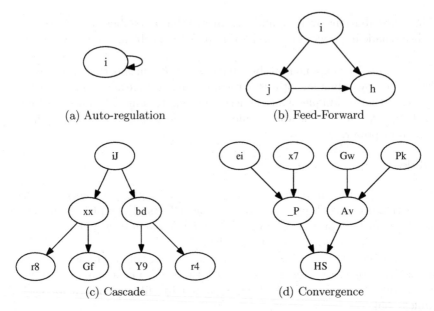

(a) Auto-regulation   (b) Feed-Forward

(c) Cascade   (d) Convergence

Fig. 1 Network motifs in genetic regulatory networks. The auto-regulatory, feed-forward, cascade and convergence motifs

2.2 Gene-Gene Interactions and Micro-characteristics

Subsection 2.1 described the graph-level statistical properties of GRN. This subsection considers individual gene-gene interactions. Crudely, i may either up-regulate (*excite*) or down-regulate (*inhibit*) j. Different formalisations model this *regulatory function* to different degrees of fidelity.

One-to-One Regulatory Functions

Imagine that i is regulated only by j. The regulatory function, $f_i(j)$, may be roughly linear, sigmoid or take some other form, and the strength of j's effect on i can range from very strong to very weak.

Also consider non-genetic influences on i, denoted ϕ_i. In this situation, $i' = f_i(j, \phi_i)$. Inter-cellular signaling is modeled in [75] and is one example of ϕ. In many circumstances we can assume that $\frac{\delta f}{\delta \phi} = 0$.

It is also possible for one gene to both up-regulate and down-regulate another gene. For example, j might actively up-regulate i when j is low, but down-regulate it otherwise. However, the chemical process underlying this is not immediately clear, and in the models inferred in [90] a previously postulated case of this was not verified. In any case, this kind of especially complex relationship is not evolutionarily robust. For that reason it will be relatively rare.

Wider properties of the organism also have an influence on the kinds of regulatory functions that are present. For example, inhibitors are more common in prokaryotes than in eukaryotes[49].

Many-to-One Regulatory Functions

If a gene is regulated by more than one gene its regulatory function is usually much more complex. In particular, eukaryotic gene regulation can be enormously complex[28]; regulatory functions may be piecewise threshold functions[18; 97; 116]. Consider the regulatory network shown in figure 1(d). If n_2 is not expressed strongly enough, n_1 may have no affect on n_5 at all.

This complexity arises because of the complex indirect, multi-level and multi-stage biological process underlying gene regulation. This regulatory process is detailed in work such as [19; 26; 116].

Some of the logically possibly regulatory relationships appear to be unlikely. For example, it appears that the *exclusive or* and *equivalent* relationships are biologically and statistically unlikely[68]. Furthermore, [57] suggests that many regulatory functions are *canalised*. A canalised[107] regulatory function is a function that is buffered and depends almost entirely on the expression level of just one other gene.

The Gene Transcription Process

Figure 2 shows how proteins transcribed by one gene may bind to several other genes as regulators and is based on figures in [19; 26]. The transcription process itself takes several steps.

First the DNA is transcribed into RNA. The creation of this fragment of RNA, known as *messenger RNA* (mRNA), is initiated when *promoters* (proteins that regulate transcription) bind ahead of the start site of the gene and cause the gene to be copied. The resulting fragment of RNA is the genetic inverse of the original DNA (i.e. an A in the DNA is transcribed as a T in the RNA). Next, the mRNA is translated into a protein.

The transcribed protein causes phenotypic effects, such as DNA repair or cellular signaling. In addition, some proteins act as promoters and bind to genes, regulating their transcriptional activity. This completes the loop.

Note that the term "motifs" is also used to refer to binding sites, to describe the kinds of promoters which can bind with and regulate a particular gene. For clarity, the term is not used in this way in this chapter. For example, [91] uses "prior knowledge of protein-binding site relationships", not "prior knowledge of motifs".

In summary, a GRN is a stochastic system of discrete components. However modeling a GRN in this way is not tractable. For that reason we do not consider stochastic systems in this chapter. Instead, a continuous model of GRN is used. In this model a GRN is a set of genes N and a set of functions F, such that there is one function for each gene. Each of these functions

Fig. 2 A protein's-eye
view of gene regulation.
The complete gene dis-
played in the bottom-left
is regulated by itself
and one other gene.
We have omitted the
mRNA → protein trans-
lation step for the sake of
clarity. A just-transcribed
protein is shown in the
figure as well. Right now
it cannot bind to the gene
which transcribed it, as
that site is occupied. It
may bind to the gene on
the right hand side of the
figure, or it may go on to
have a direct phenotypic
effect

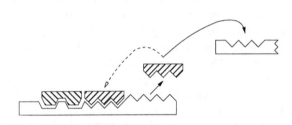

would take all or a subset of N as parameters, and ϕ_n as well. Using this
sort of model the important features of the regulatory relationships can be
inferred and represented.

3 Outstanding Problems

The key machine learning problems in gene network inference are very closely
related. Machine learning can either be used to infer *epistasis* (determine
which genes interact), or create explanatory models of the network.

Epistasis is traditionally identified through *synthetic lethality*[39; 69; 113]
and *yeast two-hybrid* (Y2H) experiments. Machine learning is necessary in
these situations because the data is often very noisy, and (as with `Per1-3`),
phenotypic changes may be invisible unless several genes are knocked out.
Two recent examples of this research are [80; 93]. Synthetic lethality and
other perturbations are discussed in more depth in subsection 4.2.

Inferring an explanatory model of the network is often better, with more
useful applications to biological understanding, genetic engineering and phar-
maceutical design. Types of model and inference techniques are discussed in
section 5. The distinction between network inference and epistatic analysis is
frequently not made clear in the research; failing to do so makes the conse-
quent publications very difficult to understand.

Interestingly, there has been very little work which has combined different
types of models. One example is [51; 112], another is [9], it is also discussed
theoretically by D'haeseleer et al. [25].

4 Available Data

To address bioinformatic network problems there are four types of data available. These are:

- Expression data
- Perturbation data
- Phylogenetic data
- Chemical and gene location data

This section describes the accessibility and utility of these for machine learning. Sometimes multiple kinds of data are used, e.g. [5; 45; 119], however this usually makes the inference more time and space complex.

4.1 Expression Data

Expression data measures how active each $n \in N$ is. As transcription activity cannot be measured directly, the concentration of mRNA (which is ephemeral) is used as a proxy.

Because regulatory or phenotypic protein interactions after transcription can consume some of the mRNA before it can regulate another gene[97] this may seem to be an inaccurate measure of gene activity[96]. Furthermore, a protein may bind to a promoter region but actually have no regulatory effect[45].

In addition, most genes are not involved in most cellular processes[55]. This means that many of the genes sampled may appear to vary randomly.

However, if the data set is comprehensive and we are just concerned with inference of the regulatory relationships then these influences are not important. Sufficient data or targeted inference obviates the problem of irrelevant genes. Non-genetic, unmodeled influences are analogous to hidden intermediate variables in a *Bayesian network*[45] (BN, subsection 5.4) whose only parent is the regulatory gene and whose only child is the regulated gene. An influence like this does not distort the overall gene-gene regulatory relationships or predictive accuracy of the model.

Tegner et al.'s [109]'s inference of a network with known post-transcription regulation and protein interactions using only (perturbed, subsection 4.2) gene expression data provides evidence of this.

Types of Expression Data

There are two kinds of expression data. They are equilibrium expression levels in a static situation and time series data that is gathered during a phenotypic process such as the cell development cycle (e.g. [105]).

Expression data is usually collected using *microarrays* or a similar technology. Time series data is gathered by using temperature- or chemical-sensitive mutants to pause the phenotypic process while a microarray is done on a sample.

A microarray is a pre-prepared slide, divided into cells. Each cell is individually coated with a chemical which fluoresces when it is mixed with the mRNA generated by just one of the genes. The brightness of each cell is used as a measurement of the level of mRNA and therefore of the gene's expression level.

Microarrays can be both *technically noisy*[60] and *biologically noisy*[85]. However, the magnitude and impact of the noise is hotly debated and dependent on the exact technology used to collect samples. Recent research[60, p. 6] (2007) argues that it has been "gravely exaggerated". New technologies[26] are also promising to deliver more and cleaner expression data in the future.

Examples of research which use equilibrium data include [2; 15; 27; 52; 58; 67; 73; 81; 96–98; 100; 105; 106; 108; 111; 112; 117; 125]. Wang et al.'s [117] work is particularly interesting as it describes how microarrays of the same gene collected in different situations can be combined into a single, larger data set.

Work that has been done using time series data includes [116], [90], [101] and [54; 103; 104; 121; 122]. Kyoda et al. [63] notes that time series data allows for more powerful and precise inference than equilibrium data, but that the data must be as noise-free as possible for the inference to be reliable.

Research on accurately simulating expression data includes [3; 7; 19; 31; 85; 103].

4.2 Perturbation Data

Perturbation data is expression data which measures what happens to the expression levels of all genes when one or more genes are artificially perturbed. Perturbation introduces a causal arrow which can lead to more efficient and accurate algorithms, e.g. [63].

Examples of experiments using perturbation data include [26; 37; 63; 109].

4.3 Phylogenetic Data

Phylogenetics is the study of species' evolutionary relationships to each other. To date, very little work has been carried out which directly uses phylogenetic conservation[17; 50] to identify regulatory relationships *de novo*. This is because phylogenetic data is not sufficiently quantified or numerous enough.

However, this sort of information can be used to validate results obtained using other methods. As [61] notes, transcriptional promoters tend to evolve phylogenetically, and as research by Pritsker et al. [91] illustrates, regulatory relationships in species of yeast are often conserved. [28] reaches similar conclusions, arguing that the "evolution of gene regulation underpins many of the differences between species".

4.4   Chemical and Gene Location Data

Along with phylogenetic data, primary chemical and gene location data can be used to validate inference from expression data or to provide an informal prior. Many types of chemical and gene location data exist; this subsection summarises some examples of recent research.

Yamanishi et al. [119] presents a technique and applies it to the yeast *Saccharomyces cerevisiae* so that the protein network could be inferred and understood in more depth than just synthetic lethality allowed. Their technique used Y2H, phylogenetics and a functional spatial model of the cell.

Hartemink et al.'s [45] work is broadly similar to Yamanishi et al.'s [119]. ChIP assays were used to identify protein interactions. Based on other forms of knowledge, in some experiments some regulatory relationships were fixed as occurring and then the results compared with the completely unfixed inference.

Harbison et al. [44] combined microarrays of the entire genome and phylogenetic insights from four related species of yeast (*Saccharomyces*). Given 203 known regulatory genes and their transcription factors, they were able to discover the genes that these factors acted as regulators for.

As these examples highlight, most research which combines multiple types of data aims to answer a specific question about a single, well known species or group of species. Although there is some work[115; 119] which investigates general and principled techniques for using multiple types of data, the general problem is open for further research. Machine learning techniques which may help maximise the value of the data include multi-classifiers[92] and fuzzy set theory[11].

5   Approaches to GRN Inference

Having discussed the relevant biology, the bioinformatic problems and the data that can be used to approach these problems, this section reviews different types of approach to network inference.

5.1   Clustering

Clustering[118] can reveal the modular structure[5; 51] of GRN, guide other experiments and be used to preprocess data before further inference.

This subsection discusses distance measures and clustering methods first. Then it gives a number of examples, including the use of clustering to preprocess data. Finally, it summarises *biclustering*.

Overview

A clustering algorithm is made up of two elements: the *method*, and the *distance measure*. The distance measure is how the similarity (difference) of any two data points is calculated, and the method determines how data points are grouped into clusters based on their similarity to (difference from) each other. Any distance measure can be used with any method.

Distance Measures

Readers are assumed to be familiar with basic distance measures such as the Euclidean distance. The *Manhattan distance*[62] is similar to the Euclidean distance. *Mutual information* (MI), closely related to the Shannon entropy[99], is also used. A gene's distance from a cluster is usually considered to be the gene's mean, maximum, median or minimum from genes in that cluster.

The *Mahalanobis distance*[20; 74] addresses a weakness in Euclidean distance measures. To understand the weakness it addresses it is important to distinguish between the real module or cluster underlying the gene expression, and the apparent cluster which an algorithm infers. Dennett [22] has a more in depth discussion of this distinction.

Imagine that we are using the Euclidean distance and that the samples we have of genes in the underlying "real" clusters C and D are biased samples of those clusters. Assume that the method is clustering the gene h, and that h is truly in D. However, because of the way that the samples of C and D are biased, h will be clustered into C. Having been clustered into C it will also bias future genes towards C even more. Because microarrays are done on genes and phenotypic situations of known interest this bias is possible and may be common.

Analysis shows that the bias comes about because naive distance measures do not consider the covariance or spread of the cluster. The Mahalanobis distance considers this; therefore it may be more likely to correctly cluster genes than measures which do not consider this factor.

Clustering Methods

Readers are assumed to be familiar with clustering methods and know that they can be *partitional* or *hierarchical* and *supervised* or *unsupervised*.

Many classic partitional algorithms, such as k-means[1; 72], work best on hyper-spherical clusters that are well separated. Further, the number of clusters must be specified in advance. For this reason they may not be ideal for gene expression clustering. We expect that *self-organising maps*[42, ch. 7] (SOM) would have similar problems, despite differences in the representation and method.

Some clustering methods which have been successfully used are *fuzzy clustering* methods. When fuzzy clustering is used a gene may be a partial member of several clusters, which is biologically accurate[6]. Fuzzy methods are also comparatively robust against noisy data.

Fuzzy (and discrete) methods that allow one gene to have total cluster membership greater than one (i.e. $\sum_j \mu_{ij} > 1$) create *covers*[73] over the data, and don't just *partition*[1] it[25]. Clustering methods which build hypergraphs[43; 79] are one kind of covering method. This is because a hyperedge can connect to any number of vertices, creating a cover.

It is important that the clustering algorithms are robust in the face of noise and missing data, [55; 114] discuss techniques for fuzzy and discrete methods.

Previous Clustering Research

Gene expression data clustering has been surveyed in several recent papers (e.g. [2; 25; 27; 98; 125] and others). Zhou et al. [125] compares a range of algorithms and focuses on combining two different distance measures (e.g. mutual information and fuzzy similarity or Euclidean distance and mutual information) into one overall distance measure. Azuaje [2] describes some freely available clustering software packages and introduces the SOTA algorithm. SOTA is a hierarchical clustering algorithm which determines the number of clusters via validity threshold.

In [125], initial membership of each gene in each fuzzy cluster is randomly assigned and cluster memberships are searched over using *simulated annealing*[78]. While searching, the fuzzy membership is swapped in the same way that discrete cluster memberships would be swapped.

GRAM[5] is a supervised clustering algorithm which finds a cover and is interesting because it combines protein-binding and gene expression data. It uses protein-binding information to group genes which are likely to share promoters together first, then other genes which match the initial members' expression profiles closely can also be included in the cluster.

[96] describes another clustering algorithm which uses a list of candidate regulators specified in advance to cluster genes into modules. This algorithm can infer more complex (Boolean AND/OR) regulatory relationships amongst genes, and its predictions have been empirically confirmed.

Multi-stage inference ([9; 25; 112] can make principled inference over larger numbers of genes tractable. Although the underlying network is directed (as described in subsection 2.2) and may have very complex regulatory relationships these factors are conditionally independent of the graphical structure and do not need to be considered simultaneously.

Horimoto and Toh [51] also found that nearly 20% of the gene pairs in a set of 2467 genes were Pearson-correlated at a 1% significance level. This emphasises the modular nature of GRN.

Mascioli et al.'s [76] hierarchical algorithm is very interesting. The validity criterion is changed smoothly, and this means that every cluster has a

---

[1] This use of the term partition is easily confused with the way a clustering method can be described as partitional. In the latter case it describes how clusters are found, in the former it describes sample membership in the clusters.

lifetime: the magnitude of the validity criterion from the point the cluster is created to the point that it splits into sub-clusters. The *dendogram* (tree) of clusters is cut at multiple levels so that the longest lived clusters are the final result.

This idea is very powerful and selects the number of clusters automatically. However a cluster's lifetime may depend on samples not in the cluster, and this is not necessarily appropriate if intra-cluster similarity is more important.

Shamir and Sharan [98] suggests not clustering genes which are distant outliers and leaving them as singletons.

Selection of the right clustering algorithm remains a challenging and demanding task, dependent on the data being used and the precise nature of any future inference.

Previous Biclustering Research

Biclustering is also known as co-clustering and direct clustering. It involves grouping subsets of the genes and subsets of the samples together at the same time. A bicluster may represent a subset of the genes which are co-regulated some of the time. Such a model generalises naturally to a cover in which each gene can be in more than one cluster. This kind of *bicovering* algorithm is described in [100] and [108].

Madeira and Oliveira's [73] article is a recent survey of the field. It tabulates and compares many biclustering algorithms.

In general, optimal biclustering is an NP-hard problem[120]. In a limited number of cases, exhaustive enumeration is possible. In other cases, heuristics such as divide-and-conquer or a greedy search may be used[73].

5.2   Logical Networks

This subsection describes research which infer Boolean or other logical networks as a representation of a GRN. Boolean networks were first described by Kauffman[56]. Prior to discussing examples of GRN inference carried out using Boolean networks we define them.

Overview

In a Boolean model of a GRN, at time t, each gene is either expressed or not. Based on a logical function over a gene's parents and their value at time t, its value can be calculated at time $t + 1$. Figure 3 is an example of a Boolean network.

Recent research[10; 12] investigates the theoretical properties of *fuzzy logic networks* (FLN) and infers biological regulatory networks using time series expression data. FLN are a fuzzy generalisation of Boolean networks.

Fig. 3 A Boolean network. For clarity each $f \in F$ has been made into a node. n and n' are connected via these function nodes. Normally the functions are implicit in the edges amongst N.
$f_i = (\neg i \vee i) \wedge j \wedge h,$
$f_j = \neg i \vee j \wedge h,$
$f_h = i \vee j \wedge h$

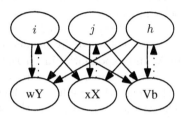

Previous Logical Network Research

Silvescu and Honavar's [101] algorithm uses time series data to find *temporal Boolean networks* (TBoN). [68] is older and uses unaugmented Boolean networks. TBoN were developed to model regulatory delays, which may come about due to missing intermediary genes and spatial or biochemical delays between transcription and regulation. An example of a temporal Boolean network has been presented in figure 4.

A temporal Boolean network is very similar to a normal Boolean network except that the functions $f \in F$ can refer to past gene expression levels. Rather than depending just on N_t to infer N_{t+1}, parameters to f_i can be annotated with an integer temporal delay.

For example: $h' = f_h(i, j, h) = i_0 \vee j_0 \wedge h_2$ means h is expressed at $t + 1$ if either i at t or j at t is expressed, so long as h also was at time $t - 2$. TBoN can also be reformulated and inferred as decision trees.

Lähdesmäki et al. [64] considered how to take into account the often contradictory and inconsistent results which are obtained from microarray data. The aim of the approach is to find not just one function f_i but a set of functions F_i for each gene i. Each member of F_i may predict the wrong value for i based on the values of N. In those situations though other $f \in F_i$ may predict correctly.

They developed a methodology which would find all functions which made less than ϵ errors on the time series training data for 799 genes. The regulatory

Fig. 4 A temporal Boolean network. Presentation and functions are as in figure 3, but delays are shown in brackets between genes and functions. The default delay if no annotation is present is assumed to be 0

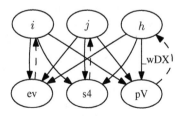

functions of only five genes were identified because the search for all consistent or all *best-fit* functions could not be done efficiently.

Boolean networks have a number of disadvantages. Compared to the underlying biology, they create such a simple model that it can only give a broad overview of the regulatory network. In addition, despite a simple model, the algorithms are usually intractable. Typically, they are polynomial or worse in N and exponential in $max(k_{in})$. Furthermore, as $max(k_{in})$ increases you need greater quantities of data to avoid overfitting.

However, the simplicity of the functional representation is a strength as well. The very low fidelity means that the models are more robust in the face of noisy data. Attractor basin analysis of Boolean networks can help provide a better understanding of the stability and causes of equilibrium gene expression levels. Such equilibria are often representative of particular phenotypes[19].

5.3   Differential Equations and Other Mathematical Formalisms

Approaches based on differential equations predict very detailed regulatory functions. This is a strength because the resulting model is more complete. It is also a weakness because it increases the complexity of the inference and there may not be enough data to reliably infer such a detailed model. In addition, the existence of a link and the precise nature of the regulatory function are two inferential steps and the regulatory function can be easily inferred given knowledge of the link.

Because there are so many different ways of doing this kind of high-fidelity inference, this subsection just presents a number of examples, as in subsection 4.4.

The NIR (*N*etwork *I*nference via multiple *R*egression) algorithm is summarised in [26; 37]. It uses gene perturbations to infer ordinary differential equations (ODEs). The method has been applied to networks containing approximately 20 genes.

Kyoda et al. [63] uses perturbations of equilibrium data and a modified version of the Floyd-Warshall algorithm[29] to infer the most parsimonious ODE model for each gene. Although noise and redundancy may make this incorrect it is arguably the most correct network which can be inferred with the training data. Different inferred networks can fit the data equally well, because GRN are cyclic[63].

Kyoda et al.'s method is very efficient $(O(N^3))$ and is not bound by arbitrary $max(k_{in})$. This is a consequence of the fact that it uses perturbation data, which is much more informative than expression data.

Toh and Horimoto [112] used *graphical Gaussian models* (GGM) to find conditional dependencies amongst gene clusters[51]. Information on the regulatory direction from primary literature was used to manually annotate the raw, undirected model.

Fig. 5 A Bayesian network and Markov blanket. Genes in the Markov blanket of n_5 are shown with a grey background. Priors for $n_{1..3}$ are denoted by incoming parentless edges

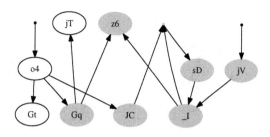

Fig. 6 A cyclic Bayesian network. Impossible to factorise

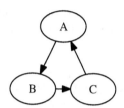

5.4 Bayesian Networks

This subsection describes and defines Bayesian networks, how they can be learnt and previous research which used them. Readers are referred to [47] for a more detailed introduction.

Bayesian Networks Described and Defined

A Bayesian network is a graphical decomposition of a joint probability distribution, such as the distribution over the state of all genes in a GRN. There is one variable for each gene.

The lack of an edge between two genes i and j means that, given i's or j's *Markov blanket*, i and j are independent: $p(i|j, mb(i)) = p(i|mb(i))$. Loosely and intuitively, the presence of an edge between i and j means that they are "directly" (causally?[89]) dependent on each other. The Markov blanket[71] consists of a variable's parents, children and children's parents as defined by the edges in and out of the variables. See figure 5 for an example.

BN must be acyclic[26]. This is a problem because auto-regulation and feedback circuits are common in GRN[110]. The reason why BN must be acyclic is that a cyclic BN cannot be factorised.

Consider the BN shown in figure 6. The value of A depends on the value of B, the value of B depends on the value of C, and the value of C depends on the value of A. Equation 1 shows what happens when we try to factorise the joint distribution by expanding the parents (π).

$$
\begin{aligned}
p(A, B, C) &= p(A|\pi(A)) \cdot p(\pi(A)) \\
&= p(A|B) \cdot p(B|\pi(B)) \cdot p(\pi(B)) \\
&= p(A|B) \cdot p(B|C) \cdot p(C|\pi(C)) \cdot p(\pi(C)) \\
&= p(A|B) \cdot p(B|C) \cdot p(C|A) \cdot p(A|\pi(A)) \cdot p(\pi(A))
\end{aligned}
\tag{1}
$$
$$\text{And so on...}$$

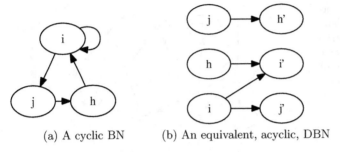

(a) A cyclic BN   (b) An equivalent, acyclic, DBN

Fig. 7 A cyclic BN and an equivalent, acyclic, DBN. The prior network[35] is not shown in this diagram

A BN is also a subtly different kind of model than a Boolean network or set of ODEs. While the latter create a definite, possibly incorrect model of the regulatory relationships, a BN model is strictly probabilistic, although it may incorrectly describe some genes as directly dependent or independent when they are not and long run frequency evaluations may suggest that it has incorrect conditional distributions for some genes.

In summary, BN are attractive (their statistical nature allows limited causal inference[89] and is robust in the face of noise and missing data) and unattractive (acyclic only). *Dynamic Bayesian networks*[35; 83] (DBN) are an elegant solution to this problem.

Dynamic Bayesian Networks
A DBN is a Bayesian network which has been temporally "unrolled". Typically we view variables as entities whose value changes over time. If we view them as constant, as they are in HMM, then we would represent i at t and i at $t + 1$ with two different variables, say i_t and i_{t+1}.

If we assume that conditional dependencies cannot point backwards or "sideways" in time this means that the graph must be acyclic, even if i auto-regulates. If we also assume that the conditional dependencies are constant over time and that the prior joint distribution[35] is the same as the temporal joint distribution then the network only needs to be unrolled for one time step. A visual illustration of this is provided in figure 7.

Modern variations of BN have also added new capabilities to them, particularly fuzzy Bayesian networks. These range from specialised techniques designed to reduce the complexity of *hybrid Bayesian network* (HBN) belief propagation with fuzzy approximations[4; 48; 86; 87] to more general formalisations which allow variables in Bayesian networks to take fuzzy states, with all of the advantages in robustness, comprehensibility and dimensionality reduction[30; 33; 88] this provides.

Learning Bayesian Networks

The problem of learning a Bayesian network can be divided into two sub-problems. The simpler problem is learning θ, the conditional distributions of the BN given its edges, η. This can be done with either full training data or training data which is partially covered.

The second and more difficult problem is inference of η and θ simultaneously. This can also be done with either full or incomplete training data.

Bayesian network inference is a large research field, and comprehensively summarising it here is impossible. For such a summary we refer interested readers to [47; 82] and [35; 41]. This subsubsection focuses on just a few algorithms. It considers the case of θ inference first, before showing how many of the same algorithms can be applied to structural inference.

θ Inference

The simplest way of performing θ inference is just to count up and categorise the examples. This is statistically valid in the case of complete data. The result of such a count is a *maximum likelihood* (ML) estimate. The desired result is the *maximum a posteriori* (MAP), which incorporates any prior information. When the prior is uniform then ML = MAP. A uniform prior is common as it also maximises the informativeness of the data.

To avoid certainty in the conditional probability distributions, *pseudo-counts* are often used. Pseudocounts were invented by Laplace[66] for the sunrise problem[2] and they can be thought of as an ad hoc adjustment of the prior distribution. Pseudocounts are invented data values, normally 1 for each entry in each conditional distribution, and their presence ensures probabilities never reach 0 or 1. This is important because $p(i|\cdot) = 0$ or $p(i|\cdot) = 1$ implies certainty, which is inferentially invalid with finite data.

Although pseudocounts are invalid if there is missing data, we speculate that if the available data is nearly complete then using pseudocounts could be accurate enough or a good starting point to search from.

If the data is too incomplete for counts to be used then there is a range of search algorithms which can be used. These include greedy hill climbing with random restarts[94], the EM algorithm[21], *simulated annealing*[78] and *Markov Chain Monte Carlo*[46; 70; 84] (MCMC).

The independence and decomposability of the conditional distributions is a crucial element in the efficiency of the algorithm, as it makes calculation of the likelihood much faster.

The EM Algorithm

The expectation maximisation (EM) algorithm[21] is shown in algorithm figure 1. A key advantage of EM is that it robust and tractably handles missing (covered) values in the training data. From an initial guess θ_0 and the observed gene expression levels we can calculate N_i, the expected gene

---

[2] *Viz:* What is the probability that the sun will rise tomorrow?

Algorithm 1. The EM algorithm[21]. p_N is a function that returns the expected expression level of all genes, N_i, given the current parameters θ and the observed training data $N \times M$. ML_N is a function that returns the θ which maximises the likelihood of some state of the genes, N. θ_{best} is the best set of parameters the search has found. If simulated annealing or random restarts are used this will be θ_{ML}

Input:
$N \times M$, the data to use in the inference
η, the edges of the graph G
Output:
$G = \langle \eta, \theta_{best} \rangle$, the maximum likelihood BN given η.
begin
    $\theta_0 \longleftarrow$ initial guess, e.g. based on counts from noisy data
    repeat
        $N_{i+1} \longleftarrow p_N(\theta_i, N \times M)$
        $\theta_{i+1} \longleftarrow ML_N(N_{i+1})$
    until $p(\theta_i) = p(\theta_{i-1})$
    return θ_{best}
end

expression levels. Holding this expectation constant and assuming that it is correct, θ_{i+1} is set so that N_i is the maximum likelihood gene expression levels.

This process is repeated and θ will converge on a local maxima. Random restarts or using simulated annealing (described next) to determine θ_{i+1} means that the algorithm can also find θ_{ML}.

Simulated Annealing

Simulated annealing [78] is a generalised Monte Carlo method which was inspired by the process of annealing metal. A Monte Carlo method is an iterative algorithm which is non-deterministic in its iterations. Metal is annealed by heating it to a very high temperature and then slowly cooling it. The resulting metal has a maximally strong structure.

Simulated annealing (SA) is very similar to hill climbing, except that the distance and direction to travel in (Δ and $grad(\theta)$ in hill climbing) are sampled from a probability distribution for each transition. Transitions to better states are always accepted, whilst transitions to worse states are accepted with a probability p, which is lower for transitions to much worse states. This probability decreases from transition to transition until only transitions to better states are accepted.

In this way simulated annealing is likely to explore a much wider part of the search space at the early stage of the search, but eventually it optimises greedily as hill climbing does[32]. If proposed transitions are compared to current positions based on their likelihood, simulated annealing finds θ_{ML}. The MAP can be found by including a prior in the calculations.

η Inference

The three search algorithms just discussed are naturally applicable to θ inference, but each of them can be used for η inference as well. For example, the EM algorithm has been generalised by Friedman *et al.*[34; 35]. *Structural EM* (SEM) is very similar to EM. The main difference is during the ML step, when SEM updates θ and also uses it to search the η-space.

Integrating the Posterior

One common weakness of these algorithms is that they all find a single solution, e.g. the MAP solution. A better technique is to integrate over the posterior distribution[45]. This is because the MAP solution may be only one of several equi-probable solutions. Ignoring these other solutions when calculating a result means that there is a greater chance it will be in error.

Analytically integrating the posterior distribution is usually impossible. Numerical integration by averaging many samples from the posterior is an alternative, and MCMC algorithms[70] are the most common way of drawing samples from the posterior.

Markov Chain Monte Carlo Algorithms

Metropolis-Hastings[46] and Gibbs sampling[38] are commonly used MCMC algorithms. Other more intricate algorithms which explore the solution space more completely (such as Hybrid Monte Carlo[84]) have also been developed.

An MCMC algorithm is very similar to simulated annealing. Starting from an initial state it probabilistically transitions through a solution space, always accepting transitions to better solutions and accepting transitions to worse states with lower probability, as in simulated annealing. Transitions in MCMC must have the *Markov property*. The probability of a transition from γ_t to γ_{t+1} which has the Markov property is independent of everything except γ_t and γ_{t+1}, including all previous states.

Higher order Markov chains can also be defined. An s'th order Markov chain is independent of all states before γ_{t-s}. Markov chains are a type of Bayesian network and are very similar to dynamic Bayesian networks.

Because of the wide range of MCMC algorithms it is difficult to give an informative list of invariant properties. We use the Metropolis-Hastings algorithm as an example instead.

Metropolis-Hastings MCMC

Assume we have a solution γ_t and that we can draw another solution conditional on it, γ_{t+1}, using a proposal distribution q. The normal distribution $N(\theta, \sigma^2)$ is frequently used as the proposal distribution q. Assume also that we can calculate the posterior probability of any solution γ (this is usually much easier than drawing samples from the posterior). The proposed transition to γ_{t+1} is accepted if and only if the *acceptance function* in equation 2 is true.

$$u < \frac{p(\gamma_{t+1})q(\gamma_t|\gamma_{t+1})}{p(\gamma_t)q(\gamma_{t+1}|\gamma_t)}, \text{ where } u \sim U(0,1) \qquad (2)$$

Metropolis-Hastings (and Gibbs sampling) converge quickest to the posterior when there are no extreme probabilities in the conditional distributions[41].

Because the probability of a transition being accepted is proportional to the posterior probability of the destination, the sequence of states will converge to the posterior distribution over time. MCMC can be used to take samples from the posterior, by giving it time to converge and by leaving a large enough number of proposed transitions between successive samples to ensure that $\Gamma_t \perp\!\!\!\perp \Gamma_{t+1}$.

The number of transitions needed to converge depends on the cragginess and dimensionality of the search space. Considering 10^5 proposed transitions is usually sufficient, and 10^4 proposed transitions between samples usually means that they are independent. Convergence and independence can be checked by re-running the MCMC algorithm. If the results are significantly different from run-to-run it indicates that one or both of the conditions was not met.

Because the number of samples that are necessary is constant as the dimensionality of the problem grows[71], MCMC are somewhat protected from the *curse of dimensionality*[23; 26].

Scoring Bayesian Networks

An important part of structural inference is comparing two models. The simplest scoring measure is the marginalised likelihood of the data, given the graph: $p(N \times M|G')p(G')$. This scoring measure is decomposable, as shown in equation 3. The log-sum is often used for pragmatic reasons. In this subsection, G refers to the set of all graphs and $G' = \langle \eta', \theta' \rangle \in G$ refers to one particular graph. η, η', θ and θ' are defined similarly.

$$p(N \times M|G') = \prod_{m \in M} \prod_{n \in N} p(n \times m|m, G') \tag{3}$$

However, the marginalised likelihood may overfit the data. This is because any edge which improves the fit will be added, and so the measure tends to make the graph too dense. A complexity penalty can be introduced if graphs are scored by their posterior probability, the *Bayesian Scoring Metric*[121] (BSM). For multinomial BN this scoring measure is commonly referred to as the BDe (*Bayesian Dirichlet equivalent*).

$$\begin{aligned} BSM(G', N \times M) &= \log p(G'|N \times M) \\ &= \log p(N \times M|G') + \log p(G') - \log p(N \times M) \end{aligned} \tag{4}$$

With the BSM, overly complex graphs can be penalised through the prior, or one can just rely on the fact that more complex graphs have more free parameters in θ. Because $p(\sum_{\theta|\eta}\theta) = 1$, the probability of any particular θ given a complex η will be relatively less. Loosely, this is because the probability must be "spread out" over a larger space[71].

The probability of the data over all possible graphs — $p(N \times M)$, expanded in equation 5 — is difficult to calculate because there are so many graphs. MCMC methods are often used to calculate it.

$$
\begin{aligned}
p(N \times M) &= \sum_{G' \in G} p(N \times M, G') \\
&= \sum_{G' \in G} p(N \times M | G')p(G') \\
&= \sum_{\eta' \in \eta} \sum_{\theta' \in \theta} p(N \times M | \theta', \eta')p(\theta'|\eta')
\end{aligned}
$$

(5)

When MCMC is too time consuming the posterior can be approximated using the *Bayesian Information Criterion*[95] (BIC, equation 6, where $|\theta_{ML}|$ is the number of parameters in the ML θ for η). This is an asymptotic approximation to the BDe and it is faster to calculate[121]. However, the BIC over penalises complex graphs when the training data is limited. Training data for GRN inference is typically very limited.

$$
\log p(N \times M | \eta') \approx BIC(N \times M, G) = \log p(N \times M | \eta', \theta') - \frac{|\theta'|}{2} \log N \quad (6)
$$

Other measures include the *minimum description length*[40; 65], the *Local Criterion*[47], the *Bayesian Nonparametric heteroscedastic Regression Criteria* (BNRC)[52; 53] and applications of computational learning theory[16]. Each of these tries to balance a better fitting graph with complexity controls.

Previous Bayesian Network Research

The project at Duke University[54; 103; 104; 121; 122] aimed to understand songbird singing. It integrates simulated neural activity data with simulated gene expression data and infers DBN models of the GRN.

Between 40–90% of the genes simulated were "distractors" and varied their expression levels according to a normal distribution. The number of genes simulated differed from experiment to experiment and was in the range 20–100 ([122] and [103], respectively). It was claimed that the simulated data showed realistic regulatory time lags, and also between gene expression and neural activity[103].

Singing/not-singing and gene expression levels were updated once each theoretical minute, and the simulated data was sampled once every 5 minutes.

It takes approximately 5 minutes for a gene to be transcribed, for the mRNA to be translated into a protein and for the protein to get back to the nucleus to regulate other genes[103].

The simulated data was continuous and a range of normalised hard and fuzzy discretisations were trialled[103; 122]. Interestingly, and contrary to information theoretic expectations[52; 102], the hard discretisations which discarded more information did better than the fuzzy ones. Linearly interpolating 5 data points between each pair of samples gave better recovery and fewer false positives[122].

[122] also developed the *influence score*, which makes the joint distribution easier to understand. If i regulates j then $-1 < I_{ij} < 1$, where the sign indicates down or up regulation and the magnitude indicates the regulatory strength.

The techniques developed could reliably infer regulatory cycles and cascade motifs. However, convergent motifs and multiple parents for a single gene were only reliably identified with more than 5000 data points. [41] and [104] discuss topological factors in more detail.

Nir Friedman and others[35; 52; 83] have also used DBNs for GRN inference. Friedman et al. [35] has extended the BIC and BDe to score graphs in the case of complete data. They have also extended SEM to incomplete data.

Friedman and others have proposed a *sparse candidate* algorithm which is optimised for problems with either a lot of data or a lot of variables. It uses MI to select candidate regulators for each gene and then only searches over networks whose regulators for each gene i come from $cand_i$. This process is iterated until convergence.

Murphy and Mian[83] discuss DBNs with continuous conditional distributions. So far our discussion has only considered multinomial BN. The value of a sample from a continuous θ_i is calculated by drawing a sample from it or by integrating over it. Continuous representations maximise the amount of information that can be extracted from the data, although they are also vulnerable to noise.

The tractability of different kinds of BN is contentious. Jarvis et al. [54], p974 claim that continuous BNs are intractable. Murphy and Mian [83], 5.1 note that exact inference in densely connected discrete BN is intractable and must be approximated. Others focus on the complexity of HBN[4; 86]. In general, Bayesian inference is NP-complete[14; 109].

There is no consensus on the most appropriate search algorithm. Hartemink et al. [45] concluded that simulated annealing was better than both greedy hill climbing and Metropolis-Hastings MCMC. Yu et al. [122] argued that greedy hill climbing with restarts was the most efficient and effective search method. Imoto et al. [52], used non-parametric regression. Because each algorithm is better for subtly different problems this variation is unsurprising, and [41, sections 3.2.1, 4.2.2] has suggestions on selecting an algorithm.

Table 1 GRN algorithmic efficiency against the number of genes N. Most of these results also require $max(k_{in}) \leq 3$. [64] found all explanatory Boolean functions with $\epsilon \lesssim 5$ for the genes it solved for

Research	$max(N)$
[68] (1998)	50
[75] (1998)	Unspecifiedly "small"
[116] (2001)	≈ 100
[109] (2003)	$\approx 10\text{--}40$
[64] (2003)	5
[103; 122] (2002–2004)	$\approx 20\text{--}100$
[8] (2004)	100, also $\overline{k}_{in} = 10$
[26] (2006)	≈ 20

6 Statistical and Computational Considerations

This section discusses the problems of tractability (subsection 6.1) and a particular statistical problem with GRN inference from microarrays (subsection 6.2).

6.1 Efficiency and Tractability

Almost all of the algorithms described above must limit N, the number of genes, and $max(k_{in})$. Unbounded $max(k_{in})$-network inference is almost always[3] $O(N^k) = O(N^N)$ or worse. Bayesian network inference is NP-hard[14; 109]; DBN inference is even harder[83].

The magnitude of this problem is clear when we consider the number of genes in, e.g., *S. cerevisiae* (approximately 6,265) and compare it to the size of the inferred networks as research has progressed. See table 1 for examples.

Two types of exception to this trend are informative. Firstly, [51; 112] and [9] used clustering to reduce the dimensionality first. However, no "de-clustering" was carried out on the inferred cluster-networks.

Kyoda et al. [63] uses perturbation data and creates a polynomial-time algorithm which is independent of $max(k_{in})$. Bernardo et al.'s [8] work uses a similar approach, and this research shows how valuable it is to use all information in the data. Note that biological analogues of the simulated data sets used in [8] are not possible with current biotechnology.

6.2 Microarrays and Inference

Chu et al. [15] identifies a statistical problem with inference from microarrays. The problem is as follows:

---

[3] [63] was better, but it used an interactive style of learning that is not practical with current biotechnology.

Table 2 A visual categorisation of GRN research. Columns denote kinds of data (sometimes simulated) which can be used and rows denote types of model. The GRN inference in [91] was secondary, and [80; 93] discuss epistatic inference. [97] also cites some work which uses phylogenetic data and ChIP assays. [45] and [44] used equilibrium microarray data as well. [13] used both equilibrium and microarray data

	ODE etc.	Boolean	BN	Neural
Time series	[24; 90; 117]	[64; 68; 101]	[36; 54; 83; 124]	[116]
Equilib.	[13; 96; 112]		[52]	
Perturb.	[8; 26; 37; 63; 109]	[123]		[80; 93]
Phylo.	[91]			
Chem./Loc.	[44]		[45]	

- Expression levels obtained from microarrays are the summed expression levels of $10^3 < x < 10^6$ cells.
- Except in limited circumstances, summed conditional dependencies may be different from individual conditional dependencies.

When the regulatory graph is singly connected (i.e. $i \rightarrow j \rightarrow h$, but not $i \rightarrow j \leftarrow h$), or if the noise is Gaussian and the regulatory relationships are all linear, then the factorisation of the sum is identical with the factorisation of the summands.

As neither of these conditions hold in GRN, the authors of [15] are concerned with the apparently successful results of machine learning using microarrays.

However, since the article was published substantially many more results have been biologically verified. This does not indicate that Chu et al. are wrong, but it does suggest that the conditional dependencies of the sum and the summands are similar enough in GRN. Further, it is important to remember that noise also blurs true conditional dependencies and that machine learning has been relatively successful anyway.

7   A Rough Map of the Field

Table 6.2 visually categorises some GRN research. It excludes results which just cluster the genes into modules and do not predict any regulatory relationships.

8   Conclusion and Directions

Research in GRN inference spans an enormous range of techniques, fields, and technologies. Techniques from other fields are used in new ways and new technology is being continuously developed. Nonetheless the problems of network inference remains open, and machine learning is essential. Fruitful avenues of research include:

- Incorporating cluster information into more detailed GRN inference.
- Combining separately learnt networks. Bayesian networks seem to be an ideal representation to use in this case. Considerations include:

  - How are two models which disagree about the edges, regulatory functions or both combined?
  - How does model agreement affect the posterior distribution?
  - What does it mean if two models agree but a third disagrees?
  - Can models inferred with different data sets be combined?

  In related research, machine inferred networks have been compared with networks manually assembled from the primary literature [49].
- Algorithmically increasing the value of different data types, as in [63]. Multi-classifiers[92] may also be fruitful.
- Incorporation of fuzzy techniques and clustering to address noise and the curse of dimensionality[33].

Little publicly available software specifically relating to GRN inference exists, but a wide range of general machine learning software is available to buy or download. Authors are usually happy to share their work when contacted directly. GRN simulators include `GeneSim`[122] and `GreenSim`[31]. `GreenSim` has been released under the GPL and is available online.

Acknowledgements. This research is funded by the Commonwealth Scholarship Commission in the United Kingdom and also supported by the Oxford-Man Institute, and the authors gratefully thank both institutions for their support and assistance.

References

[1] Arthur, D., Vassilvitskii, S.: k-means++: The advantages of careful seeding. Technical Report 2006-13, Stanford University (2006)

[2] Azuaje, F.: Clustering-based approaches to discovering and visualing microarray data patterns. Brief. in Bioinf. 4(1), 31–42 (2003)

[3] Balagurunathan, Y., et al.: Noise factor analysis for cDNA microarrays. J. Biomed. Optics 9(4), 663–678 (2004)

[4] Baldwin, J.F., Di Tomaso, E.: Inference and learning in fuzzy Bayesian networks. In: FUZZ 2003: The 12th IEEE Int'l Conf. on Fuzzy Sys., vol. 1, pp. 630–635 (May 2003)

[5] Bar-Joseph, Z., et al.: Computational discovery of gene modules and regulatory networks. Nat. Biotech. 21(11), 1337–1342 (2003)

[6] Barabasi, A.-L., Oltvai, Z.N.: Network biology: Understanding the cell's functional organisation. Nat. Rev. Genetics 5(2), 101–113 (2004)

[7] Ben-Dor, A., et al.: Clustering gene expression patterns. J. Comp. Bio. 6(3/4), 281–297 (1999)

[8] Di Bernardo, D., et al.: Robust identification of large genetic networks. In: Pacific Symp. on Biocomp., pp. 486–497 (2004)

[9] Bonneau, R., et al.: The inferelator: An algorithm for learning parsimonious regulatory networks from systems-biology data sets de novo. Genome Bio. 7(R36) (2006)

[10] Cao, Y., et al.: Reverse engineering of NK boolean network and its extensions — fuzzy logic network (FLN). New Mathematics and Natural Computation 3(1), 68–87 (2007)

[11] Cao, Y.: Fuzzy Logic Network Theory with Applications to Gene Regulatory Sys. PhD thesis, Department of Electrical and Computer Engineering, Duke University (2006)

[12] Cao, Y., et al.: Pombe gene regulatory network inference using the fuzzy logic network. New Mathematics and Natural Computation

[13] Zeke, S., Chan, H., et al.: Bayesian learning of sparse gene regulatory networks. Biosystems 87(5), 299–306 (2007)

[14] Chickering, D.M.: Learning Bayesian networks is NP-Complete. In: Fisher, D., Lenz, H.J. (eds.) Learning from Data: Artificial Intelligence and Statistics, pp. 121–130. Springer, Heidelberg (1996)

[15] Chu, T., et al.: A statistical problem for inference to regulatory structure from associations of gene expression measurements with microarrays. Bioinf. 19(9), 1147–1152 (2003)

[16] Cohen, I., et al.: Learning Bayesian network classifiers for facial expression recognition using both labeled and unlabeled data. CVPR 1, 595–601 (2003)

[17] Conant, G.C., Wagner, A.: Convergent evolution of gene circuits. Nat. Genetics 34(3), 264–266 (2003)

[18] Cui, Q., et al.: Characterizing the dynamic connectivity between genes by variable parameter regression and kalman filtering based on temporal gene expression data. Bioinf. 21(8), 1538–1541 (2005)

[19] de Jong, H.: Modeling and simulation of genetic regulatory systems: A literature review. J. Comp. Bio. 9(1), 67–103 (2002)

[20] de Leon, A.R., Carriere, K.C.: A generalized Mahalanobis distance for mixed data. J. Multivariate Analysis 92(1), 174–185 (2005)

[21] Dempster, A.P., et al.: Maximum likelihood from incomplete data via the EM algorithm. J. the Royal Statistical Society. Series B (Methodological) 39(1), 1–38 (1977)

[22] Dennett, D.C.: Real patterns. J. Philosophy 88, 27–51 (1991)

[23] D'haeseleer, P.: Resconstructing Gene Networks from Large Scale Gene Expression Data. PhD thesis, University of New Mexico, Albuquerque, New Mexico (December 2000)

[24] D'haeseleer, P., Fuhrman, S.: Gene network inference using a linear, additive regulation model. Bioinf. (submitted, 1999)

[25] D'haeseleer, P., et al.: Genetic network inference: From co-expression clustering to reverse engineering. Bioinf. 18(8), 707–726 (2000)

[26] Driscoll, M.E., Gardner, T.S.: Identification and control of gene networks in living organisms via supervised and unsupervised learning. J. Process Control 16(3), 303–311 (2006)

[27] Eisen, M.B., et al.: Cluster analysis and display of genome-wide expression patterns. Proc. of the National Academy of Sciences USA 95(25), 14863–14868 (1998)

[28] FitzGerald, P.C., et al.: Comparative genomics of drosophila and human core promoters. Genome Bio. 7, R53+ (2006)

[29] Floyd, R.W.: Algorithm 97: Shortest path. Communications of the ACM 5(6), 345 (1962)

[30] Fogelberg, C.: Belief propagation in fuzzy Bayesian networks: A worked example. In: Faily, S., Zivny, S. (eds.) Proc. 2008 Comlab. Student Conference (October 2008)

[31] Fogelberg, C., Palade, V.: GreenSim: A genetic regulatory network simulator. Tech. Report PRG-RR-08-07, Computing Laboratory, Oxford University, Wolfson Building, Parks Road, Oxford, OX1-3QD (May 2008), http://syntilect.com/cgf/pubs:greensimtr

[32] Fogelberg, C., Zhang, M.: Linear genetic programming for multi-class object classification. In: Zhang, S., Jarvis, R. (eds.) AI 2005. LNCS (LNAI), vol. 3809, pp. 369–379. Springer, Heidelberg (2005)

[33] Fogelberg, C., et al.: Belief propagation in fuzzy bayesian networks. In: Hatzilygeroudis, I. (ed.) 1st Int'l Workshop on Combinations of Intelligent Methods and Applications(CIMA) at ECAI 2008, University of Patras, Greece, July 21–22 (2008)

[34] Friedman, N.: Learning belief networks in the presence of missing values and hidden variables. In: Proc. of the 14th Int'l Conf. on Machine Learning, pp. 125–133. Morgan Kaufmann, San Francisco (1997)

[35] Friedman, N., et al.: Learning the structure of dynamic probabilistic networks. In: Proc. of the 14th Annual Conf. on Uncertainty in Artificial Intelligence (UAI 1998), vol. 14, pp. 139–147. Morgan Kaufmann, San Francisco (1998)

[36] Friedman, N., et al.: Using Bayesian networks to analyze expression data. J. Comp. Bio. 7(3), 601–620 (2000)

[37] Gardner, T.S., et al.: Inferring microbial genetic networks. ASM News 70(3), 121–126 (2004)

[38] Geman, S., Geman, D.: Stochastic relaxation, Gibbs distributions and the Bayesian restoration of images. IEEE Transactions on Pattern Analysis and Machine Intelligence 6, 721–742 (1984)

[39] Giaever, G., et al.: Functional profiling of the Saccharomyces cerevisiae genome. Nat. 418(6896), 387–391 (2002)

[40] Grünwald, P.: The minimum description length principle and non-deductive inference. In: Flach, P. (ed.) Proc. of the IJCAI Workshop on Abduction and Induction in AI, Japan (1997)

[41] Guo, H., Hsu, W.: A survey of algorithms for real-time Bayesian network inference. In: Joint AAAI 2002/KDD 2002/UAI 2002 workshop on Real-Time Decision Support and Diagnosis Sys. (2002)

[42] Gurney, K.: An Introduction to Neural Networks. Taylor & Francis, Inc., Bristol (1997)

[43] Han, E.-H., et al.: Clustering based on association rule hypergraphs. In: Research Issues on Data Mining and Knowledge Discovery, TODO (1997)

[44] Harbison, C.T., et al.: Transcriptional regulatory code of a eukaryotic genome. Nat. 431(7004), 99–104 (2004)

[45] Hartemink, A.J., et al.: Combining location and expression data for principled discovery of genetic regulatory network models. In: Pacific Symp. on Biocomp, pp. 437–449 (2002)

[46] Hastings, W.K.: Monte Carlo sampling methods using Markov chains and their applications. Biometrika 57(1), 97–109 (1970)

[47] Heckerman, D.: A tutorial on learning with Bayesian networks. Technical report, Microsoft Research, Redmond, Washington (1995)

[48] Heng, X.-C., Qin, Z.: Fpbn: A new formalism for evaluating hybrid Bayesian networks using fuzzy sets and partial least-squares. In: Huang, D.-S., Zhang, X.-P., Huang, G.-B. (eds.) ICIC 2005. LNCS, vol. 3645, pp. 209–217. Springer, Heidelberg (2005)

[49] Herrgard, M.J., et al.: Reconciling gene expression data with known genome-scale regulatory network structures. Genome Research 13(11), 2423–2434 (2003)

[50] Hinman, V.F., et al.: Developmental gene regulatory network architecture across 500 million years of echinoderm evolution. Proc. of the National Academcy of Sciences, USA 100(23), 13356–13361 (2003)

[51] Horimoto, K., Toh, H.: Statistical estimation of cluster boundaries in gene expression profile data. Bioinf. 17(12), 1143–1151 (2001)

[52] Imoto, S., et al.: Estimation of genetic networks and functional structures between genes by using Bayesian networks and nonparametric regression. In: Pacific Symp. on Biocomp., vol. 7, pp. 175–186 (2002)

[53] Imoto, S., et al.: Bayesian network and nonparametric heteroscedastic regression for nonlinear modeling of genetic network. J. Bioinf. and Comp. Bio. 1(2), 231–252 (2003)

[54] Jarvis, E.D., et al.: A framework for integrating the songbird brain. J. Comp. Physiology A 188, 961–980 (2002)

[55] Jiang, D., et al.: Cluster analysis for gene expression data: A survey. IEEE Transactions on Knowledge and Data Engineering 16(11), 1370–1386 (2004)

[56] Kauffman, S.A.: The Origins of Order: Self-Organization and Selection in Evolution. Oxford University Press, Oxford (1993)

[57] Kauffman, S.A.: Antichaos and adaptation. Scientific American 265(2), 78–84 (1991)

[58] Kim, S., et al.: Dynamic Bayesian network and nonparametric regression for nonlinear modeling of gene networks from time series gene expression data. Biosys. 75(1-3), 57–65 (2004)

[59] Kitano, H.: Computational systems biology. Nat. 420(6912), 206–210 (2002)

[60] Klebanov, L., Yakovlev, A.: How high is the level of technical noise in microarray data? Bio. Direct 2, 9+ (2007)

[61] Koch, M.A., et al.: Comparative genomics and regulatory evolution: conservation and function of the chs and apetala3 promoters. Mol. Bio. and Evolution 18(10), 1882–1891 (2001)

[62] Krause, E.F.: Taxicab Geometry. Dover Publications (1987)

[63] Kyoda, K.M., et al.: A gene network inference method from continuous-value gene expression data of wild-type and mutants. Genome Informatics 11, 196–204 (2000)

[64] Lähdesmäki, H., et al.: On learning gene regulatory networks under the Boolean network model. Machine Learning 52(1–2), 147–167 (2003)

[65] Lam, W., Bacchus, F.: Learning Bayesian belief networks: An approach based on the MDL principle. In: Comp. Intelligence, vol. 10, pp. 269–293 (1994)

[66] Laplace, P.-S.: Essai philosophique sur les probabilités. Mme. Ve. Courcier (1814)

[67] Le, P.P., et al.: Using prior knowledge to improve genetic network reconstruction from microarray data. Silico Bio. 4 (2004)

[68] Liang, S., et al.: REVEAL: a general reverse enginerring algorithm for inference of genetic network architectures. In: Pacific Symp. on Biocomp, pp. 18–29 (1998)

[69] Lum, P.Y., et al.: Discovering modes of action for therapeutic compounds using a genome-wide screen of yeast heterozygotes. Cell 116(1), 121–137 (2004)

[70] MacKay, D.J.C.: Introduction to Monte Carlo methods. In: Jordan, M.I. (ed.) Learning in Graphical Models. NATO Science Series, pp. 175–204. Kluwer, Dordrecht (1998)

[71] MacKay, D.J.C.: Information Theory, Inference, and Learning Algorithms. Cambridge University Press, Cambridge (2003)

[72] MacQueen, J.B.: Some methods for classification and analysis of multivariate observations. In: Proc. of the 5th Berkeley Symp. on Mathematical Statistics and Probability, pp. 281–297. University of California Press (1967)

[73] Madeira, S.C., Oliveira, A.L.: Biclustering algorithms for biological data analysis: a survey. IEEE/ACM Transactions on Comp. Bio. and Bioinf. 1(1), 24–45 (2004)

[74] Mahalanobis, P.C.: On the generalised distance in statistics. In: Proc. of the National Institute of Science of India, vol. 12, pp. 49–55 (1936)

[75] Marnellos, G., Mjolsness, E.: A gene network approach to modeling early neurogenesis in drosophila. In: Pacific Symp. on Biocomp., vol. 3, pp. 30–41 (1998)

[76] Massimo, F., Mascioli, F., et al.: Scale-based approach to hierarchical fuzzy clustering. Signal Processing 80(6), 1001–1016 (2000)

[77] McShan, D.C., et al.: Symbolic inference of xenobiotic metabolism. In: Altman, R.B., et al. (eds.) Pacific Symp. on Biocomp., pp. 545–556. World Scientific, Singapore (2004)

[78] Metropolis, N.A., et al.: Equation of state calculations by fast computing machines. J. Chemical Physics 21, 1087–1092 (1956)

[79] Mjolsness, E., et al.: Multi-parent clustering algorithms from stochastic grammar data models. Technical Report JPL-ICTR-99-5, JPL (1999)

[80] Motsinger, A.A., et al.: GPNN: Power studies and applications of a neural network method for detecting gene-gene interactions in studies of human disease. BMC Bioinf. 7, 39 (2006)

[81] Murali, T.M., Kasif, S.: Extracting conserved gene expression motifs from gene expression data. In: Pacific Symp. on Biocomp., pp. 77–88 (2003)

[82] Murphy, K.: Learning Bayes net structure from sparse data sets. Technical report, Comp. Sci. Div., UC Berkeley (2001)

[83] Murphy, K., Mian, S.: Modelling gene expression data using dynamic Bayesian networks. Technical report, Computer Science Division, University of California, Berkeley, CA (1999)

[84] Neal, R.M.: Probabilistic inference using Markov chain Monte Carlo methods. Technical Report CRG-TR-93-1, University of Toronto (1993)

[85] Nykter, M., et al.: Simulation of microarray data with realistic characteristics. Bioinf. 7, 349 (2006)

[86] Pan, H., Liu, L.: Fuzzy Bayesian networks - a general formalism for representation, inference and learning with hybrid Bayesian networks. IJPRAI 14(7), 941–962 (2000)

[87] Pan, H., McMichael, D.: Fuzzy causal probabilistic networks - a new ideal and practical inference engine. In: Proc. of the 1st Int'l Conf. on Multisource-Multisensor Information Fusion (July 1998)

[88] Park, H.-S., et al.: A context-aware music recommendation system using fuzzy Bayesian networks with utility theory. In: Wang, L., Jiao, L., Shi, G., Li, X., Liu, J. (eds.) FSKD 2006. LNCS, vol. 4223, pp. 970–979. Springer, Heidelberg (2006)

[89] Pearl, J.: Causal diagrams for empirical research. Biometrika 82(4), 669–709 (1995)

[90] Perkins, T.J., et al.: Reverse engineering the gap gene network of drosophila melanogaster. PLoS Comp. Bio. 2(5), e51+ (2006)

[91] Pritsker, M., et al.: Whole-genome discovery of transcription factor binding sites by network-level conservation. Genome Research 14(1), 99–108 (2004)

[92] Ranawana, R., Palade, V.: Multi-classifier systems: Review and a roadmap for developers. Int'l J. Hybrid Intelligent Sys. 3(1), 35–61 (2006)

[93] Ritchie, M.D., et al.: Optimization of neural network architecture using genetic programming improves detection and modeling of gene-gene interactions in studies of human diseases. BMC Bioinf. 4, 28 (2003)

[94] Russell, S.J., Norvig, P.: Artificial Intelligence: A Modern Approach, 2nd edn. Prentice Hall, Englewood Cliffs (2002)

[95] Schwarz, G.: Estimating the dimension of a model. The Annals of Statistics 6(2), 461–464 (1978)

[96] Segal, E., et al.: Module networks: identifying regulatory modules and their condition-specific regulators from gene expression data. Nat. Genetics 34(2), 166–176 (2003)

[97] Segal, E., et al.: From signatures to models: Understanding cancer using microarrays. Nat. Genetics 37, S38–S45 (2005) (By invitation)

[98] Shamir, R., Sharan, R.: Algorithmic approaches to clustering gene expression data. In: Jiang, T., Smith, T., Xu, Y., Zhang, M.Q. (eds.) Current Topics in Comp. Bio., pp. 269–300. MIT press, Cambridge (2002)

[99] Shannon, C.E.: A mathematical theory of communication. Bell System Technical Journal 27, 379–423, 623–656 (1948)

[100] Sheng, Q., et al.: Biclustering microarray data by Gibbs sampling. Bioinf. 19, ii196–ii205 (2003)

[101] Silvescu, A., Honavar, V.: Temporal Boolean network models of genetic networks and their inference from gene expression time series. Complex Sys. 13, 54–70 (2001)

[102] Sivia, D.S.: Data Analysis: A Bayesian Tutorial. Clarendon Press, Oxford (1996)

[103] Smith, V.A., et al.: Evaluating functional network inference using simulations of complex biological systems. Bioinf. 18, S216–S224 (2002)

[104] Smith, V.A., et al.: Influence of network topology and data collection on network inference. In: Pacific Symp. on Biocomp., pp. 164–175 (2003)

[105] Spellman, P.T., et al.: Comprehensive identification of cell cycle-regulated genes of the yeast Saccharomyces cerevisiae by microarray hybridization. Mol. Bio. of the Cell 9(12), 3273–3297 (1998)

[106] Spirtes, P., et al.: Constructing Bayesian network models of gene expression networks from microarray data. In: Proc. of the Atlantic Symp. on Comp. Bio., Genome Information Sys. and Technology (2000)

[107] Sterelny, K., Griffiths, P.E.: Sex and Death: An Introduction to Philosophy of Bio. Science and Its Conceptual Foundations series. University Of Chicago Press (June 1999) ISBN 0226773043

[108] Tang, C., et al.: Interrelated two-way clustering: An unsupervised approach for gene expression data analysis. In: Proc. of the IEEE 2nd Int'l Symp. on Bioinf. and Bioeng. Conf., 2001, November 4–6, pp. 41–48 (2001)

[109] Tegner, J., et al.: Reverse engineering gene networks: integrating genetic perturbations with dynamical modeling. Proc. of the National Academy of Sciences, USA 100(10), 5944–5949 (2003)

[110] Thomas, R.: Laws for the dynamics of regulatory networks. Int'l J. Developmental Bio. 42, 479–485 (1998)

[111] Tibshirani, R., et al.: Clustering methods for the analysis of DNA microarray data. Technical report, Stanford University (October 1999)

[112] Toh, H., Horimoto, K.: Inference of a genetic network by a combined approach of cluster analysis and graphical Gaussian modeling. Bioinf. 18(2), 287–297 (2002)

[113] Tong, A.H., et al.: Systematic genetic analysis with ordered arrays of yeast deletion mutants. Science 294(5550), 2364–2368 (2001)

[114] Troyanskaya, O., et al.: Missing value estimation methods for DNA microarrays. Bioinf. 17(6), 520–525 (2001)

[115] Vert, J.-P., Yamanishi, Y.: Supervised graph inference. In: Saul, L.K., et al. (eds.) Advances in Neural Information Processing Sys., vol. 17, pp. 1433–1440. MIT Press, Cambridge (2005)

[116] Vohradský, J.: Neural network model of gene expression. FASEB Journal 15, 846–854 (2001)

[117] Wang, Y., et al.: Inferring gene regulatory networks from multiple microarray datasets. Bioinf. 22(19), 2413–2420 (2006)

[118] Xu, R., Wunsch II, D.: Survey of clustering algorithms. IEEE Transactions on Neural Networks 16(3), 645–678 (2005)

[119] Yamanishi, Y., et al.: Protein network inference from multiple genomic data: a supervised approach. Bioinf. 20(1), 363–370 (2004)

[120] Yang, E., et al.: A novel non-overlapping bi-clustering algorithm for network generation using living cell array data. Bioinf. 23(17), 2306–2313 (2007)

[121] Yu, J., et al.: Using Bayesian network inference algorithms to recover molecular genetic regulatory networks. In: Int'l Conf. on Sys. Bio. (ICSB 2002) (December 2002)

[122] Yu, J., et al.: Advances to Bayesian network inference for generating causal networks from observational biological data. Bioinf. 20(18), 3594–3603 (2004)

[123] Yuh, C.H., et al.: Genomic cis-regulatory logic: Experimental and computational analysis of a sea urchin gene. Science 279, 1896–1902 (1998)

[124] Zhang, Y., et al.: Dynamic Bayesian network (DBN) with structure expectation maximization (SEM) for modeling of gene network from time series gene expression data. In: Arabnia, H.R., Valafar, H. (eds.) BIOCOMP, pp. 41–47. CSREA Press (2006)

[125] Zhou, X., et al.: Gene clustering based on clusterwide mutual information. J. Comp. Bio. 11(1), 147–161 (2004)

Automatic Approximation of Expensive Functions with Active Learning

Dirk Gorissen, Karel Crombecq, Ivo Couckuyt, and Tom Dhaene

Abstract. The use of computer simulations has become a viable alternative for real-life controlled experiments. Due to the computational cost of these high fidelity simulations, surrogate models are often employed as an approximation for the original simulator. Because of their compact formulation and negligible evaluation time, global surrogate models are very useful tools for exploring the design space, optimization, visualization and sensitivity analysis. Additionally, multiple surrogate models can be chained together to model large scale systems where the direct use of the original expensive simulators would be too cumbersome. Many surrogate model types, such as neural networks, support vector machines and rational models have been proposed, and many more techniques have been developed to minimize the number of expensive simulations required to train a sufficiently accurate surrogate model. In this chapter, we present a fully automated and integrated global surrogate modeling methodology for regression modeling and active learning that readily enables the adoption of advanced global surrogate modeling methods by application scientists. The work brings together insights from distributed systems, artificial intelligence, and modeling & simulation, and has applications in a very wide range of fields. The merits of this approach are illustrated with several examples, and several surrogate model types.

1 Introduction

Regardless of the advances in computational power, the the simulation of complex systems with multiple input and output parameters continues to be a time-consuming process. This is especially evident for routine tasks such as

Dirk Gorissen and Tom Dhaene
INTEC-IBBT, Ghent University, Gaston Crommenlaan 8, 9050 Belgium
e-mail: {dirk.gorissen,tom.dhaene}@ugent.be

Karel Crombecq and Ivo Couckuyt
University of Antwerp, Middelheimlaan 1, 2020 Antwerp, Belgium
e-mail: {karel.crombecq,ivo.couckuyt}@ua.ac.be

A.-E. Hassanien et al. (Eds.): Foundations of Comput. Intel. Vol. 1, SCI 201, pp. 35–62.
springerlink.com

prototyping, high dimensional visualization, optimization, sensitivity analysis and design space exploration [1, 2, 3]. For example, Ford Motor Company reported on a crash simulation for a full passenger car that takes 36 to 160 hours to compute [4]. Furthermore, it is not unsual for the system under study to be a black box, with little or no additional information available about its inner working outside of the output it generates. Consequently, different approximation methods have been developed that mimic the behavior of the 'true' simulation model as closely as possible and are computationally cheap(er) to evaluate. Many methods exist, their relative strengths depending on the final goal (optimization, design space eploration, conservation of physical laws, etc.). Popular choices are polynomial and rational functions [5, 6], Kriging (or DACE) models [7], neural networks and RBF models [8]. This work concentrates on the automatic generation of such data-driven, global surrogate models (also known as emulators, metamodels or response surface models (RSM)).

It is very important that we first stress the difference between local and global surrogate models, for the two are often confused. Yet the motivation and philosophy are quite distinct. Local surrogate modeling involves building small, relatively low fidelity surrogates for use in optimization. Local surrogates are used as rough approximators of the (costly) optimization surface and guide the optimization algorithm towards good extrema while minimizing the number of simulations [9]. Once the optimum is found, the surrogate is discarded. In general the theory is referred to as Surrogate Based Optimization (SBO) or Metamodel Assisted Optimization (MAO). A good overview reference is given by [10]. In contrast, in global surrogate modeling the surrogate model *itself* is the goal. The objective is to construct a high fidelity approximation model that is as accurate as possible over the *complete* design space of interest using as little simulation points as possible. Once constructed, the global surrogate model (also referred to as a replacement metamodel) is reused in other stages of the computational science and engineering pipeline (see section 3.1). Optimization is not the goal, but rather a useful (optional) post processing step.

The scientific goal of global surrogate modeling is the generation of a surrogate that is as accurate as possible (according to some, possibly physics-based, measure), using as few simulation evaluations as possible, and with as little overhead as possible. If we regard previous work in *global* surrogate modeling and the techniques in use in industry today, our experience is that most work has focused on a particular subset of the global surrogate modeling process (hyperparameter optimization, model selection, etc.), with particular assumptions on the other sub problems. See for example [11, 12, 13].

The authors argue that a more holistic, automated, and integrated approach is needed to deal with the increasing diversity and computational complexity of current simulation codes. Fully exploiting the close interplay between active learning and model management. This has, of course, been remarked before (see [14] for example, or the discussion in [15]). Various powerful and novel sequential metamodeling techniques have been proposed in literature: e.g., [16, 17, 18, 15, 19, 20]. However, there is still room for a more extensible, flexible, and automated approach to global surrogate modeling, that does not mandate assumptions (but does not preclude them either) about the problem, model type, sampling algorithm, etc. Particularly, if such assumptions are neither applicable nor practical, or if not enough information is available to make any assumptions. For example, [16] present a powerful model generation framework based on Gaussian Process (GP) models. However, GP models are not the best choise for all problems (e.g., they are outperformed Support Vector Regression (SVR) on various benchmarks [21]). In such cases it would be useful if the model type could easily be swapped out, or parts of the framework be re-used to ensure a wide applicability with a minimum of overhead.

This work brings together advanced solutions to the different sub problems of global surrogate modeling in a unified, pluggable, adaptive framework (complementing the results in [22]). The motivation stems from our observation that the primary concern of an application scientist is obtaining an accurate replacement metamodel (with a predefined accuracy) for their problem as fast as possible (i.e., with a minimal number of simulations) and with minimal user interaction. Model (type) selection, model parameter optimization, active learning, etc. are of lesser or no interest to them. Thus the contribution of this work is in the general methodology, the various extensions (secions 4.4 to 4.8) and the available software implementation. This framework will be useful in any domain (biology, economics, electronics, ...) where a cheap, accurate approximation is needed for an expensive reference model. In addition the work makes advances in computaional intelligence (kernel methods, feature selection, hyperparameter optimization, active learning strategies, etc.) more easily available to application scientists. Thus lowering the barrier of entry and bridging the gap between theory and application.

2   Related Work

Research efforts related to the work presented here can be described on different levels. On the conceptual and software engineering level our work is similar to popular machine learning toolkits like Rapidminer (formerly Yale), Spider, Shogun, Weka, and Plearn. They try to bring together multiple machine learning techniques and data mining algorithms together into one coherent, easy to use, software package. The difference is that they are primarily concerned with classification and data mining and typically assume all data is available and cheap (there is no expensive simulation code that must be driven). Different active learning strategies (including the tie-in with the hyperparameter optimization) are often missing as is the link with distributed computing (to allow efficient scheduling of simulations), automatic model type selection, and the ability to deal with complex-valued data directly.

On a problem domain level, there are many projects that tackle surrogate modeling for optimization, also known as Metamodel Assisted Optimization (MAO) (see also section 1). These projects are similar in that they also have to deal with computationally expensive simulation data. The difference is that they are heavily focused on optimization, the surrogate model itself is not the main goal but used to approximate the fitness landscape. The surrogate models in use are often quite simple (cfr. traditional Response Surface Methodology (RSM)), with polynomial regression being the most popular among practitioners [23]. This, of course, need not be the case. For example, complicated ensemble methods combining different surrogate model types have been used [24], as have been innovative trust region methods [25] and sampling techniques [26, 27]. A good overview reference of SBO is given by Eldred and Dunlavy in [10]. Well known examples in this category are the proprietary tools developed by LMS/Noesis (Optimus, Virual.Lab) and Vanderplaats R&D (VisualDOC). From academia, the more prominent projects are Geodise [28] from the University of Southampton, and the DAKOTA toolkit from Sandia National Labs [29].

In sum, the differences between our work and toolkits like DAKOTA, VisualDOC, and Geodise, is that we place much more emphasis on the surrogate model construction process itself, trying to generate a high fidelity replacement metamodel over the complete design space, using as little data points as possible while minimizing a priory knowledge and user interaction. In contrast the aforementioned projects put most effort into efficiently seeking through the design space in search of regions of good design (optimization). (Global) surrogates are useful as a rough guide but not the final objective. Thus the approaches are complementary, some features are shared but the philosophy and focus is distinct (see also section 4).

Finally, the last level of related work are research efforts that come closest to what we try to achieve. These include the Hierarchical Nonlinear Approximation algorithm of Busby et al. [16], the Treed GP method of [17], the Bayesian sequential strategy of Farhang-Mehr et al. [18], the Efficient Robust Concept Exploration Method (ERCEM) of [15], the *NeuroModeler* tool by [30], and the proprietary *Datascape* [31], and *ClearVu Meta-Modeling* [20] tools developed by Third Millennium Productions (originally by Lockheed Martin) and NuTech respectively. All these efforts have as a goal to approximate the complete design space as accurately as possible while minimizing the number of data points needed. The contribution of this paper is to take these efforts one step further, further automating, integrating, and improving the global surrogate modeling process and removing explicit dependencies on model types or sampling algorithms.

[16] present a sequential metamodeling approach based on the characteristics of GP models. [18] do the same but then for Bayesian models, and [30] do so for ANN models (specifically for the RF and microwave domain). [17] use GP models as well but adopt a divide an conquer approach to better capture variations of the design space. [20] do not make assumptions about

the model type but fix the model hyperparameter optimization algorithm to an Evolutionary Strategy (ES). In addition they do not use sequential design but try to predict the number of samples needed from previous runs based on different fixed experimental designs. The work by [15] is similar to aforementioned references, but takes a more engineering oriented view. [15] gives an excellent discussion on the need for a close integration between the different surrogate modeling sub-problems but does not take the integration as far as could be done. Hyperparameter optimization and integration with distributed computing are also not considered. Finally, Datascape uses a proprietary, private method for constructing surrogates based on fuzzy logic, non-linear regression and numerical optimization [31]. Also, the number of samples needed must be chosen upfront, there is no sequential design.

In contrast we make no assumptions about the model type, sampling algorithm, model selection measure, or any other component. Instead we, like DAKOTA and Spider, focus on engineering a well thought out object oriented framework where different components can easily be plugged in, benchmarked, and used together without any a priory assumptions. This results in a generic framework that can be applied to any problem, while not excluding the use of problem specific techniques that *do* make assumptions about the underlying problem, model type or data distribution. Besides developing the common infrastructure we have also developed new algorithms for the components themselves (e.g., the gradient active learning algorithm described in [32]). Also, an engineer not familiar with the intricacies of surrogate modeling should not be overwhelmed with options, therefore we place a very strong focus on sensible defaults, adaptivity, robustness, self-healing and easy integration into the engineering design process. This focus has led to a level of automation not found in other projects (see sections 4.4 to 4.7). A final key point is that support for grid computing is inherently built into the framework (cfr section 4.7). Knowledge of the surrogate modeling process is used to automatically and optimally schedule simulations on a cluster or grid.

3   Global Surrogate Modeling

3.1   Motivation

This chapter is concerned with data-driven modeling based on global surrogate models. Again we stress that, while optimization can be performed as a post-processing step, it is not the main goal.

Global surrogate models are particularly useful for design space exploration, sensitivity analysis, prototyping, (high dimensional) visualization, and *what-if* analysis. They are also widely used to build model libraries for use in controllers or engineering design software packages (for example, see the global rational models in use in ADS Momentum [33]). In addition, they can cope with varying boundary conditions. This enables them to be chained together in a model cascade in order to approximate large scale systems [34]. A classic example is the full-wave simulation of an electronic circuit board. Electro-magnetic modeling of the whole board in one run is almost intractable. Instead the board is modeled as a collection of small, compact, accurate surrogates that represent the different functional components (capacitors, transmission lines, resistors, etc.) on the board. Finally, if optimization is the goal, one could argue that a global model is less useful since significant time savings could be achieved if more effort were directed at finding the optimum rather than modeling regions of poor designs. However, this is the logic of purely local models, but they forgo any wider exploration of radical designs [35].

The principal reason driving the use of surrogate models is that the simulator is too time consuming to run for a large number of simulations [36]. Nevertheless, one could argue that in order to construct an accurate global surrogate, one still needs to perform a large number of simulations (to obtain a sufficiently high accuracy), thus running into the same problem. However, this is not the case since: (1) building a global surrogate is a one-time, up-front investment (assuming the simulation code stays the same) and (2) adaptive modeling and active learning can drastically decrease the required number of data points to produce an accurate model[1]. In addition, the active learning process can be integrated with distributed computing to ensure simulations are run in parallel and scheduled optimally. Thus reducing the computation time further.

Surrogate modeling has found its way into many fields where it is used to approximate some complex and/or expensive reference model. Some examples of applications are economic validation of capital projects [37] (Economics), prediction of fibrinogen absorption onto polymer surfaces [38] (Chemistry), electromagnetic simulations (Electronics) [33] and modeling colon coloration [39] (Medicine).

3.2 Sub-problems

Accurate global surrogate modeling of an expensive simulation code involves a number of sub-problems that need to be addressed.

Sampling Strategy

As a preliminary remark, it is important to remind the reader that we are typically dealing with computer experiments, thus data is deterministic and noise free. There certainly are simulators using stochastic codes, but the main

---

[1] Obviously, due to the 'curse of dimensionality' this can not continue to hold for a high number of dimensions (>6). As the dimensionality increases the accuracy requirements must be relaxed. Unless, of course, there is no sampling (fixed data set). In this case there is no such restriction.

focus of computer experimentation has been on deterministic codes. Thus, in computer experiments there is no need to perform replication (as opposed to physical experiments) and the classic theory of optimal designs can not be applied. Instead, a major concern is to create an experimental design which can sample the complete design space in a representative way with a minimum number of samples [40]. Data points must be selected iteratively, at points where the information gain will be the greatest. Mathematically this means defining a sampling function

$$\phi(X_{i-1}) = X_i \, , \, i = 1, .., L \tag{1}$$

that constructs a data hierarchy

$$X_0 \subset X_1 \subset X_2 \subset ... \subset X_L \subset X \tag{2}$$

of nested subsets of $X = \{x_1, ..., x_k\}$, where L is the number of levels. X_0 is referred to as the *initial experimental design* and is constructed using one of the many algorithms available from the theory of Design and Analysis of Computer Experiments (DACE) (see the work by Kleijnen et al. [41]). Once the initial design X_0 is available it can be used to seed the the sampling function ϕ. An important requirement of ϕ is to minimize the number of sample points $|X_i| - |X_{i-1}|$ selected each iteration (f is expensive to compute), yet maximize the information gain of each successive data level. This process is called active learning [42], but is also known as adaptive sampling [43], reflective exploration [33], Optimal Experimental Design (OED) [44] and sequential design [45]. In the context of computer experiments, a large variety of active learning methods have been developed. Some methods are developed specifically with one model type in mind, using the properties of the model to select new sample locations. Examples of this approach can be found in [16, 46, 17]. Other sampling strategies work independent of the model type, using only information provided by previously evaluated samples. This approach is used in [47, 13, 33]. Remember that we are only concerned with sampling for global surrogate modeling. Adaptive sampling can also be used in the context of optimization, to help guide the optimizer to a good local extrema (e.g., [26]). While we have also tackled that case, it is not the focus of this work.

Modeling Strategy

A second sub-problem is selecting the surrogate model type and the model parameter (hyperparameter) optimization method. Popular surrogate model types include Radial Basis Function (RBF) models, Rational Functions, Artificial Neural Networks (ANN), Support Vector Machines (SVM), Multivariate Adaptive Regression Splines (MARS) and Kriging models. Different model types are preferred in different domains. For example rational functions are widely used by the Electro-Magnetic (EM) community [48, 49], while ANNs are preferred for hydrological modeling [50, 51]. Differences in model type usage are mainly due to practical reasons (available expertise, tradition, computation time, etc.). In general there is no hard theory that can be used as an a priori guide (cfr. the *No-Free-Lunch* theorems [52]). Of course this is not always the case. In some cases knowledge of the physics of the underlying system can make a particular model type to be preferred. For example, rational functions are popular for all kinds of Linear Time-Invariant (LTI) systems since theory is available that can be used to prove that rational pole-residue models conserve certain physical quantities (see [53]).

Selecting the model type is only part of the problem, since the complexity of the model needs to be chosen as well. Each model has a set of hyperparameters θ that control the complexity of the model M and thus the bias-variance trade-off. In general, finding the optimal bias-variance trade-off, i.e., s^* as defined in section 3.1, is hard. All too often this is done through a trial-and-error procedure (see for example [36, 21]). A better approach is to use an optimization algorithm guided by a performance metric (e.g., external validation set, leave-one-out error, an approximation of the posterior $p(\theta, M | data)$ such as AIC/BIC/MDL, etc.). In this way a successive set of approximation models $M_1, M_2, ..., M_m$ are generated that converge towards a local minimum of the optimization landscape, as determined by the performance metric. The challenge here is to converge to a good local optimum, since the landscape can be expected to be highly multi-modal, deceptive and epistatic.

This makes the choice of the performance metric crucial, leading us to the well known problem of model selection and generalization estimation (for which there is also no "best" solution, but much depends on the the model type and the amount and distribution of data points [15]). There is a vast body of research available on these topics, particularly in the fields of neural networks and kernel methods. Examples from literature include ANN structure optimization with pattern search [54], genetic SVM hyperparameter optimization [55] and automatic trimming of redundant polynomial terms [56].

Note, though, that in the context of this chapter the hyperparameter optimization problem is more difficult than is typically the case. Since data points are costly, the use of active learning is unavoidable. This implies that the data distribution (from which models must be constructed) is not constant, and consequently that the hyperparameter optimization surface is dynamic instead of static (as is often assumed).

Sample Evaluation

More of a practical consideration is the sample evaluation strategy to use. When constructing a global surrogate model for an expensive simulation engine the largest computational bottleneck is performing the necessary simulations. A natural step is to harness the power of a grid or cluster to relieve this cost. Integration with a distributed system can occur on multiple levels. This will be explained further in section 4.7.

Others

Besides the three major sub-problems listed above surrogate modeling also involves making informed choices with respect to: feature selection, data pre/post-processing, termination criteria and model validation. In particular, feature selection is gaining increasing importance as the number of dimensions considered by engineers continues to increase.

3.3 Integration

In general, there has been much research on the different sub-problems of *global* surrogate modeling, but less on the combination of different solutions in a generic way (see the next section for an overview of related work). For the majority of real world problems, data is expensive and its use should be optimized. It is in these situations that a close integration of sampling, modeling and sample evaluation strategies is really needed (an good discussion of these aspects is given by [15]).

Secondly, the primary users of global surrogate modeling methods are application scientists, few of which will be experts in the intricacies of efficient sampling and modeling strategies. Their primary concern is obtaining an accurate replacement metamodel as fast as possible and with minimal overhead. Model (type) selection, hyperparameter optimization, sampling strategy, etc. are of lesser or no interest. At the same time every discipline or engineer has their preferred technique and approximation method and there is no such thing as a "one size fits all". Thus any efforts to integrate the different sub problems must do so in a generic and extensible way without excluding the use of specialized algorithms and techniques. As a result surrogate modeling algorithms should be as adaptive, self-tuning, and as robust as possible, and easily integrate with the other components of the Computer Science and Engineering (CS&E) pipeline. This is the focus of this work.

3.4 Adaptive Global Surrogate Modeling

The algorithm that lies at the core of our approach is presented in Algorithm 1. The algorithm is conceptually very simple, but in its simplicity lies its power and flexibility.

There are two main parts to the algorithm: an outer, active learning loop and an inner adaptive modeling loop. Given a set of sample points, the inner loop will optimize the model hyperparameters until no further improvement is possible on the given set of samples. Having reached a local minimum of the hyperparameter space, the algorithm then signals the sample selection loop to determine a new, maximally informative set of simulation points. These are then evaluated, intelligently merged with the existing sample points, and the adaptive modeling process is allowed to resume. This whole process continues

Algorithm 1. Adaptive global surrogate modeling algorithm (pseudo-code)

01. $target$ = getAccuracyTarget();
02. X = initialExperimentalDesign();
03. $f|_X$ = evaluateSamples(X);
04. M = [];

05. while($target$ not reached)
06.　　M = buildModels($X, f|_X$);
07.　　while(improving(M)) do
08.　　　M = optimizeHyperparameters(M);
09.　　end
10.　　X_{new} = sampleSelection($X, f|_X, M$);
11.　　$f|_{X_{new}}$ = evaluateSamples(X_{new});
12.　　$[X, f|_X]$ = merge($X, f|_X, X_{new}, f|_{X_{new}}$);
13. end

14. return bestModel(M);

until the user-defined accuracy has been reached or some other timeout occurs. The final surrogate model is returned to the user.

The real usefulness of this algorithm comes from an implementation that provides swappable components for each of the different steps. Such an implementation is presented in the next section: the **SU**rrogate **MO**delling toolbox (SUMO Toolbox).

4 SUMO Toolbox

4.1 Introduction

The SUMO Toolbox is an adaptive tool that integrates different modeling approaches and implements a fully automated, global surrogate model construction algorithm. Given a simulation engine, the toolbox automatically generates a surrogate model within the predefined accuracy and time limits set by the user. At the same time, taking into account that there is no such thing as a 'one size fits all', as different problems need to be modeled differently. Therefore the toolbox was designed to be modular and extensible but not be too cumbersome to use or configure.

Given this design philosophy, the toolbox caters to both the scientists working on novel approximation or active learning methods as well as to the engineers who need the surrogate model as part of their design process. For the former, the toolbox provides a common platform on which to deploy, test, and compare new modeling algorithms and sampling techniques. For the latter, the software functions as a highly configurable and flexible component to which surrogate model construction can be delegated, enhancing productivity.

Finally, while the SUMO Toolbox is still under development, stable releases for all platforms are available for download at
http://www.sumo.intec.ugent.be.

4.2 Core Algorithm

The control flow of the toolbox is driven by the algorithm presented in subsection 3.4, and illustrated in Fig. 1. It is instructive to walk through the control flow. Initially, a small initial set of samples is chosen according to some experimental design (e.g., Latin hypercube, Box-Behnken, etc.). Based on this initial set, one or more surrogate models are constructed and their parameters optimized according to a chosen hyperparameter optimization algorithm (e.g., BFGS, Particle Swarm Optimization (PSO), Genetic Algorithm (GA), DIRECT, etc.). Models are assigned a score based on one or more measures (e.g., cross validation, AIC, etc.) and the optimization continues until no further improvement is possible. The models are then ranked according to their score and new samples are selected based on the top k models, as described in section 4.5. The hyperparameter optimization process is restarted intelligently (cfr. section 4.4) and the whole process repeats itself until one of the following three conditions is satisfied: (1) the maximum number of samples has been reached, (2) the maximum allowed time has been exceeded, or (3) the user required accuracy has been met. Note that the sample evaluation component runs in parallel with the other components (non blocking) and not sequentially as is often the case.

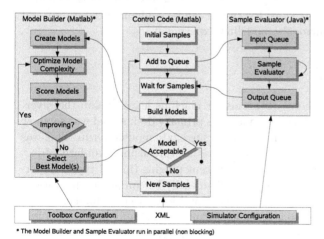

* The Model Builder and Sample Evaluator run in parallel (non blocking)

Fig. 1 The SUMO Toolbox control flow

Fig. 2 SUMO Toolbox
Plugins

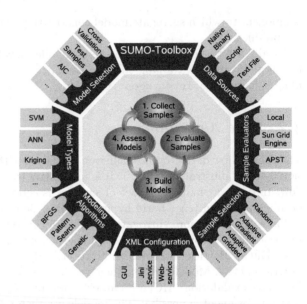

Selection of what algorithms to use, which simulator outputs to model, if noise should be added, if certain inputs should be clamped, whether complex data should be split into its real and imaginary components, etc. is completely determined by two XML configuration files. The first defines the interface to the simulation code: number of input and output parameters, type of each parameter (real, discrete or complex), simulator executables & dependencies and/or one or more data sets. The second XML file contains the configuration of the toolbox itself: which outputs to model, model type to use, whether sample evaluation should occur on a grid or cluster, etc. This file also allows the user to specify multiple *runs*. Every run can be configured separately and represents a different surrogate modeling experiment.

4.3 Extensibility

In light of the *No-Free-Lunch* theorems for machine learning and the diversity of techniques in use among engineers, a primary design goal was to allow maximum flexibility in composing a global surrogate modeling run. This was achieved by designing a plugin-based infrastructure using standard object oriented design patterns (see Fig. 2). Different plugins (model types, hyperparameter optimization algorithms, etc.) can easily be composed into various configurations or replaced by custom, more problem specific plugins.

The plugins currently available include (but are certainly not limited to):

- *Data sources:* native executable, Java class, Matlab script, database, user defined

- *Model types:* Polynomials, Rational functions, Multi Layer Perceptrons, RBF models, RBF Neural Networks, Support Vector Machines ($\epsilon - SVM$, $\nu - SVM$, LS-SVM), GP models, splines, hybrid (ensembles), automatic model type selection
- *Hyperparameter optimization algorithms:* hill climbing, BFGS, Pattern Search, GA, PSO, DIRECT, simulated annealing, NSGA-II, Differential Evolution
- *Initial experimental design:* random, Central Composite, Box-Behnken, Optimal Latin Hypercube (following [57]), full factorial, user defined DOE
- *Active learning:* error-based, density-based, gradient based, random, optimization-driven infill sampling[2]
- *Model selection:* cross validation, validation set, Leave-one-out, AIC, MDL, model difference

In addition, the toolbox includes built-in support for high performance computing. Sample evaluation can occur locally (with the option to take advantage of multi-CPU or multi-core architectures) or through a distributed middleware (possibly accessed through a remote head-node). Currently the Sun Grid Engine (SGE), A Parameter Sweep Tool (APST) and the Globus based LHC Computing Project (LCG) middlewares are supported, though other interfaces (e.g., for Condor) may be added (DRMAA and CoG backends are under development). All interfacing with the grid middleware (submission, job monitoring, rescheduling of failed/lost simulation points, etc.) is handled completely transparently and automatically (see [58] for details).

4.4   Pareto Optimal Model Scoring

The optimization of the model parameters is driven by a metric that assigns a score to a model. A novel feature of the SUMO Toolbox is that more than one metric is permitted. In this case a multi-objective optimization algorithm is used to drive the model parameter optimization process (note that this has nothing to do with the optimization of the simulator or surrogate model itself). Currently a plugin using NSGA-II [59] is available and a second one (based on SCEM-UA [60]) will be available in the near future. This feature is useful if one wishes to combine traditional model selection criteria with domain specific constraints, or penalties. A simple example is a Min-Max constraint that penalizes a model if its response exceeds physically acceptable ranges. A second example is to minimize a maximum and global (average) approximation error simultaneously.

---

[2] This is an extended version of the EGO algorithm originally developed by Jones et. al. [26]. It includes extensions discussed by [27] and provides a configurable algorithm, supporting constraints, that balances global sampling and optimization driven sampling.

Of course it is always still possible to transform the multi-objective problem to a single objective one. In this case the the objectives are combined by a user-defined function. For example, the geometric mean as done in [23].

4.5 Intelligent Active Learning Strategies

When the hyperparameter optimization algorithm has reached a local optimum, the active learning algorithm is triggered to select a certain number of new samples. Since evaluating a new sample is potentially a very costly operation, care must be taken that the new sample locations are maximally informative. This means that, ideally, samples should be selected where (1) data is sparse, (2) the response is non-linear, (3) the current model is uncertain or inaccurate. Model types which depend heavily on the euclidean distance between sample points, such as GP models, need a proper distribution of samples over the entire design space in order to guarantee an acceptable accuracy. Artificial neural networks, for example, can cope better with clustering, allowing the active learning algorithm to zoom in on regions of high variance. Thus, it is clear that the ideal sample distribution for a given problem depends on many different factors, and that there exists no single optimal distribution.

In order to balance these different factors, several active learning algorithms were developed and integrated in the SUMO Toolbox. Space filling methods try to cover the complete design space evenly. Error based methods on the other hand use an error measure to guide sampling to areas which need more exploration. These may be highly non-linear areas, areas containing (local) optima or areas in which the model fails to approximate the response accurately. Finally, a gradient based method was developped and implemented which tries to combine these two extremes into one robust algorithm that performs a trade-off between exploration and exploitation [32]. Finally, a sampling strategy based on the EGO [26] approach was developed to cater for those problems where an accurate model of the optima is most important.

4.6 Automatic Model Type Selection

Arguably the biggest difficulty in applying surrogate modeling is identifying the most adequate surrogate model type. We have implemented a novel algorithm to tackle this problem automatically. This is achieved through the use of heterogeneous evolution. The island model with speciation is used to evolve multiple surrogate model types together in different sub-populations. Migration between populations causes competition between surrogate model types to fit the data, with more accurate models having a higher chance of propagating to the next generation. In this way automatic model type selection is implemented while at the same time allowing the model type to vary dynamically and enabling hybrid solutions through the use of ensembles. Details can be found in [61] and an application to a RF-circuit block modeling problem is described in [62].

Note that this approach is different than that described in [12, 63, 64] and others. In contrast to [63] for example, the type of the ensemble members is not fixed in any way but varies dynamically. Also, [63] is concerned with optimization, not the model itself. Surrogates are used together with a trust region framework to quickly and robustly identify the optimum. In contrast to this work, the model parameters are taken fixed, and they make only a *"mild assumption on the accuracy of the metamodeling technique"*. The work of Sanchez et. al [12] and Goel et. al [64] is more useful in our context since they provide new algorithms for generating an optimal set of ensemble members for a fixed set of data points (no sampling). Unfortunately, though, the parameters of the models involved must still be chosen manually. Nevertheless, their approaches are interesting, and can be used to further enhance the current approach. For example, instead of returning the single final best model, an optimal ensemble member selection algorithm can be used to return a potentially much better model based on the final population or pareto front (cfr. section 4.4).

4.7 Integration with Distributed Computing

Integration with distributed computing can occur on two main levels[3], each of which is explained briefly below. For the details please see [58].

Resource Level

The most obvious approach is to simply use a cluster or grid to run simulations in parallel, reducing the overall 'wall clock' execution time. This can be achieved by using a specific Sample Evaluator (SE) (cfr. Figs. 1 and 2). The SE can be seen as a kind of Application Aware meta-Scheduler (AAS) that forms the glue between the modeler and the grid middleware. It is responsible for translating modeler requests (i.e., evaluations of data points) into middleware specific jobs (e.g., jdl files in the case of LCG), polling for results, and returning them to the modeler. SE was designed to be middleware agnostic, besides a few configuration options, the user should not have to worry about the intricacies of each job authentication or submission system.

Scheduling Level

Performance can be improved further if one considers that not all data points are equally important, a partial ordering exists. For example, data points

----

[3] Actually, there is a third level of integration: at the service level. This would however involve a discussion of Service Oriented Architectures which would lead us too far. Details can be found in [58].

lying close to interesting features (e.g., extrema, domain boundaries), or far from other data points have a higher priority. These priorities are assigned by the sample selection algorithm and dynamically managed by the input queue (cfr. Fig. 1) which is a priority queue. Consequently, the priorities are reflected in the scheduling decisions made by the SE and distributed backend (simulations with higher priority are done first). The priority queue can have different management policies. For example, a policy can be to let the priority decrease with time unless interest in a sample point is renewed. Additionally, the number of simulations to run (as requested by the modeler) is not fixed but changes dynamically. This number is determined based on average time needed for modeling, the average duration of a single simulation, and the number of compute nodes available at that point in time. In this way the underlying resources are always used optimally. Work is underway to extend this scheme so to directly integrate the SE with the grid information system used (e.g., Ganglia). Combining knowledge on real-time system load, network traffic, etc. with data point priorities would allow the SE to achieve an optimal job-host mapping (i.e., the data points with the highest priority should be scheduled on the fastest nodes).

4.8   Configuration

Each of the components available in the toolbox has its own set of parameters and options. In order for power-users to retain full flexibility these options should not be hidden but at the same time should not confuse more casual users. An extensive configuration framework based on XML is available. XML was chosen since it is the defacto (platform independent) standard format for machine readable structured data. Sensible defaults are provided but every modeling aspect can be adjusted as needed. The use of XML also allows all the different components to be combined in endless different ways. Each output of a simulation code can be modeled by separate methods, multiple outputs can be combined into one model, different active learning strategies can be used (also per output) or combined into a hierarchy, multiple experiments can be combined in different *runs* which allows for easy benchmarking and comparison of techniques, etc. For a concrete example see [22] or the SUMO implementation.

5   Examples

In this section, two different examples from different application domains will be used to demonstrate the flexibility of our approach. The first problem is a tapered transmission line that is terminated with a matched load. The third problem is related to airfoil design and models the drag given a particular airfoil geometry.

5.1 Tapered Transmission Line

Experimental Setup

The first example is a high-speed 2-port microwave structure. The goal is to model the complex electro-magnetic behavior of this system. This behavior can be characterized by the scattering parameters or S-parameters, which describe the response of an N-port system to signals at each port. These S-parameters are a function of the geometrical parameters of the system (length, substrate parameters, dielectric constants, ...) as well as the frequency.

The frequency is treated as a special parameter since, for any fixed combination of geometrical parameters, the output can be computed efficiently for a large set of different frequencies using only a single simulation. Because of this special property, the frequency dimension is sampled more densely than the other dimensions. This means that one simulation will return m samples instead of one, with m the number of frequency samples selected by the simulator.

Mathematically, this can be defined as follows. We are to model a function $f : \mathbb{R}^{d+1} \mapsto \mathbb{C}^n$, where d dimensions are sampled normally, and one additional frequency parameter is sampled automatically on a dense grid but modeled as a normal parameter. Because of this property, new samples should be selected in a lower-dimensional subspace of the design space. This poses some problems for the adaptive sampling algorithm, as it has to remove one dimension from the data points before performing selecting new samples. This is done by fixing the frequency to the maximum value, and sampling in this slice only, resulting in d-dimensional

In this particular example we model the (complex) reflection coefficient S_{11} of a lossless exponential tapered transmission line terminated with a matched load, as described in [65, 6]. This structure is displayed in Fig. 3, where Z_0 and Z_L represent the reference impedance and the load impedance respectively. The reflection coefficient S_{11} is modeled as a function of the relative dielectric constant $\epsilon_r \in [3,5]$ and the line length $L \in [1cm, 10cm]$, as well as the frequency $f \in [1kHz, 3GHz]$. The simulator, however, only accepts 2 inputs: ϵ_r and L. For each combination of these two inputs, the output for a complete set of frequency ranges (20, linearly distributed over the frequency range) is computed and returned for modeling.

This system was modeled with ANN models using a genetic algorithm to optimize the topology and the initial weights (see [61] for details). The GA was run for 10 generations between each sampling iteration with a population size of 10. The network itself was trained using Levenberg-Marquardt backpropagation in conjunction with Bayesian regularization [66] for 300 epochs. The intial experimental design is a Latin hypercube design of 9 points in the 2-dimensional geometric parameter space. For each of these points, 30 frequencies values were returned, resulting in a total of 270 initial data points. From then on, each sampling iteration selected 2 additional sample locations, resulting in 60 more data points, up to a total of 1000.

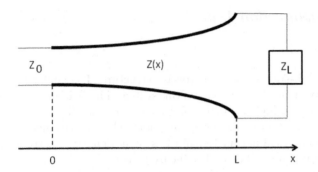

Fig. 3 An exponential tapered microstrip transmission line. $Z_0 = 50\Omega$ and $Z_L = 100\Omega$

Because the ANN implementation that is used does not support complex data directly, real and imaginary parts were modeled separately. Two different approaches were tried. In the first approach, a separate model was trained for each component of the complex output. In the second approach a single model was trained to model both components (as two separate real numbers), resulting in a considerable gain in speed at the (possible) cost of some accuracy. At the end, these two real outputs are then combined again to produce a model of the original complex output parameter.

Results

The results are summarized in Table 1. A plot of the modulus of the best model produced using separate neural networks for the real and imaginary part can be found in 4. The accuracy of the models was calculated by testing the models on a dense validation dataset of 6750 points which was not used for training. Two measures were calculated: the root relative square error and the maximum absolute error. The root relative square error is defined as:

$$RRSE = \sqrt{\frac{\sum_{i=1}^{n}(y_i - \tilde{y}_i)^2}{\sum_{i=1}^{n}(y_i - \bar{y})^2}}$$

where $y_i, \tilde{y}_i, \bar{y}$ are the true, predicted and mean true response values respectively. The maximum absolute error is defined as:

$$MAE = Max_i\|y_i - \tilde{y}_i\|$$

As expected, by training the real and imaginary part with a separate neural network, a small accuracy improvement can be obtained at the cost of a longer run time. However, both approaches performed very well, producing models with a root relative square error of less than 1% after only 570 samples. By

Table 1 The error of the models on a external validation set of 6750 points. RRSE is the root relative square error, MAE is the maximum absolute error. The results from training with separate networks are on the left, the results from training with only one neural network are on the right

	ANN Split			ANN Combined		
	Real	Imaginary	Complex	Real	Imaginary	Complex
RRSE	3.38E-3	3.19E-3	3.31E-3	4.60E-3	6.18E-3	5.19E-3
MAE	1.85E-4	1.42E-4	2.57E-4	3.00E-4	3.13E-4	4.73E-4

Fig. 4 A plot of the modulus of the output produced by training the real and imaginary part with separate neural networks

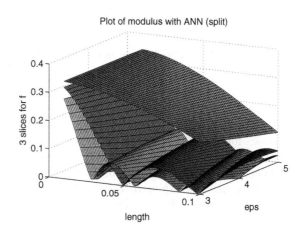

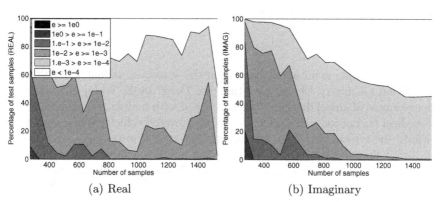

(a) Real   (b) Imaginary

Fig. 5 A plot of the evolution of the error on an external validation set as a function of the number of samples

adding another 1000 samples, the accuracy can be further improved by almost a factor 10. This can be seen in Fig. 5. The left plot shows the evolution of the error on the real part, while the right plot shows the evolution of the imaginary part. This plot represents the percentage of samples in the

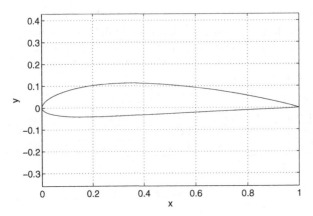

Fig. 6 The NACA 4415: A wheel fairing airfoil [68]

validation set that have an error in the a range corresponding to a particular shade of grey on the plot.

This plot shows that the error lowers steadily as the number of samples increases for the imaginary part, but oscillates slightly for the real part. When more samples are evaluated, the model might have to find new optima for its hyperparameters, thus temporarily causing a drop in accuracy while the parameter space is explored. Eventually, a new optimum is found and the accuracy increases substantially.

5.2 Airfoil Design Example

Context

The next example is an application from the aerospace industry. It is based on "Airfoil Geometry Design for Minimum Drag" by Z. Wang [67] and is a nice example of airfoil design. [67] is concerned with finding an optimal design of a wheel fairing (or wheel pant) on a solar powered vehicle. When designing airfoils, the typical goal is to optimize the lift-to-drag ratio $\frac{Lift}{Drag}$, i.e., design an airfoil that has a high lift while having not a too high drag coefficient. However for a wheel fairing for a vehicle or airplane the main aim is to have a low drag coefficient C_d. There are several published standard airfoils, for example the NACA series of the, now dissolved, National Advisory Committee for Aeronautics [68] who created series of airfoils for different purposes using analytical equations (see Fig. 6 for an example profile). Nonetheless, in many cases it is useful and more efficient to create a custom-made design.

In this use case Xfoil [69] is used to evaluate the performance of custom airfoils, while Matlab is used to construct different airfoil geometries searching for the optimum design, and calculating the objective function. The original optimization problem is defined as:

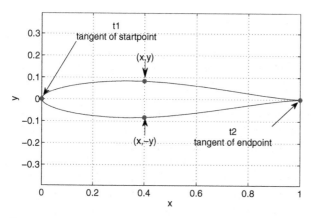

Fig. 7 Optimal airfoil geometry

$$min_{Airfoil\,Geometry} Drag(Airfoil\,Geometry)$$

subject to
$$Desired\,Thickness - Thickness(Airfoil\,Geometry) \leq 0$$

where the actual objective is constructed as $Drag(AirFoil\,Geometry) = C_d^0 + C_d^3 - k\,min(pressure^3)$. C_d^i denotes the drag coefficient at degree of attack i, $min(pressure^3)$ is the minimum pressure at degree of attack 3 and k is a weighting constant set to 0.00200. The original goal of [67] was to design a wheel fairing airfoil that would perform better than a NACA four digit airfoil. The following Xfoil options were used:

- Viscous mode is on
- Reynold number is 10^6
- Maximum number of iterations is 50 (viscous-solution iteration limit)
- There is auto point accumulation to active polar

Technically it works as follows. Initially, an airfoil geometry is generated with the spline toolbox of Matlab (using 4 control points). A discretization of this spline is saved to a file. Subsequently, this file, along with xfoil instructions, is fed into xfoil which simulates wind flow (computational fluid dynamics) and returns several performance metrics saved in a file. This file is easily read into Matlab which is able to combine the drag coefficient and maximum pressure into the aggregated objective mentioned above.

Summarizing the whole setup, there are 4 inputs (x, y, t_1, t_2) and 1 output $Drag$. The first 2 input parameters implicitly define 2 control points with coordinates (x, y) and $(x, -y)$. The other two control points are endpoints of the airfoil and are fixed at $(0, 0)$ and $(0, 1)$ respectively. The last 2 parameters are the tangents of these endpoints. This is illustrated in Fig. 7.

Experimental Setup

However, in some cases knowing the optimum is not sufficient. Instead global information (relationships between parameters, sensitivities, ...) is needed. Therefore it is useful to apply global surrogate modeling to this problem. Using the surrogate model a designer is able to explore the design domain fast and efficiently, directly locating interesting designs, and gaining more insight into the behavior of the system.

The airfoil geometry problem has been modeled with the SUMO toolbox using the same XFoil setup and objective function as [67]. The additional SUMO toolbox configuration was chosen as follows: An inital set of 38 samples were generated (20 by a Latin Hypercube design and 18 corner points). New samples are adaptively chosen by the gradient sample selection method (up to a maximum of 500) and an ANN model (using the same settings as the previous example) is used to approximate the data. The hyperparameter optimization is driven by the RRSE on a max-min validation set of 20% which grows adaptively as more samples become available. Outliers caused by failed simulations (complete failure or wrong results due to failed convergence of the solver) were removed during sampling.

Results

Figure 8 shows the plot of the final ANN model. The final model is a 4-14-2-1 network with a RRSE of 1% on the validation set and (when the final model is trained on all the data) a 10-fold cross validation error of 0.17%. Thus the attained accuracy is very high. This is understandable, as Fig. 8 shows, the response is very smooth, almost parabolic, and thus easy to fit (this was further confirmed by using the SUMO model browser GUI). The model also shows the impact of the tangent t_1, t_2 parameters to be small. The three slices for t_1 almost coincide, so we can safely say that the tangent at the start point of the airfoil has virually no effect on the combined drag. Similarly, the second tangent parameter t_2 only influences the drag for 'thick' airfoils (large values of x).

Besides a global model, the optimum is still of interest. Therefore, as a post-processing, the model was optimized (this can now be done very cheaply) using the DIviding REctangles (DIRECT) [70] algorithm. This resulted in a *Drag* value of 0.0166 at $x = 0.4, y = 0.0833, t_1 = 0.9167, t_2 = -0.05$. A plot of the airfoil geometry at this optimal point is shown in Fig. 7.

6 Conclusion and Discussion

In this work we presented an extensible, adaptive approach to global surrogate modeling. As such, this work brings together insights from distributed

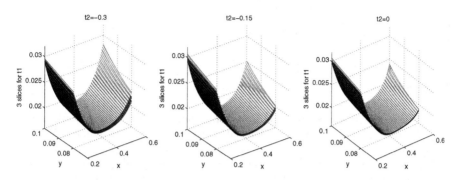

Fig. 8 Final 4-14-2-1 ANN model of $Drag$ (t_1, t_2 are plotted at 3 fixed values: min, max and $\frac{min+max}{2}$)

systems, modeling & simulation, and computational intelligence. Good performance has been demonstrated on two complex, real world problems.

In addition, a strong point of this contribution is that all the algorithms and examples described (including many others) are readily available in the form of the platform independent SUMO Toolbox that is available for download at http://www.sumo.intec.ugent.be. The SUMO Toolbox can be used as a standalone modeling tool or it can be integrated as a modeling backend in existing tools such as DAKOTA and VisualDOC. There is no inherent bias towards a specific application or field, thus the approach is useful whenever a cheap, accurate approximation for a complex system or dataset is needed.

Future efforts will involve work on a divide and conquer strategy with automatic domain decomposition (taking ideas from [17, 12]), more extensive use of hybrid models (ensembles), extend constraint support to other components so that non-rectangular domains are better handled, inclusion of standard methods for feature selection, and continuing work on a higher level of adaptivity (e.g., dynamically changing the model selection metric depending on the data distribution). For the latter the use of Fuzzy theory is also under investigation. In addition work is underway to further improve the scheduling process when running simulations on a cluster or grid (tighter integration with the grid resource information system), and reducing the barrier of entry for users by focusing on ease of use and easy integration (through the use of Jini and/or web services) into the engineering design process.

Of course, we will also continue to work with engineers and industrial partners to fine tune the existing algorithms, add new plugins (e.g., support advanced regression techniques such as projection pursuit and recursive partitioning), and perform extensive benchmarking on new and challenging problems.

Acknowledgements. The authors would like to thank Adam Lamecki from the Technical University of Gdansk, Poland, for providing the simulation code for the electromagnetic example. This research was supported by the Fund for Scientific Research (FWO) Flanders.

References

1. Wang, G.G., Shan, S.: Review of metamodeling techniques in support of engineering design optimization. Journal of Mechanical Design 129(4), 370–380 (2007)
2. Gu, L.: A comparison of polynomial based regression models in vehicle safety analysis. In: Diaz, A. (ed.) 2001 ASME Design Automation Conference, ASME, Pittsburgh, PA (2001)
3. Giunta, A., McFarland, J., Swiler, L., Eldred, M.: The promise and peril of uncertainty quantification using response surface approximations. Structure & Infrastructure Engineering 2, 175–189 (2006)
4. Gorissen, D., Crombecq, K., Hendrickx, W., Dhaene, T.: Grid enabled metamodeling. In: Daydé, M., Palma, J.M.L.M., Coutinho, Á.L.G.A., Pacitti, E., Lopes, J.C. (eds.) VECPAR 2006. LNCS, vol. 4395. Springer, Heidelberg (2007)
5. Hendrickx, W., Dhaene, T.: Sequential design and rational metamodelling. In: Kuhl, M., Steiger, N.M., Armstrong, F.B., Joines, J.A. (eds.) Proceedings of the 2005 Winter Simulation Conference, pp. 290–298 (2005)
6. Deschrijver, D., Dhaene, T., Zutter, D.D.: Robust parametric macromodeling using multivariate orthonormal vector fitting. IEEE Transactions on Microwave Theory and Techniques 56(7), 1661–1667 (2008)
7. Simpson, T.W., Poplinski, J.D., Koch, P.N., Allen, J.K.: Metamodels for computer-based engineering design: Survey and recommendations. Eng. Comput (Lond.) 17(2), 129–150 (2001)
8. Balewski, L., Mrozowski, M.: Creating neural models using an adaptive algorithm for optimal size of neural network and training set. In: 15th International Conference on Microwaves, Radar and Wireless Communications, MIKON 2004, Conference proceedings of MIKON 2004, vol. 2, pp. 543–546 (2004)
9. Booker, A.J., Dennis, J.E., Frank, P.D., Serafini, D.B., Torczon, V., Trosset, M.W.: A rigorous framework for optimization of expensive functions by surrogate. Structural and Multidisciplinary Optimization 17(1), 1–13 (1999)
10. Eldred, M.S., Dunlavy, D.M.: Formulations for surrogate-based optimization wiht data fit, multifidelity, and reduced-order models. In: 11th AIAA/ISSMO Multidisciplinary Analysis and Optimization Conference, Protsmouth, Virginia (2006)
11. Anguita, D., Ridella, S., Rivieccio, F., Zunino, R.: Automatic hyperparameter tuning for support vector machines. In: Dorronsoro, J.R. (ed.) ICANN 2002. LNCS, vol. 2415, pp. 1345–1350. Springer, Heidelberg (2002)
12. Sanchez, E., Pintos, S., Queipo, N.: Toward an optimal ensemble of kernel-based approximations with engineering applications. In: Proceedings of the International Joint Conference on Neural Networks, 2006. IJCNN 2006, pp. 2152–2158 (2006)
13. Turner, C.J., Crawford, R.H., Campbell, M.I.: Multidimensional sequential sampling for nurbs-based metamodel development. Eng. with Comput. 23(3), 155–174 (2007)

14. Sugiyama, M., Ogawa, H.: Release from active learning/model selection dilemma: optimizing sample points and models at the same time. In: Neural Networks, 2002. IJCNN 2002. Proceedings of the 2002 International Joint Conference on Neural Networks, 2002, vol. 3, pp. 2917–2922 (2002)
15. Lin, Y.: An Efficient Robust Concept Exploration Method and Sequential Exploratory Experimental Design. PhD thesis, Georgia Institute of Technology (2004)
16. Busby, D., Farmer, C.L., Iske, A.: Hierarchical nonlinear approximation for experimental design and statistical data fitting. SIAM Journal on Scientific Computing 29(1), 49–69 (2007)
17. Gramacy, R.B., Lee, H.K.H., Macready, W.G.: Parameter space exploration with gaussian process trees. In: ICML 2004: Proceedings of the twenty-first international conference on Machine learning, p. 45. ACM Press, New York (2004)
18. Farhang-Mehr, A., Azarm, S.: Bayesian meta-modelling of engineering design simulations: a sequential approach with adaptation to irregularities in the response behaviour. International Journal for Numerical Methods in Engineering 62(15), 2104–2126 (2005)
19. Devabhaktuni, V., Chattaraj, B., Yagoub, M., Zhang, Q.J.: Advanced microwave modeling framework exploiting automatic model generation, knowledge neural networks, and space mapping. IEEE Tran. on Microwave Theory and Techniques 51(7), 1822–1833 (2003)
20. Ganser, M., Grossenbacher, K., Schutz, M., Willmes, L., Back, T.: Simulation meta-models in the early phases of the product development process. In: Proceedings of Efficient Methods for Robust Design and Optimization (EUROMECH 2007) (2007)
21. Clarke, S.M., Griebsch, J.H., Simpson, T.W.: Analysis of support vector regression for approximation of complex engineering analyses. In: Proceedings of the 29th Design Automation Conference (ASME Design Engineering Technical Conferences) (DAC/DETC 2003) (2003)
22. Gorissen, D., Crombecq, K., Couckuyt, I., Dhaene, T.: Automatic approximation of expensive functions with active learning. Technical Report TR-10-08, University of Antwerp, Middelheimlaan 1, 2020 Antwerp, Belgium (2008)
23. Goel, T., Haftka, R., Shyy, W.: Comparing error estimation measures for polynomial and kriging approximation of noise-free functions. Journal of Structural and Multidisciplinary Optimization (published online) (2008)
24. Ong, Y.S., Nair, P.B., Keane, A.J., Wong, K.W.: Surrogate-Assisted Evolutionary Optimization Frameworks for High-Fidelity Engineering Design Problems, Knowledge Incorporation in Evolutionary Computation. Studies in Fuzziness and Soft Computing Series, pp. 307–331. Springer, Heidelberg (2004)
25. Giunta, A., Eldred, M.: Implementation of a trust region model management strategy in the DAKOTA optimization toolkit. In: Proceedings of the 8th AIAA/USAF/NASA/ISSMO Symposium on Multidisciplinary Analysis and Optimization, Long Beach, CA (2000)
26. Jones, D.R., Schonlau, M., Welch, W.J.: Efficient global optimization of expensive black-box functions. J. of Global Optimization 13(4), 455–492 (1998)
27. Sasena, M.J., Papalambros, P.Y., Goovaerts, P.: Metamodeling sampling criteria in a global optimization framework. In: 8th AIAA/ USAF/NASA/ISSMO Symposium on Multidisciplinary Analysis and Optimization, Long Beach, CA, AIAA Paper, 2000–4921 (2000)

28. Parmee, I., Abraham, J., Shackelford, M., Rana, O.F., Shaikhali, A.: Towards autonomous evolutionary design systems via grid-based technologies. In: Proceedings of ASCE Computing in Civil Engineering, Cancun, Mexico (2005)

29. Eldred, M., Outka, D., Fulcher, C., Bohnhoff, W.: Optimization of complex mechanics simulations with object-oriented software design. In: Proceedings of the 36th IAA/ASME/ASCE/AHS/ASC Structures, Structural Dynamics, and Materials Conference, New Orleans, LA, pp. 2406–2415 (1995)

30. Zhang, Q.J., Gupta, K.C.: Neural Networks for RF and Microwave Design (Book + Neuromodeler Disk). Artech House, Inc., Norwood (2000)

31. Gano, S., Kim, H., Brown, D.: Comparison of three surrogate modeling techniques: Datascape, kriging, and second order regression. In: Proceedings of the 11th AIAA/ISSMO Multidisciplinary Analysis and Optimization Conference, AIAA-2006-7048, Portsmouth, Virginia (2006)

32. Crombecq, K.: A gradient based approach to adaptive metamodeling. Technical report, University of Antwerp (2007)

33. De Geest, J., Dhaene, T., Faché, N., De Zutter, D.: Adaptive CAD-model building algorithm for general planar microwave structures. IEEE Transactions on Microwave Theory and Techniques 47(9), 1801–1809 (1999)

34. Barton, R.R.: Design of experiments for fitting subsystem metamodels. In: WSC 1997: Proceedings of the 29th conference on Winter simulation, pp. 303–310. ACM Press, New York (1997)

35. Forrester, A.I.J., Bressloff, N.W., Keane, A.J.: Optimization using surrogate models and partially converged computational fluid dynamics simulations. Proceedings of the Royal Society 462, 2177–2204 (2006)

36. Yesilyurt, S., Ghaddar, C.K., Cruz, M.E., Patera, A.T.: Bayesian-validated surrogates for noisy computer simulations; application to random media. SIAM Journal on Scientific Computing 17(4), 973–992 (1996)

37. Chaveesuk, R., Smith, A.: Economic valuation of capital projects using neural network metamodels. The Engineering Economist 48(1), 1–30 (2003)

38. Knight, D., Kohn, J., Rasheed, K., Weber, N., Kholodovych, V., Welsh, W., Smith, J.: Using surrogate modeling in the prediction of fibrinogen adsorption onto polymer surfaces. Journal of chemical information and computer science 55, 1088–1097 (2004)

39. Hidovic, D., Rowe, J.: Validating a model of colon colouration using an evolution strategy with adaptive approximations. In: Deb, K., et al. (eds.) GECCO 2004. LNCS, vol. 3103, pp. 1005–1016. Springer, Heidelberg (2004)

40. Brown, M., Adams, S., Dunlavy, B., Gay, D., Swiler, D., Giunta, L., Hart, A., Watson, W., Eddy, J.P., Griffin, J., Hough, J., Kolda, P., Martinez-Canales, T., Eldred, M., Williams, P.: Dakota, a multilevel parallel object-oriented framework for design optimization, parameter estimation, uncertainty quantification, and sensitivity analysis: Version 4.1 users manual. Technical Report SAND2006-6337, Sandia Labs (2007)

41. Kleijnen, J.P., Sanchez, S.M., Lucas, T.W., Cioppa, T.M.: State-of-the-art review: A user's guide to the brave new world of designing simulation experiments. INFORMS Journal on Computing 17(3), 263–289 (2005)

42. Ding, M., Vemur, R.: An active learning scheme using support vector machines for analog circuit feasibility classification. In: 18th International Conference on VLSI Design, pp. 528–534 (2005)

43. Devabhaktuni, V.K., Zhang, Q.J.: Neural network training-driven adaptive sampling algorithm. In: Proceedings of 30th European Microwave Conference, Paris, France, vol. 3, pp. 222–225 (2000)
44. Robertazzi, T.G., Schwartz, S.C.: An accelerated sequential algorithm for producing d-optimal designs. SIAM Journal on scientific Computing 10, 341–358 (1989)
45. Keys, A.C., Rees, L.P.: A sequential-design metamodeling strategy for simulation optimization. Comput. Oper. Res. 31(11), 1911–1932 (2004)
46. Sasena, M.: Flexibility and Efficiency Enhancements For Constrainted Global Design Optimization with Kriging Approximations. PhD thesis, University of Michigan (2002)
47. Jin, R., Chen, W., Sudjianto, A.: On sequential sampling for global metamodeling in engineering design, detc-dac34092. In: ASME Design Automation Conference, Montreal, Canada (2002)
48. Lehmensiek, R., Meyer, P., Muller, M.: Adaptive sampling applied to multivariate, multiple output rational interpolation models with applications to microwave circuits. International Journal of RF and microwave computer aided engineering 12(4), 332–340 (2002)
49. Deschrijver, D., Dhaene, T.: Rational modeling of spectral data using orthonormal vector fitting. In: Proceedings of 9th IEEE Workshop on Signal Propagation on Interconnects, 2005, pp. 111–114 (2005)
50. Kingston, G., Maier, H., Lambert, M.: Calibration and validation of neural networks to ensure physically plausible hydrological modeling. Journal of Hydrology 314, 158–176 (2005)
51. Srinivasulu, S., Jain, A.: A comparative analysis of training methods for artificial neural network rainfall-runoff models. Appl. Soft Comput. 6(3), 295–306 (2006)
52. Wolpert, D.: The supervised learning no-free-lunch theorems. In: Proceedings of the 6th Online World Conference on Soft Computing in Industrial Applications (2001)
53. Triverio, P., Grivet-Talocia, S., Nakhla, M., Canavero, F.G., Achar, R.: Stability, causality, and passivity in electrical interconnect models. IEEE Transactions on Advanced Packaging 30(4), 795–808 (2007)
54. Ihme, M., Marsden, A.L., Pitsch, H.: Generation of optimal artificial neural networks using a pattern search algorithm: Application to approximation of chemical systems. Neural Computation 20, 573–601 (2008)
55. Lessmann, S., Stahlbock, R., Crone, S.: Genetic algorithms for support vector machine model selection. In: International Joint Conference on Neural Networks, 2006. IJCNN 2006, pp. 3063–3069 (2006)
56. JiGuan, G.L.: Modeling test responses by multivariable polynomials of higher degrees. SIAM Journal on Scientific Computing 28(3), 832–867 (2006)
57. Ye, K., Li, W., Sudjianto, A.: Algorithmic construction of optimal symmetric latin hypercube designs. Journal of Statistical Planning and Inference 90, 145–159 (2000)
58. Gorissen, D., Dhaene, T., Demeester, P., Broeckhove, J.: Grid enabled surrogate modeling. In: The Encyclopedia of Grid Computing Technologies and Applications (in press) (2008)
59. Deb, K., Pratap, A., Agarwal, S., Meyarivan, T.: A fast and elitist multiobjective genetic algorithm: Nsga-ii. IEEE Transactions on Evolutionary Computation 6(2), 182–197 (2002)

60. Vrugt, J.A., Gupta, H.V., Bouten, W., Sorooshian, S.: A shuffled complex evolution metropolis algorithm for optimization and uncertainty assessment of hydrologic model parameters. Water Resources Research 39(8), 1214–1233 (2003)

61. Gorissen, D.: Heterogeneous evolution of surrogate models. Master's thesis, Master of AI, Katholieke Universiteit Leuven, KUL (2007)

62. Gorissen, D., Tommasi, L.D., Croon, J., Dhaene, T.: Automatic model type selection with heterogeneous evolution: An application to rf circuit block modeling. In: Proceedings of the IEEE Congress on Evolutionary Computation, WCCI 2008, Hong Kong (2008)

63. Lim, D., Ong, Y.S., Jin, Y., Sendhoff, B.: A study on metamodeling techniques, ensembles, and multi-surrogates in evolutionary computation. In: GECCO 2007: Proceedings of the 9th annual conference on Genetic and evolutionary computation, pp. 1288–1295. ACM, New York (2007)

64. Goel, T., Haftka, R., Shyy, W., Queipo, N.: Ensemble of surrogates. Structural and Multidisciplinary Optimization 33, 199–216 (2007)

65. Pozar, D.M.: Microwave Engineering, 2nd edn. John Wiley and Sons, Chichester (1998)

66. Foresee, F., Hagan, M.: Gauss-newton approximation to bayesian regularization. In: Proceedings of the 1997 International Joint Conference on Neural Networks, pp. 1930–1935 (1997)

67. Wang, Z.: Airfoil geometry design for minimum drag. Technical Report AAE 550, Purdue University (2005)

68. UIUC Airfoil Coordinates Database (2008),
    `http://www.ae.uiuc.edu/m-selig/ads/coord_database.html`

69. Design and analysis of subsonic isolated airfoils (2008),
    `http://web.mit.edu/drela/public/web/xfoil/`

70. Finkel, D.E., Kelley, C.T.: Additive scaling and the direct algorithm. J. of Global Optimization 36(4), 597–608 (2006)

New Multi-Objective Algorithms for Neural Network Training Applied to Genomic Classification Data

Marcelo Costa, Thiago Rodrigues, Euler Horta, Antônio Braga, Carmen Pataro, René Natowicz, Roberto Incitti, Roman Rouzier, and Arben Çela

Abstract. One of the most used Artificial Neural Networks models is the Multi-Layer Perceptron, which is capable to fit any function as long as they have enough number of neurons and network layers. The process of obtaining a properly trained Artificial Neural Network usually requires a great effort in determining the parameters that will make it to learn. Currently there are a variety of algorithms for Artificial Neural Networks's training working, simply, in order to minimize the sum of mean square error. However, even if the network reaches the global minimum error, it does not imply that the model response is optimal. Basically, a network with large number of weights but with small amplitudes behaves as an underfitted model that gradually overfits data during training. Solutions that have been overfitting are unnecessary complexity solutions. Moreover, solutions with low norm of the weights are those that present underfitting, with low complexity. The

Marcelo Costa
Universidade Federal de Minas Gerais, Depto. de Estatística, Brazil
e-mail: azevedo@est.ufmg.br

Thiago Rodrigues
Universidade Federal de Lavras, Depto. Ciência da Computação, Brazil
e-mail: thiago@dcc.ufla.br

Euler Horta, Antônio Braga, and Carmen Pataro
Universidade Federal de Minas Gerais, Depto. Engenharia Eletrônica, Brazil
e-mail: {apbraga,eulerhorta,cdmp}@cpdee.ufmg.br

René Natowicz and Arben Çela
Université Paris-Est, ESIEE-Paris, France
e-mail: {r.natowicz,a.cela}@esiee.fr

Roberto Incitti
Institut Mondor de Médecine Moléculaire, Créteil, France
e-mail: roberto.incitti@inserm.fr

Roman Rouzier
Hôpital Tenon, department of Gynecology, Paris, France
e-mail: roman.rouzier@tnn.aphp.fr

A.-E. Hassanien et al. (Eds.): Foundations of Comput. Intel. Vol. 1, SCI 201, pp. 63–82.
springerlink.com
© Springer-Verlag Berlin Heidelberg 2009

Multi-Objective Algorithm controls the weights amplitude by optimizing two objective functions: the error function and norm function. The high generalization capability of the Multi-Objective Algorithm and an automatic weight selection is aggregated by the LASSO approach, which generates networks with reduced number of weights when compared with Multi-Objective Algorithm solutions. Four data sets were chosen in order to compare and evaluate MOBJ, LASSO and Early-Stopping solutions. One generated from a function and tree available from a Machine Learning Repository. Additionally, the MOBJ and LASSO algorithms are applied to a microarray data set, which samples correspond to a genetic expression profile from DNA microarray technology of neoadjuvant chemotherapy (treatment given prior to surgery) for patients with breast cancer. Originally, the dataset is composed of 133 samples with 22283 attributes. By applying e probe section method described in the literature, 30 attributes were selected and used to train the Artificial Neural Networks. In average, the MOBJ and LASSO solutions were the same, the main difference is the simplified topology achieve by LASSO training method.

1   Introduction

Artificial Neural Networks (ANN) are data driven models capable to fit any function as long as they have enough number of neurons and network layers. One of the most used neural network model is the Multi-Layer Perceptron (MLP) [18]. A MLP with one hidden layer with sigmoid activation function and linear output is capable to approximate any continuous function [19]. However, in practical situations, the data is corrupted with noise and the ANN learn the true relationship among the input and output patterns minimizing the noise effect.

The process of obtaining a properly trained Artificial Neural Network is not a simple task and usually requires a great effort in determining the parameters that will make it to learn a given task

A network properly trained do not only responds adequately to the samples used in the training process but also to others samples which was not presented to the artificial neural network. This ability is called generalization.

Several factors may influence the ANN's ability of generalization such as the selection of optimal parameters for the process of training and the selection of the appropriate topology. The choice of the topology of a Multilayer Perceptron (MLP) neural network [18] is a task of the designer. It consists, basically, in determining the number of layers of the network and the number of neurons in each layer. The process of training consists on finding a suitable solution in a large universe of possible solutions and the choice between these solutions must be made on completion of a specific set of training [17].

Currently there are a variety of algorithms for ANN's training working, simply, in order to minimize the sum of mean square error. As the case of Backpropagation [18] and its variations, such as Quickprop [16], Rprop [15], Levenberg Marquardt [14], and sliding model training [13].

In gradient descent learning the algorithm approximates locally the error surface by calculating the first derivative of the error. However, the algorithm does not guarantee convergence to the global minimum of error. Moreover, even if the network reaches the global minimum error, it does not imply that the model response is optimal. An optimal solution represents a model whose response is very close to the function that generated the data. This model has the ability to provide reliable outputs for patterns which were not presented to the networks in the training process. For neural networks, the number of weights and their magnitudes are related to its fittness to the data [21].

Basically, a network with large number of weights but with small amplitudes behaves as an underfitted model that gradually overfits data during training. Based on this principle, the early-stopping algorithm [22] splits the data set into training and validation set. The training set is used to calculate the weights update, the validation set is used as a goodness of fit measure in the meanwhile. The algorithm stops when the validation error starts to increase. The weight-decay algorithm [23] adds a penalty term to the error cost function that confines the norm of the weights to certain proportions. This procedure smoothes the model output but it is also sensitive to initial conditions. Pruning algorithms [24] aim at extracting weights after training process as a way to reduce the network complexity. In general, those algorithms try to find an unknown condition for the weight's amplitude that improves the network generalization capability [21].

Solutions that have been overfitting are unnecessary complexity solutions. Moreover, solutions with low norm of the weights are those that present underfitting, with low complexity. Furthermore, considering the norm of the weights as an ANN's measure of complexity, the idea is to use this measure as a function to be minimized in conjunction with the mean square error. So you can formalize the problem of multi-objective optimization which considers two goals to be minimized.

The Multi-Objective Algorithm (MOBJ) [25] controls the weights amplitude by optimizing two objective functions: the error function and norm function. The algorithm is able to reach a high generalization solution and avoid over and undefitting. A comparison with standard algorithms can be found in [25]. The LASSO approach aggregates the MOBJ high generalization capability and an automatic weight selection through the specification of an alternative objective function: the sum of the absolute weights. Therefore, it generates networks with reduced number of weights when compared with MOBJ solutions but with similar prediction performance.

2   The Multi-Objective Algorithm

The original Multi-Objective algorithm (MOBJ) [25] optimizes the error and norm cost functions. Both functions have different minima which means that a solution with both minimum error and norm functions is non-feasible. The Pareto set represents the set of solutions between two extreme solutions: the solution with minimum error (and large norm) and the solution with minimum norm (and large error). Figure 1 shows the Pareto set shape.

The Pareto is obtained by first obtaining its two extremes, which is formed by underfitted and overfitted solutions, represented, in Figure 1, by the axes x and y respectively. The minimum mean square error solution, overfitted solution, is obtained by training a network with a standard training algorithm, such as Backpropagation [18]. The minimum norm solution, underfitted solution, is obtained simply making all network weights equal to zero. Intermediate solutions obtained by the Multi-Objective algorithm are called Pareto-optimal (Figure 1). Solutions belong to the Pareto-optimal set cannot be improved considering both objective functions simultaneously. This is in fact the definition of the Pareto-optimal set, which is used here to obtain a good compromise between the two conflicting objectives, error and norm.

The Figure 1 represents the boundary between feasible solutions and non-feasible region. Moreover, the Pareto set represents solutions with minimum error subject to specific norm constraints. Based on this concept the overall multi-objective optimization problem can then be written as a constrained mono-objective function:

$$\mathbf{w}^* = arg \ min \ \frac{1}{N} \sum_{j=1}^{N} \left(d_j - y(\mathbf{w}, \mathbf{x}_j) \right)^2$$

$$subject\ to: \ \|\mathbf{w}\| \le \eta_i \tag{1}$$

where $\{\mathbf{x}_j, d_j\}_{j=1}^{N}$ are the input-output data set, N is the sample size, $\|\mathbf{w}\| = \sqrt{\sum_t w_t^2}$ and η_j is the norm constraint value. For each constraint value η_j, there is one solution associated within the Pareto set.

The multi-objective problem can be solved by a constrained optimization method as the *ellipsoidal algorithm* [26]. Alternative multi-objective algorithms include Sliding Mode Multi-Objective algorithm [27], Levenberg-Marquardt based Multi-Objective [28] and Multi-Objective Evolutionary Algorithms [29, 30].

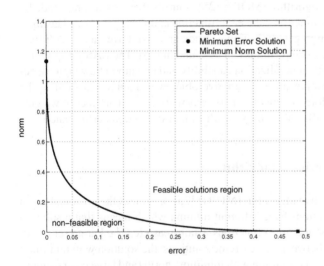

Fig. 1 Pareto Set and the two objective functions minima

3   The LASSO Method

The *least absolute shrinkage and selection operator* (LASSO) [32] minimizes the residual sum of squares (error) subject to the sum of the absolute weights being less than a constant t.

$$\mathbf{w}^* = arg \ min \ \tfrac{1}{N} \sum_{j=1}^{N} \left(d_j - y(\mathbf{w}, \mathbf{x}_j)\right)^2$$

$$subject \ to : \ \sum_i |w_i| \le t \tag{2}$$

The method is very similar to the norm constrain approach with subtle but important differences. To illustrate the differences between the MOBJ and LASSO methods, we present a single perceptron with hyperbolic tangent activation function, one input and two weights: the input weight w and the bias weight, b. The perceptron's output equation is: $y(x_i) = tanh(w.x_i + b)$. The following patterns: $(x_i, y_i) = \{(-3, -0.4), (2, -0.9)\}$ define the training set. The error surface is shown in Figure 2.

A perceptron with linear activation function has an elliptical error surface centered at the full least squares estimates. However, the non-linear activation function turns the surface irregular but with a distinct minimum at $w_o = -0.207$ and $b_o = -1.045$. Both norm and sum of the absolute weights functions have their minimum at the origin ($w = 0$, $b = 0$). The associated Pareto sets for norm and absolute weights functions are sets of solutions that start from origin and end at the minimum error point. To compare the solutions conditioned to the previous constraints, the error contours as well as the norm and the LASSO contours are shown in Figure 3. The constraint region for the norm is the disk $w^2 + b^2 \le \eta$ while that for LASSO is the diamond $|w| + |b| \le t$. Both methods find the first point where the error contours hit the constraint regions which represents a solution with minimum error conditioned to the respective constraint. Unlike the disk, the diamond has corners, if the solution

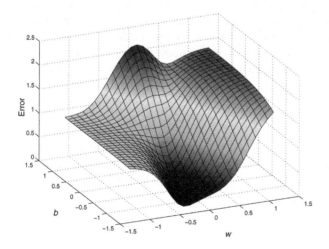

Fig. 2 Perceptron's Error Surface for the training set: $(x_i, y_i) = \{(-3, -0.4), (2, -0.9)\}$

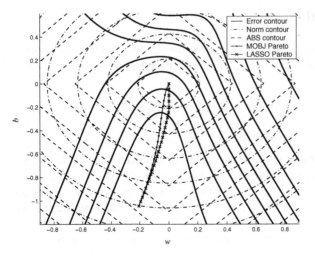

Fig. 3 Error, Norm and LASSO contours

occurs at a corner then it has one weight equal to zero. When the number of weights is larger than 2, the diamond becomes a rhomboid, and has many corners, flat edges and faces, there are many more opportunities for the estimated weights to be zero[31].

Although MOBJ and LASSO methods have common solutions the path between origin and minimum error are quite different. Figure 3 shows that the norm solutions are non-zero for any constraint except at the origin. The LASSO approach have a sub-set of solutions where w is null. As the solutions reach the minimum error, they become closer. The main advantages of the LASSO constraint are the capability to control generalization just as the norm constraint does but with an automatic weight selection procedure.

For multi-layer perceptron (MLP) neural networks as long as the topology is larger than the unknown optimal number of hidden nodes, the LASSO approach can reduce or eliminate some weights and network inputs on the final solution. This may result on a reduced topology and improved network performance. Moreover, this information can be used as a data underlying function complexity measure.

4   Methodology

The LASSO method for neural networks can be implemented with the *ellipsoidal algorithm* [26], the same algorithm used by MOBJ [25] to search the norm constrained solutions. First, the multi-objective problem can be written as a constrained mono-objective optimization:

$$\mathbf{w}^* = arg \min_{w} J(w)$$

$$subject\,to: \; g(w) \leq 0$$

(3)

where $w \in \Re^p$, $J(w) = \sum_{j=1}^{N} (y_j - y(\mathbf{w}, \mathbf{x}_j))^2$, $g(w) = \left(\sum_j |w_j| \right) - \varepsilon_i$, ε_i is the limit of equality for the sum of the absolute weights' function $\left(\sum_j |w_j| \leq \varepsilon \right)$ and $y(x_i, w)$ is the neural network response.

The conventional ellipsoidal method can be written as:

$$m(w) = \begin{cases} \nabla g(w), & \text{if } g(w) > 0 \\ \nabla J(w), & \text{if } g(w) \leq 0 \end{cases} \qquad (4)$$

where $\nabla(.)$ means the gradient or any subgradient of the argument. The weight update equation is:

$$w_{k+1} = w_k - \beta_1 \frac{Q_k m_k}{\left(m_k^T Q_k m_k \right)^{1/2}} \qquad (5)$$

where:

$$Q_{k+1} = \beta_2 \frac{Q_k - \beta_3 (Q_k m_k)(Q_k m_k)^T}{m_k^T Q_k m_k} \qquad (6)$$

where w_0 is a random initial vector and Q_0 is a symetric positive matrix, in general, the identity matrix, $\beta_1 = \frac{1}{N+1}$, $\beta_2 = \frac{N^2}{N^2-1}$ and $\beta_3 = \frac{2}{N+1}$.

A subgroup of efficient solutions can be generated from a sequence of normally equidistant constraints $\varepsilon_1, \varepsilon_2, ..., \varepsilon_{max}$ such that, $\varepsilon_1 < \varepsilon_2 < ... < \varepsilon_{max}$, where for each constraint, the ellipsoidal method is used to generate a single solution being the final solution selected as the one that has the least validation error.

The algorithm needs the gradient vector of the error and the sum of absolute weights. The following equations were used to calculate the absolute weight function derivative:

$$|w_j| = \begin{cases} +w_j, & \text{if } w_j \geq 0 \\ -w_j, & \text{if } w_j < 0 \end{cases} \qquad (7)$$

$$\frac{\partial |w_j|}{\partial w_j} = \begin{cases} +1, & \text{if } w_j \geq 0 \\ -1, & \text{if } w_j < 0 \end{cases} \qquad (8)$$

As described in Equation 8 the absolute weight derivative is discontinuous at the origin. In addition to this, the ellipsoidal algorithm performs an iterative search. Consequently, the generated solutions will not achieve some exactly null weights but it will generate some weights with very small amplitudes, closer to zero. In order to detect and eliminate those weights we use a very simple procedure that was previous proposed to simplify MOBJ solutions [33]. We randomly select weights in the final neural network solution, if the validation error decreases when the weight is set to zero then it is definitely extracted, otherwise it is restored and a different weight is randomly selected. The procedure is repeated until every weight in the network has been tested. This approach provides a fare comparison of the effective number of weights between MOBJ and LASSO solutions.

Four data sets were chosen in order to compare and evaluate MOBJ, LASSO and Early-Stopping solutions. The algorithms were set with the same initial random weights and the training stop was based on validation criterion. The Data sets were divided into training, validation and test sets with 50%, 25% and 25%, respectively and 30 simulations were performed for each algorithm. For each simulation the data sets were randomly divided and at the end of the training process the networks were pruned with the random selection procedure.

The first data set was generated from the following function:

$$f(x) = \frac{(x-2)(2x+1)}{1+x^2} + \epsilon \tag{9}$$

where x were generated form a uniform distribution within the range $[-8; 12]$ and ϵ is a random variable normally distributed with 0 mean and standard deviation $\sigma = 0.2$. The data set consists of 200 samples. The neural network topology used has one input, one output, 15 hidden nodes enumerating 46 weights. The number of nodes was chosen empirically.

The next data sets used are available at Machine Learning Repository[1]. The Boston data set [34] represents a regression problem where the output is the median value of owner-occupied homes in $1000's and the inputs are 13 variables related to physical characteristics of the house and neighborhood quality. It has 506 instances. The cancer data set consists of 9 input attributes, 2 possible outputs and 699 instances. The inputs are numerical attributes of cancer cells. The outputs associate each input to one of the two following classes: benign or malignant (tumor). The card data set is concerned with credit card applications, with 51 input attributes, 2 possible outputs and 690 instances. The inputs correspond to a mix of continuous and nominal attributes and the outputs associates the acceptance or not for personal credit. Both Cancer and Card data sets represent classification problems. Similarly to the function regression problem, all networks were created with 15 hidden nodes. The respective numbers of weights in the networks for each problem are: 226 for Boston, 151 for Cancer and 781 for Card.

5 Results

The main difference between MOBJ and LASSO constrained solutions is demonstrated in Figure 4 for the Boston data set In this case, no pruning was performed. The MOBJ approach reduces the norm of the weights and consequently theirs amplitudes but it does not aggregate the functionality of selecting weights as demonstrated by LASSO. As described in Section 3, the LASSO solutions have many more opportunities for the estimated weights to be zero.

Tables 1, 2 and 3 present some descriptive statistics (Stat.): mean (ave), standard deviation (std) and trained network solutions with minimum number of weights (w^*) for each data set. For $f(x)$ and Boston data sets, *Train.*, *Valid.* and *Test* represent training error (residual sum of squares), validation error and test error. For Cancer and Card data sets they represent the percentage of correct classification patterns for

---

[1] http://www.ics.uci.edu/ mlearn/MLRepository.html

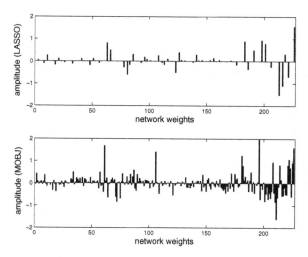

Fig. 4 Amplitude of the weights for Boston data set trained with MOBJ and LASSO

training, validation and test sets. The norm of the weights ($\|\mathbf{w}\|$) and the sum of the absolute weights ($\sum_j |w_j|$) are also included. n_w represents the number weights after pruning. CPU$_{time}$ describes the training time (in seconds) on a Pentium IV, 2.4GHz, 1GB (RAM) using the software MATLAB. *Train.** *Valid.** and *Test** are the error results after pruning.

Figure 5 shows the approximations performed by MOBJ, LASSO and Early-Stopping for the non-linear regression problem ($f(x)$). In general both methods (MOBJ and LASSO) were able to provide good solutions and on average both mean squared error for training (*Train.*) and validation (*Valid.*) sets are close to the known error variance ($\sigma^2 = 0.04$) which emphasizes that the solutions have achieved good generalization. The Early-Stopping method presented, on average a mean training error close to $\sigma^2 = 0.04$ but with higher validation and test errors. The LASSO best solution is a network with 16 weights that represents a fully connected network with 5 hidden nodes against 27 weights found by MOBJ that is associated to a 10 hidden nodes network and 28 weights found by Early-Stopping.

On average, MOBJ, LASSO and Early-Stopping achieved solutions with closer errors and classification rates for the training set. For validation and test sets, MOBJ and LASSO methods obtained the best results. As expected, LASSO solutions reached least sum of the absolute weights and least number of parameters after pruning. MOBJ solutions achieved least norm function values but very close to the LASSO ones. On the contrary, LASSO solutions have lower sum of the absolute weights in comparison to MOBJ. The Early-Stopping solutions present the highest values of norm and sum of the absolute weights and also increased error for the test set. However, it presented the least computational time followed by MOBJ and LASSO algorithms. Methods MOBJ and LASSO generate multiple solutions being the final one selected based on validation error. The Early-Stopping algorithm

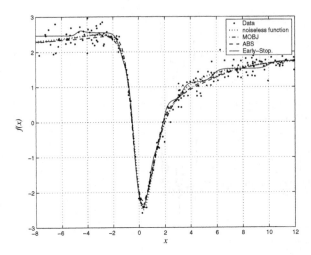

Fig. 5 Approximated function with MOBJ and LASSO

only generates a single solution. This factor must be considered when evaluating processing time of each method.

For the $f(x)$ function, the Early-Stopping algorithm achieved the highest error for the test set. LASSO method provided the least training, validation and test errors for Boston data set after pruning. For CANCER and CARD database, the LASSO method presented slightly superior classification rates when compared to MOBJ and Early-Stopping results.

The random pruning technique provided a remarkable simplification of the networks. The number of extracted weights were 30 weights for $f(x)$ (65.2%), 204 weights for Boston (90.27%), 145 weights for Cancer (96.03%) and 774 weights for Card (99.1%). For some cases like Cancer and Card the solutions with minimum number of weights were also generated by MOBJ. However, on average, LASSO provided solutions with both minimum number of weights and standard deviation. It is important to notice that the random pruning technique is able to generate MOBJ solutions with reduced number of weights but this feature is not as frequent as for LASSO solutions.

Based on validation and training criterion, the network found by MOBJ with minimum training and validation errors has 14 hidden nodes against 6 hidden nodes found by LASSO technique for Boston data set. Figure 6 shows the simplified topology found with best generalization performance. In this case: $E_{train} = 0.1324$, $E_{val} = 0.1418$ and $E_{test} = 0.1198$ are the respective training, validation and test errors for MOBJ and $E_{train} = 0.0525$, $E_{val} = 0.1340$ and $E_{test} = 0.1156$ for LASSO. Although it was not evident, the method's capability to exclude inputs, only one input has been eliminated from the trained network with LASSO. Figure 7 emphasizes the input selection property of LASSO. From 51 initial inputs only 13 were left with the LASSO algorithm against 16 with MOBJ for Cancer data set. Classification rates are 97.43% (train.), 97.71% (valid.) and 96.55% (test.) for

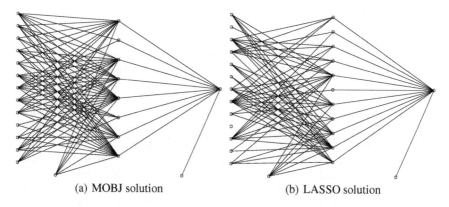

(a) MOBJ solution   (b) LASSO solution

Fig. 6 Pruned Networks for Boston data set

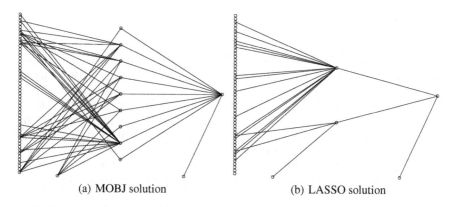

(a) MOBJ solution   (b) LASSO solution

Fig. 7 Pruned Networks for Cancer data set

MOBJ and 97.43% (train.), 97.71% (val.) and 98.85% (test.) for LASSO. Quite the same classification rates with a slight positive difference for LASSO. Early-Stopping results were not considered here because they presented smaller classification rates.

6 Analysis of Microarray Data with Multi-Objective Neural Networks

The multi-objective training algorithms MOBJ and LASSO have great performance to select and to estimate the neural network parameters. Both are able to generate solutions with high generalization performance but the LASSO algorithm is also able to optimize the number of weights, neurons and inputs of the neural network through the training process. In this section, the properties of the MOBJ

Table 1 MOBJ results

Data	Stat.	Train.	Valid.	Test	$\|\mathbf{w}\|$	$\Sigma_j \|w_j\|$	n_w	CPU$_{time}$	Train.*	Valid.*	Test*
	w^*	0.0335	0.0490	0.0507	6.01	22.39	27	47.77	0.0334	0.0490	0.0508
$f(x)$	ave	0.0398	0.0404	0.0471	5.87	23.61	36	39.52	0.0406	0.0403	0.0482
	std	0.0062	0.0086	0.0112	1.10	4.87	8	13.53	0.0068	0.0086	0.0113
	w^*	0.1654	0.1852	0.1886	2.31	8.90	28	77.33	0.2889	0.1806	0.1914
Boston	ave	0.0865	0.1578	0.1970	4.44	30.15	75	88.32	0.1665	0.1524	0.2303
	std	0.0665	0.0462	0.0597	1.60	16.62	36	16.93	0.0886	0.0469	0.0603
	w^*	0.9800	0.9486	0.9828	1.49	3.24	6	28.69	0.9714	0.9543	0.9770
Cancer	ave	0.9781	0.9741	0.9575	3.53	18.22	55	33.76	0.9670	0.9758	0.9546
	std	0.0093	0.0121	0.0184	2.33	11.44	41	8.26	0.0119	0.0100	0.0176
	w^*	0.9246	0.8728	0.8488	2.04	4.04	7	354.58	0.8638	0.8902	0.8430
Card	ave	0.8923	0.8798	0.8543	1.89	8.40	62	346.57	0.8562	0.8815	0.8463
	std	0.0264	0.0163	0.0267	1.71	6.46	48	55.98	0.0181	0.0178	0.0220

Table 2 LASSO results

Data	Stat.	Train.	Valid.	Test	$\|\mathbf{w}\|$	$\Sigma_j \|w_j\|$	n_w	CPU$_{time}$	Train.*	Valid.*	Test*
	w^*	0.0406	0.0364	0.0411	9.46	26.17	16	326.11	0.0405	0.0360	0.0406
$f(x)$	ave	0.0416	0.0399	0.0481	6.38	19.52	26	165.11	0.0425	0.0397	0.0490
	std	0.0071	0.0080	0.0094	1.18	4.32	9	109.76	0.0084	0.0080	0.0100
	w^*	0.1075	0.1137	0.3761	2.53	8.08	22	187.95	0.1123	0.1135	0.3736
Boston	ave	0.0924	0.1532	0.1993	4.39	20.23	37	160.42	0.1358	0.1466	0.2130
	std	0.0438	0.0433	0.0740	1.33	9.32	15	17.20	0.0465	0.0409	0.0646
	w^*	0.9829	0.9600	0.9598	5.16	11.49	6	68.73	0.9686	0.9657	0.9540
Cancer	ave	0.9741	0.9758	0.9625	5.49	13.78	10	76.56	0.9704	0.9768	0.9602
	std	0.0087	0.0118	0.0138	4.34	11.44	3	17.99	0.0100	0.0110	0.0142
	w^*	0.8899	0.8960	0.8430	2.49	4.81	7	330.61	0.8551	0.8960	0.8430
Card	ave	0.8781	0.8802	0.8539	2.04	4.83	13	304.18	0.8589	0.8819	0.8473
	std	0.0237	0.0174	0.0201	1.30	3.41	4	42.29	0.0155	0.0177	0.0194

algorithms are applied to a microarray data set. The samples correspond to a genetic expression profile from DNA microarray technology of neoadjuvant chemotherapy (treatment given prior to surgery) for patients with breast cancer. These expression profiles obtained from biopsies or fine needle aspirations can then be correlated with traditional tumor characteristics (size, grade) and behavior (recurrence, sensitivity to treatment). In breast cancer, neoadjuvant chemotherapy, which is treatment provided prior to surgery, allows breast tumor chemosensitivity to be tested in vivo [9, 10, 11, 12].

An accurate prediction of patient sensitivity to preoperative chemotherapy is an important issue in the treatment of breast cancer: patients who would present a complete pathologic response (PCR) patient (cured after chemotherapy and operation) if the treatment were prescribed to them, must be predicted with the highest possible accuracy. So are patients who would not present a complete pathologic response (NoPCR patients, i.e with residual disease after treatment), in order to avoid the

Table 3 Early Stopping results

Data	Stat.	Train.	Valid.	Test	$\|\mathbf{w}\|$	$\Sigma_j\|w_j\|$	n_w	CPU_{time}	Train.*	Valid.*	Test*
	w^*	0.0323	0.0741	0.0469	53.90	214.54	28	237.41	0.0453	0.0656	0.0391
$f(x)$	ave	0.0370	0.0565	0.0586	54.72	219.25	39	57.52	0.0508	0.0522	0.0667
	std	0.0061	0.0118	0.0179	1.06	3.14	4	70.38	0.0253	0.0085	0.0252
	w^*	0.3216	0.2237	0.2851	13.18	112.68	27	1.33	0.4945	0.1982	0.4065
Boston	ave	0.0877	0.1607	0.1886	12.37	108.66	76	52.63	0.1876	0.1488	0.2346
	std	0.0517	0.0452	0.0592	0.84	5.63	21	47.51	0.0758	0.0434	0.0804
	w^*	0.9886	0.9600	0.9310	19.63	203.74	6	49.14	0.9600	0.9657	0.9253
Cancer	ave	0.9741	0.9690	0.9630	19.53	205.19	16	45.27	0.9591	0.9726	0.9588
	std	0.0092	0.0137	0.0150	1.04	7.02	9	46.21	0.0089	0.0126	0.0168
	w^*	0.9014	0.8671	0.8198	3386.84	23437.99	3	12.47	0.8522	0.8902	0.8256
Card	ave	0.8977	0.8516	0.8498	2875.94	17936.89	15	21.65	0.8523	0.8697	0.8469
	std	0.0192	0.0219	0.0250	514.93	6059.30	10	15.02	0.0202	0.0179	0.0220

prescription of a treatment that would be inefficient, and to allow the prescription of other treatments. Therefore, such a predictor could give the possibility to prescribe individualized treatments [10, 8].

The dataset is composed, a priori, of 133 samples with 22283 attributes, called probes. Each probe corresponds to one gene expression value, so, the goal here is to build a multigenic predictor of the outcome of the neoadjuvant chemotherapy treatment. Microarrays can be used as multigenic predictors of patient response, but probe selection remains problematic. The most commonly used methods for selecting a subset of DNA probe sets identify probes that deviate most from a random distribution of expression levels or that are the most differentially expressed between the PCR and NoPCR learning cases. In the former approach, statistical analysis is used to rank genes based on the calculated p-values of the probe sets, and this ranking provides the basis for gene selection [4, 5, 6, 7, 8].

To select the relevant genic probes set we use the method proposed in [3]. It consider a hypothetic probe s^*, that would be ideal for classifying learning cases into PCR and NoPCR ones. For each patient cases, knowing the expression level of this single probe would be enough to decide the patient membership to either PCR or NoPCR class. Let $I_p(s^*)$ be the expression level interval containing the expression levels of the PCR patient learning cases, and let $I_n(s^*)$ be the minimum interval containing the expression levels of the NoPCR patient learning cases. As the probe s^* is supposed to be ideal, these two intervals are disjoint: $I_p(s^*) \cap I_n(s^*) = \emptyset$. Hence, this ideal probe would be correctly expressed by any patient learning case, i.e. the expression level of any PCR learning case would be in the interval $I_p(s^*)$ and that of any NoPCR one would be in the interval $I_n(s^*)$.

In the general case of non ideal probes, expression level intervals are not disjoint for PCR and NoPCR learning cases, and expression levels are blurred with noise. Therefore, we decide to attach two minimum sets of expression levels to any probe s, sets $E_p(s)$ et $E_n(s)$, computed from the learning data as follows:

- Let $m_p(s)$ et $sd_p(s)$ be the mean and standard deviation of probe s expression levels, on the PCR learning cases. The interval of positive expression levels of probe s, denoted $I_p(s)$, is that of radius $sd_p(s)$, centered on the mean $m_p(s)$:

$$I_p(s) = [m_p(s) - sd_p(s), m_p(s) + sd_p(s)];$$

- in the same way, let $m_n(s)$ and $sd_n(s)$ be the mean and standard deviation of probe s expression levels on the subset of NoPCR patient learning cases. We define the interval $I_n(s)$ of probe s negative expressions as:

$$I_n(s) = [m_n(s) - sd_n(s), m_n(s) + sd_n(s)].$$

The two expression level intervals $I_p(s)$ and $I_n(s)$ are not disjoint in general. So, we define the minimum set of positive expression levels of probe s, denoted $E_p(s)$, as the interval of positive expressions $I_p(s)$ minus its intersection with interval $I_n(s)$, and conversely, its minimum set of negative expression levels, denoted $E_n(s)$, as the interval of negative expression levels $I_n(s)$ deprived of its intersection with interval $I_p(s)$:

- probe s minimum set of positive expression levels:

$$E_p(s) = I_p(s) \backslash (I_p(s) \cap I_n(s));$$

- probe s minimum set of negative expression levels:

$$E_n(s) = I_n(s) \backslash (I_p(s) \cap I_n(s)).$$

From the two minimum sets of expression levels attached to any probe s, we define the valuation $v(s)$ of probe s. Let p be a PCR patient case of the learning data set. The patient p is *positively* expresses probe s if the expression level of s for this patient, denoted $e(s, p)$, is in the minimum set of positive expression levels of probe s: $e(s, p) \in Ep(s)$. In the same way, we say that patient p negatively expresses probe s if the expression level $e(s, p)$ is in the minimum set of negative expression levels of probe s : $e(s, p) \in E_p(s)$.

Let $p(s)$ be the number of PCR patient cases for which probe s is positively expressed:

$$p(s) = |p \in PCR; e(s, p) \in E_p(s)|$$

Let $n(s)$ be the number of NoPCR patient cases for which probe s is negatively expressed:

$$n(s) = |p \in NoPCR; e(s, p) \in E_n(s)|$$

We define the probesˊvaluation function v(s), whose values are in the real interval $[0, 1]$, as a measure of the proximity between probe s and the ideal probe s^*. Let P be the number of PCR learning cases and let N bet that of NoPCR learning cases, let $C = P + N$ be the size of the learning set (for the data at hand, one has $P = 21$ and $N = 61$), and let $c(s) = p(s) + n(s)$ be the number of patients that correctly express the probe. We say that an actual probe is close to the ideal one if its number of correct expressions $c(s)$ is close to the total number C of learning cases and if the ratio $r(s) = p(s)/n(s)$ of its correct positive and negative expressions is close to the ratio $R = P/N$ of PCR and NoPCR cases of the learning set. In other words, proximity between an actual probe and the ideal one depends upon the number of patients that correctly express the probe, and depends upon how well this last set of patients is representative of whole set of learning cases. This leads to a simple valuation function that satisfies both requirements:

$$v(s) = 0.5 \times \left(\frac{p(s)}{P} + \frac{n(s)}{N}\right)$$

the coefficient value 0.5 that appears in this definition is only for function's values to be in the unit interval.

With this definition, the valuation of the ideal probe s^* would be $v(s^*) = 1$, and any probe which is neither positively nor negatively expressed by patients of the learning set has a valuation $v(s) = 0$. In the general case, due to the valuation function's definition, of two probes s and s' having the same correct expressions numbers, $c(s) = c(s')$, probe s will have a higher valuation if its ratio $r(s)$ of positive and negative expressions is closer to the total ratio R of PCR and NoPCR learning cases. On the other hand, of two probes s and s' having the same ratio $r(s) = r(s')$ of positive and negative expressions, probe s valuation will be higher if $c(s) > c(s')$.

In the following, probes are ranked in decreasing order according to their valuations (the probe at first rank is one of the probes which have the highest valuation). We define a k-probes predictor (in short a k-predictor) as the set S_k of k first ranked probes, together with a classification decision criterion which is the majority voting [35]: let p be a patient case, let $pcr(k,p)$ be the number of probes of set S_k that are positively expressed by patient p and let $nopcr(k,p)$ be the number of probes of set S_k that are negatively expressed by patient p. The prediction of patient's treatment outcome is:

- if $pcr(k,p) > nopcr(k,p)$, k-predictor's prediction of the outcome of the treatment for patient case p is 'PCR';
- if $pcr(k,p) < nopcr(k,p)$, k-predictor's prediction is 'NoPCR';
- if $pcr(k,p) = nopcr(k,p)$, i.e tie, no prediction can be done.

When evaluating the performances of the predictor, any PCR patient case predicted as NoPCR or not predicted will account for a *False Negative* (FN), and any NoPCR patient case either predicted as PCR or not predicted at all will account for a *False Positive* (FP).

According to [3], the best classifier is composed of 30 probes, showed in Table 4. The data set is composed of treatment outcomes and expression levels of 22283 DNA probes corresponding to 133 patients. The set of learning cases is composed of 82 patient data, where 21 patient cases have a PCR treatment outcome and 61 a NoPCR outcome. The set of test cases is composed of 51 patient data. Among this test set, 13 patient cases have a PCR treatment outcome and 38 a NoPCR outcome.

Table 4 The 30 first ranked probes computed on the learning data set composed of 21 PCR cases and 61 NoPCR cases

Probe	Gene Symbol	$v(s)$	$p(s)$	$n(s)$	$c(s)$
213134_x_at	BTG3	0.61	12	40	52
205548_s_at	BTG3	0.61	12	40	52
209604_s_at	GATA3	0.59	15	29	48
209603_at	GATA3	0.49	12	26	38
212207_at	THRAP2	0.46	8	34	42
201826_s_at	SCCPDH	0.46	12	22	34
205339_at	SIL	0.45	10	27	37
209016_s_at	KRT7	0.45	6	38	44
201755_at	MCM5	0.45	7	35	42
204862_s_at	NME3	0.44	10	25	35
219051_x_at	METRN	0.44	11	22	33
211302_s_at	PDE4B	0.43	9	27	36
212660_at	PHF15	0.42	7	32	39
200891_s_at	SSR1	0.42	7	32	39
202392_s_at	PISD	0.42	11	20	31
204825_at	MELK	0.41	8	28	36
215867_x_at	CA12	0.41	10	22	32
214164_x_at	CA12	0.41	10	22	32
212046_x_at	MAPK3	0.41	10	22	32
209602_s_at	GATA3	0.41	13	13	26
212745_s_at	BBS4	0.41	3	42	45
203139_at	DAPK1	0.41	9	24	33
203226_s_at	SAS	0.40	7	29	36
219044_at	FLJ10916	0.40	8	26	34
203693_s_at	E2F3	0.40	8	26	34
220016_at	AHNAK	0.40	9	23	32
214383_x_at	KLHDC3	0.40	9	23	32
212721_at	SFRS12	0.40	9	23	32
202200_s_at	SRPK1	0.39	6	31	37
217028_at	CXCR4	0.39	8	25	33

For the training data set, the 30 Majority Voting predictor misclassifies a total of 13 cases ($accuracy = 13/82 = 0.84$) with 4 false negative ($specificity = 4/21 = 0.80$) and 9 false positives ($sensitivity = 9/61 = 0.85$) For the validation data set, the predictor misclassifies a total of 7 cases ($accuracy = 7/51 = 0.86$) with 1 false negative ($specificity = 1/13 = 0.92$) and 6 false positives ($sensitivity = 6/38 = 0.84$), as described in [3].

The 30 probes selected were used to train the Artificial Neural Network in other to compare the MOBJ and LASSO results on both generalization and probe selection properties The *Early Stopping* method was not used in the comparison because require data availability, what is not the case for the current data set. The solutions

Table 5 False Positive (FP), False Negative (FN), Accuracy (Acc), Sensitivity (Sens) and Specificity (Spec) for Majority Voting, MOBJ and LASSO classifiers

Model	Training					Test				
	FP	FN	Acc.	Sens.	Spec.	FP	FN	Acc.	Sens.	Spec.
Majority Voting	9	4	0.84	0.85	0.80	6	1	0.86	0.84	0.92
MOBJ	1	4	0.94	0.98	0.80	6	1	0.86	0.84	0.92
LASSO	3	5	0.90	0.95	0.86	7	2	0.82	0.81	0.84

Fig. 8 Pruned Network for MOBJ solution

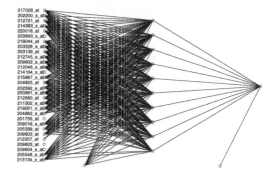

Fig. 9 Pruned Network for LASSO solution

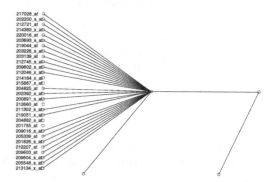

were all obtained for a neural network with 10 neurons in the hidden layer and the algorithms were set with the same initial random weights.

After optimization, the Pareto set for MOBJ and LASSO contains the solutions that can not be improved in one of the objectives without degrading the other. The simplest selection approach is to minimize the error of a validation set, but this requires data availability. The model in the intersection between the error and the norm curves given by solutions in the Pareto set was selected for the MOBJ and the LASSO solutions. This strategy was used because do not depend on a validation set. Other selection strategies that explore Pareto set properties have been applied successfully to classification and regression problems [1, 2].

The Table 5 presents the *Majority Voting* and the *Trained Networks* solutions for the *Gene Expression* data set. On average, MOBJ, LASSO achieved solutions with closer accuracy, sensitivity and specificity for the training set. The Majority Voting method achieved worst result for accuracy and sensitivity. For the test set the Majority Voting, MOBJ and LASSO methods obtained the the same solutions, in average, for the accuracy and the sensitivity. For the specificity the Majority Voting and MOBJ methods achieved the same solution, which is better than the LASSO solution. This similarity of the classification results indicates that the problem can be solved with a linear separation surface as the one used by the single neuron model generated by the LASSO algorithm.

Although, in average, the MOBJ and LASSO solutions were the same, the main difference between this solutions is the simplified topology achieve by LASSO training method. Figures 8 and 9 show the MLP topology after training and random pruning for MOBJ and LASSO, respectively. Remarkably, the LASSO algorithm was capable to reduce significantly the number of neurons in the hidden layer and to prune probes. From 10 initial hidden neurons, 09 were eliminated with the LASSO against none elimination with MOBJ. Both methods left 25 form the 30 initial input neurons.

7 Discussion and Conclusions

This chapter presented the LASSO multi-objective algorithm. The algorithm is able to reach high generalization solutions for regression and classification problems. In addition, the sum of the absolute weights objective function aggregates topology selection into the training process. The method was first compared with the original Multi-Objective and Early-Stopping algorithms and it achieved better results with high generalization responses and reduced topology. To measure the effective number of weights for each algorithm a random selection procedure were applied to both MOBJ and Early-Stopping solutions. Although the pruning approach is also able to simplify MOBJ and Early-Stopping solutions, it performed better with LASSO trained networks. A second application with genomic data was provided to allow the network to adjust its parameters in order to find the topology which better fits the data and with proper generalization outputs. The final solution provided by the LASSO algorithm was able to select probes and to reduce the topology to a single perceptron model with no loss in classification. This result illustrates the potential of the algorithm to train and to select compact models.

A preprocessing tool was also used to rank and to select a subset of probes prior to the neural network fitting. The procedures applies cross-validation to select a maximum number of probes needed to obtain the maximum classification results on the validation set.

References

1. Medeiros, T., Braga, A.P.: A new decision strategy in multi-objective training of artificial neural networks. In: European Symposium on Neural Networks (ESANN 2007), pp. 555–560 (2007)
2. Kokshenev, I., Braga, A.P.: Complexity bounds for radial basis functions and multi- objective learning. In: European Symposium on Neural Networks (ESANN 2007), pp. 73–78 (2007)
3. Natowicz, R., Incitti, R., Horta, E.G., Charles, B., Guinot, P., Yan, K., Coutant, C., Andre, F., Pusztai, L., Rouzier, R.: Prediction of the outcome of preoperative chemotherapy in breast cancer by DNA probes that convey information on both complete and non complete responses. BMC Bioinformatics 9, 149 (2008)
4. Simon, R.: Development and validation of therapeutically relevant multi-gene biomarker classifiers. J. Natl. Cancer Inst. 97, 866–867 (2005)
5. Maglietta, R., D'Addabbo, A., Piepoli, A., Perri, F., Liuni, S., Pesole, G., Ancona, N.: Selection of relevant genes in cancer diagnosis based on their prediction accuracy. Artif. Intell. Med. 40, 29–44 (2007)
6. Verducci, J.S., Melfi, V.F., Lin, S., Wang, Z., Roy, S., Sen, C.K.: Microarray analysis of gene expression: Considerations in data mining and statistical treatment. Physiol. Genomics 25, 355–363 (2006)
7. Ancona, N., Maglietta, R., Piepoli, A., D'Addabbo, A., Cotugno, R., Savino, M., Liuni, S., Carella, M., Pesole, G., Perri, F.: On the statistical assessment of classifiers using DNA microarray data. BMC Bioinformatics 7, 387 (2006)
8. Chen, J.J., Wang, S.J., Tsai, C.A., Lin, C.J.: Selection of differentially expressed genes in microarray data analysis. Pharmacogenomics 7, 212–220 (2007)
9. Tozlu, S., Girault, I., Vacher, S., Vendrell, J., Andrieu, C., Spyratos, F., Cohen, P., Lidereau, R., Bieche, I.: Identification of novel genes that co-cluster with estrogen receptor alpha in breast tumor biopsy specimens, using a large-scale real-time reverse transcription-PCR approach. Endocr. Relat. Cancer 13, 1109–1120 (2006)
10. Rouzier, R., Pusztai, L., Delaloge, S., Gonzalez-Angulo, A.M., Andre, F., Hess, K.R., Buzdar, A.U., Garbay, J.R., Spielmann, M., Mathieu, M.C., Symmans, W.F., Wagner, P., Atallah, D., Valero, V., Berry, D.A., Hortobagyi, G.N.: Nomograms to predict pathologic complete response and metastasis-free survival after preoperative chemotherapy for breast cancer. J. Clin. Oncol. 23, 8331–8339 (2005)
11. Rouzier, R., Rajan, R., Wagner, P., Hess, K.R., Gold, D.L., Stec, J., Ayers, M., Ross, J.S., Zhang, P., Buchholz, T.A., Kuerer, H., Green, M., Arun, B., Hortobagyi, G.N., Symmans, W.F., Pusztai, L.: Microtubule-associated protein tau: A marker of paclitaxel sensitivity in breast cancer. Proc. Natl. Acad. Sci. USA 102, 8315–8320 (2005)
12. Chang, J.C., Hilsenbeck, S.G., Fuqua, S.A.: The promise of microarrays in the management and treatment of breast cancer. Breast Cancer Res. 7, 100–104 (2005)
13. Parma, G.G., Menezes, B.R., Braga, A.P.: Sliding mode algorithm for training multilayer neural network. IEEE Letters 38(1), 97–98 (1998)
14. Hagan, M.T., Menhaj, M.B.: Training feedforward networks with Marquardt algorithm. IEEE Transactions on Neural Networks 5(6), 989–993 (1994)
15. Riedmiller, M., Braun, H.: A direct adaptive method for faster backpropagation learning: The RPROP algorthm. In: Proceddings of the IEEE Intl. Conf. on Neural Networks, San Francisco, pp. 586–591 (1993)
16. Fahlman, S.E.: Faster-learning variations on backpropagation: an empirical study. In: Proceddings of a 1988 Connectionist Models Summer School, Pittsburg, pp. 38–51 (1988)

17. Hinton, G.E.: Connectionist learning procedures. Artificial Inteligence 40, 185–234 (1989)
18. Rumelhart, D., Hinton, G., Williams, R.: Learning representations by back-propagating errors. Nature 323, 533–536 (1986)
19. Haykin, S.: Neural networks: A comprehensive foundation. Macmillan, New York (1994)
20. Rumelhart, D., McClelland, J.: Parallel Distributed Processing, vol. 1. MIT Press, Cambridge (1986)
21. Bartlett, P.L.: For valid generalization, the size of the weights is more important than the size of the network. In: Proceedings of NIPS 9, pp. 134–140 (1997)
22. Prechelt, L.: Automatic early stopping using cross validation: quantifying the criteria. Neural Networks 11(4), 761–767 (1998)
23. Krogh, A., Hertz, J.A.: A Simple Weight Decay Can Improve Generalization. In: Proceedings of NIPS, vol. 4, pp. 950–957 (1991)
24. Reed, R.: Pruning algorithms - a survey. IEEE Transactions on Neural Networks 4(5), 740–746 (1993)
25. Teixeira, R.A., Braga, A.P., Takahashi, R.H.C., Saldanha, R.R.: Recent Advances in the MOBJ Algorithm for training Artificial Neural Networks. International Journal of Neural Systems 11, 265–270 (2001)
26. Bland, R.G., Goldfarb, D., Todd, M.J.: The ellipsoid method: a survey. Operations Research 29(6), 1039–1091 (1981)
27. Costa, M.A., Braga, A.P., Menezes, B.R., Teixeira, R.A., Parma, G.G.: Training neural networls with a multi-objective sliding mode control algorithm. Neurocomputing 51, 467–473 (2003)
28. Costa, M.A., Braga, A.P., Menezes, B.R.: Improved Generalization Learning with Sliding Mode Control and the Levenberg-Marquadt Algorithm. In: Proceedings of VII Brazilian Symposium on Neural Networks (2002)
29. Jin, Y., Okabe, T., Sendhoff, B.: Neural network regularization and ensembling using multi-objective algorithms. In: Proceedings of the 2004 IEEE Congress on Evolutionary Computation, pp. 1–8 (2004)
30. Fieldsend, J.E., Singh, S.: Optimizing forecast model complexity using multi-objective evolutionary algorithms. In: Applications of Multi-Objective Evolutionary Algorithms, pp. 675–700. World Scientific, Singapore (2004)
31. Hastie, T., Tibshirani, R., Friedman, J.: The Elements of Statistical Learning: Data Mining, Inference and Prediction. Springer Series in Statistics (2001)
32. Tibshirani, R.: Regression Shrinkage and Selection via the Lasso. Journal of the Royal Statistical Society. Series B (Methodological) 58(1), 267–288 (1996)
33. Costa, M.A., Braga, A.P., Menezes, B.R.: Improving neural networks generalization with new constructive and pruning methods. Journal of intelligent & Fuzzy Systems 13, 73–83 (2003)
34. Harrison, D., Rubinfeld, D.L.: Hedonic prices and the demand for clean air. J. of Eviron. Economics & Management 5, 81–102 (1978)
35. May, K.O.: A Set of Independent Necessary and Sufficient Conditions for Simple Majority Decision. Econometrica 20(4), 680–684 (1952)

An Evolutionary Approximation for the Coefficients of Decision Functions within a Support Vector Machine Learning Strategy

Ruxandra Stoean, Mike Preuss, Catalin Stoean, Elia El-Darzi, and D. Dumitrescu

Abstract. Support vector machines represent a state-of-the-art paradigm, which has nevertheless been tackled by a number of other approaches in view of the development of a superior hybridized technique. It is also the proposal of present chapter to bring support vector machines together with evolutionary computation, with the aim to offer a simplified solving version for the central optimization problem of determining the equation of the hyperplane deriving from support vector learning. The evolutionary approach suggested in this chapter resolves the complexity of the optimizer, opens the 'black-box' of support vector training and breaks the limits of the canonical solving component.

1 Introduction

This chapter puts forward a hybrid approach which embraces the geometrical consideration of learning within support vector machines (SVMs) while it considers the estimation for the coefficients of the decision surface through the direct search capabilities of evolutionary algorithms (EAs).

Ruxandra Stoean and Catalin Stoean
Department of Computer Science, University of Craiova, Romania
e-mail: {ruxandra.stoean,catalin.stoean}@inf.ucv.ro

Mike Preuss
Department of Computer Science, University of Dortmund, Germany
e-mail: mike.preuss@cs.uni-dortmund.de

Elia El-Darzi
Department of Computer Science, University of Westminster, UK
e-mail: eldarze@westminster.ac.uk

D. Dumitrescu
Department of Computer Science, University of Cluj-Napoca, Romania
e-mail: ddumitr@cs.ubbcluj.ro

A.-E. Hassanien et al. (Eds.): Foundations of Comput. Intel. Vol. 1, SCI 201, pp. 83 114.
springerlink.com   © Springer-Verlag Berlin Heidelberg 2009

The SVM framework views artificial learning from an interesting perception: A hyperplane geometrically discriminates between training samples and the coefficients of its equation have to be determined, with respect to both the particular prediction ability and the generalization capacity. On the other hand, EAs are universal optimizers that generate solutions based on abstract principles of evolution and heredity. The aim of this work thus becomes to approximate the coefficients of the decision hyperplane through a canonical EA.

The motivation for the emergence of this combined technique resulted from several findings. SVMs are a top performing tool for data mining, however, the inner-workings of the optimization component are rather constrained and very complex. On the other hand, the adaptable EAs achieve learning relatively difficult from a standalone perspective. Taking advantage of the original interpretation of learning within SVMs and the flexible optimization nature of EAs, hybridization aims to accomplish an improved methodology. The novel approach augments support vector learning to become more 'white-box' and to be able to converge independent of the properties of the underlying kernel for a potential decision function. Apart from the straightforward evolution of hyperplane coefficients, an additional aim of the chapter is to investigate the treatment of several other variables involved in the learning process. Furthermore, it is demonstrated that the transparent evolutionary alternative is performed at no additional effort as regards the parametrization of the EA. Last but not least, on a different level from the theoretical reasons, the hybridized technique offers a simple and efficient tool for solving practical problems. Several real-world test cases served not only as benchmark, but also for application of the proposed architecture, and results bear out the initial assumption that an evolutionary approach is useful in terms of deriving the coefficients of such learning structures.

The research objectives and aims of this chapter will be carried out through the following original aspects:

- The hybrid technique will consider the learning task as in SVMs but use an EA to solve the optimization problem of determining the decision function.
- Classification and regression particularities will be treated separately. The optimization problem will be tackled through two possible EAs: One will allow for a more relaxed, adaptive evolutionary learning condition, while the second will be more similar to support vector training.
- Validation will be achieved by considering five diverse real-world learning tasks.
- Besides comparing results, the potential of the utilized, simplistic EA through parametrization is to be investigated.
- To enable handling large data sets, the first adaptive EA approach will be enhanced by the use of a chunking technique, with the purpose of resulting in a more versatile approach.
- The behavior of a crowding-based EA on preserving the performance of the technique will be examined with the purpose of a future employment for the coevolution of nonstandard kernels.

- The second methodology, which is more straightforward, will be generalized through the additional evolution of internal parameters within SVMs; a very general method of practical importance is therefore desired to be achieved.

The chapter contributes some key elements to both EAs and SVMs:

- The hybrid approach combines the strong characteristics of the two important artificial intelligence fields, namely: The original learning concept of SVMs and the flexibility of the direct search and optimization power of EAs.
- The novel alternative approach simplifies the support vector training.
- The proposed hybridization offers the possibility of a general evolutionary solution to all SVM components.
- The novel technique opens the direction towards the evolution and employment of nonstandard kernels.

The remainder of this chapter is organized as follows: Section 2 outlines the primary concepts and mechanisms underlying SVMs. Section 3 illustrates the means to achieve the evolution of the coefficients for the learning hyperplane. Section 4 describes the insides of the technique and its application to real-world problems. The chapter closes with several conclusions and ideas for future enhancement.

2   The SVM Learning Scheme

SVMs are a powerful approach to data mining tasks. Their originality and performance emerge as a result of the inner learning methodology, which is based on the geometrical relative position of training samples.

2.1   A Viewpoint on Learning

Given $\{(x_i, y_i)\}_{i=1,2,...,m}$, a training set where every $x_i \in R^n$ represents a data sample and each y_i corresponds to a target, a learning task is concerned with the discovery of the optimal function that minimizes the discrepancy between the given targets of data samples and the predicted ones; the outcome of previously "unknown" samples, $\{(x_i', y_i')\}_{i=1,2,...,p}$, is then tested.

The SVM technique is equally suited for classification and regression problems. The task for classification is to achieve an optimal separation of given samples into classes. SVMs assume the existence of a separating surface between every two classes labelled as -1 and 1. The aim then becomes the discovery of the appropriate decision hyperplane.

The standard assignment of SVMs for regression is to find the optimal function to be fitted to the data such that it achieves at most ϵ deviation from the actual targets of samples; the aim is thus to estimate the optimal regression coefficients of such a function.

2.2 SVM Separating Hyperplanes

If training examples are known to be linearly separable, then there exists a linear hyperplane of equation (1), which separates the samples according to classes. In (1), w and b are the coefficients of the hyperplane and $\langle\rangle$ denotes the scalar product.

$$\langle w, x_i \rangle - b = 0, w \in \Re^n, b \in \Re, x_i \in R^n, i = 1, 2, ..., m \ . \tag{1}$$

The positive data samples lie on the corresponding side of the hyperplane and their negative counterparts on the opposite side. As a stronger statement for linear separability [1], each of the positive and negative samples lies on the corresponding side of a matching supporting hyperplane for the respective class (denoted by y_i) (2).

$$y_i(\langle w, x_i \rangle - b) > 1, i = 1, 2, ..., m \ . \tag{2}$$

SVMs must determine the optimal values for the coefficients of the decision hyperplane that separates the training data with as few exceptions as possible. In addition, according to the principle of Structural Risk Minimization [2], separation must be performed with a maximal margin between classes. This high generalization ability implies a minimal $\|w\|$. In summary, the SVM classification of linear separable data with a linear hyperplane leads to the optimization problem (3).

$$\begin{cases} \min_{w,b} \|w\|^2 \\ \text{subject to } y_i(\langle w, x_i \rangle - b) \geq 1, i = 1, 2, ..., m \ . \end{cases} \tag{3}$$

Generally, the training samples are not linearly separable. In the nonseparable case, it is obvious that a linear separating hyperplane is not able to build a partition without any errors. However, a linear separation that minimizes training error can be applied to derive a solution to the classification problem [3]. The idea is to relax the separability statement through the introduction of slack variables, denoted by ξ_i for every training example. This relaxation can be achieved by observing the deviations of data samples from the corresponding supporting hyperplane, i.e. from the ideal condition of data separability. Such a deviation corresponds to a value of $\frac{\pm\xi_i}{\|w\|}$, $\xi_i \geq 0$ [4]. These values may indicate different nuanced digressions, but only a ξ_i higher than unity signals a classification error (Fig. 1). Minimization of training error is achieved by adding the indicator of error for every data sample into the separability statement while minimizing their sum. Hence, the SVM classification of linear nonseparable data with a linear hyperplane leads to the primal

Fig. 1 The separating
and supporting linear hy-
perplanes for the nonsep-
arable training subsets
(*squares* denote positive
samples, while *circles*
stand for the negative
ones). The support vec-
tors are *circled* and the
misclassified data point
is *highlighted*

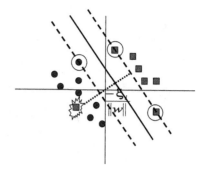

optimization problem (4), where C represents the penalty for errors and is what is called a hyperparameter (free parameter) of the SVM method.

$$\begin{cases} \min_{w,b} \|w\|^2 + C \sum_{i=1}^{m} \xi_i, C > 0 \\ \text{subject to } y_i(\langle w, x_i \rangle - b) \geq 1 - \xi_i, \xi_i \geq 0, i = 1, 2, ..., m . \end{cases} \quad (4)$$

If a linear hyperplane does not provide satisfactory results for the classifi-cation task, then a nonlinear decision surface can be formulated. The initial space of training data samples can be nonlinearly mapped into a higher di-mensional one, called the feature space and further denoted by H, where a linear decision hyperplane can be subsequently built. The separating hy-perplane can achieve an accurate classification in the feature space which corresponds to a nonlinear decision function in the initial space (Fig. 2). The procedure therefore leads to the creation of a linear separating hyperplane that would, as before, minimize training error; however, in this case, it will perform in the feature space. Accordingly, a nonlinear map $\Phi : R^n \rightarrow H$ is considered and data samples from the initial space are mapped into H.

As in the classical SVM solving procedure, vectors appear only as part of scalar products, the issue can be further simplified by substituting the scalar product by what is referred to as kernel.

A kernel is defined as a function with the property given by (5).

$$K(x_i, x_j) = \langle \Phi(x_i), \Phi(x_j) \rangle, x_i, x_j \in R^n . \quad (5)$$

The kernel can be perceived as to express the similarity between samples. SVMs require the kernel to be a positive (semi-)definite function in order for the standard approach to find a solution to the optimization problem [5]. Such a kernel satisfies Mercer's theorem below and is, therefore, a scalar product in some space [6].

Theorem 1. *(Mercer) [3], [7], [8], [9] Let $K(x,y)$ be a continuous symmetric kernel that is defined in the closed interval $a \leq x \leq b$ and likewise for y. The kernel $K(x,y)$ can be expanded in the series*

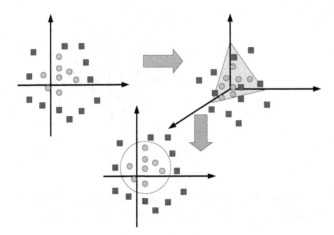

Fig. 2 Initial data space (*left*), nonlinear map into the higher dimension where the objects are linearly separable/the linear separation (*right*), and corresponding nonlinear surface (*bottom*)

$$K(x,y) = \sum_{i=1}^{\infty} \lambda_i \Phi(x)_i \Phi(y)_i$$

with positive coefficients, $\lambda_i > 0$ for all i. For this expansion to be valid and for it to converge absolutely and uniformly, it is necessary that the condition

$$\int_a^b \int_a^b K(x,y)\psi(x)\psi(y)dxdy \geq 0$$

holds for all $\psi(\cdot)$ for which

$$\int_a^b \psi^2(x)dx < \infty$$

The problem with this restriction is twofold [5]. Firstly, Mercer's condition is very difficult to check for a newly constructed kernel. Secondly, kernels that fail the theorem could prove to achieve a better separation of the training samples. Applied SVMs consequently use a couple of classical kernels that had been demonstrated to meet Mercer's condition:

- the polynomial classifier of degree p: $K(x,y) = \langle x, y \rangle^p$
- the radial basis function classifier: $K(x,y) = e^{\frac{\|x-y\|^2}{\sigma}}$, where p and σ are also hyperparameters of SVMs.

However, as a substitute for the original problem solving, a direct search algorithm does not depend on the condition whether the kernel is positive (semi-)definite or not.

After the optimization problem is solved, the class of every test sample is calculated: The side of the decision boundary on which every new data example lies is determined (6).

$$class(x_i') = sgn(\langle w, \Phi(x_i') \rangle - b), i = 1, 2, ..., p .$$ (6)

As it is not always possible to determine the map Φ and, as a consequence of the standard training methodology, either to explicitly obtain the coefficients, the class follows from further artifices.

The classification accuracy is then defined as the number of correctly labelled cases over the total number of test samples.

2.3 Addressing Multi-class Problems through SVMs

k-class SVMs build several two-class classifiers that separately solve the matching tasks. The translation from multi-class to two-class is performed through different systems, among which one-against-all, one-against-one or decision directed acyclic graph are the most commonly employed.

One-against-all Approach

The one-against-all (1aa) technique [10] builds k classifiers. Every j^{th} SVM considers all training samples labelled with j as positive and all the remaining as negative.

Consequently, by placing the problem in the initial space, the aim of every j^{th} SVM is to determine the optimal coefficients w^j and b^j of the decision hyperplane which best separates the samples with outcome j from all the other samples in the training set, such that (7) holds.

$$\begin{cases} \min_{w^j, b^j} \frac{\|w^j\|^2}{2} + C \sum_{j=1}^{m} \xi_i^j, \\ \text{subject to } y_i(\langle w^j, x_i \rangle - b^j) \geq 1 - \xi_i^j, \\ \xi_i^j \geq 0, i = 1, 2, ..., m, j = 1, 2, ..., k . \end{cases}$$ (7)

Once all hyperplanes are determined following the classical SVM training, the label for a test sample x_i' is given by the class that has the maximum value for the learning function, as in (8).

$$class(x_i') = argmax_{j=1,2,...,k}(\langle w^j, \Phi(x_i') \rangle) - b^j), i = 1, 2, ..., p .$$ (8)

One-against-one Approach

The one-against-one (1a1) technique [10] builds $\frac{k(k-1)}{2}$ SVMs. Every machine is trained on data from every two classes, l and j, where samples labelled with l are considered positive while those in class j are taken as negative.

Accordingly, the aim of every SVM is to determine the optimal coefficients w^{lj} and b^{lj} of the decision hyperplane which best separates the samples with outcome l from the samples with outcome j, such that (9).

$$
\begin{cases}
\min_{w^{lj}, b^{lj}} \frac{\|w^{lj}\|^2}{2} + C \sum_{i=1}^m \xi_i^{lj}, \\
\text{subject to } y_i(\langle w^{lj}, x_i \rangle - b) \geq 1 - \xi_i^{lj}, \\
\xi_i^{lj} \geq 0, i = 1, 2, ..., m, l, j = 1, 2, ..., k, l \neq j .
\end{cases}
\tag{9}
$$

Once the hyperplanes of the $\frac{k(k-1)}{2}$ SVMs are found, a *voting* method is used to determine the class for a test sample $x_i', i = 1, 2, ..., p$. For every SVM, the label is computed following the sign of the corresponding decision function applied to x_i'. Subsequently, if the sign says x_i' is in class l, the vote for the l-th class is incremented by one; conversely, the vote for class j is increased by unity. Finally, x_i' is taken to belong to the class with the largest vote. In case two classes have an identical number of votes, the one with the smaller index is selected.

Decision Directed Acyclic Graph

Learning within the decision directed acyclic graph (DDAG) technique [11] follows the same procedure as in 1a1. After the hyperplanes of the $\frac{k(k-1)}{2}$ SVMs are discovered, a graph system is used to determine the class for a test sample $x_i', i = 1, 2, ..., p$. Each node of the graph has a list of classes attached and considers the first and last elements of the list. The list that corresponds to the root node contains all k classes. When a test instance x_i' is evaluated, it is descended from node to node, in other words, one class is eliminated from each corresponding list, until the leaves are reached. The mechanism starts at the root node which considers the first and last classes. At each node, l vs j, we refer to the SVM that was trained on data from classes l and j. The class of x is computed by following the sign of the corresponding decision function applied to x_i'. Subsequently, if the sign says x is in class l, the node is exited via the right edge while, conversely, through the left edge. The wrong class is thus eliminated from the list and it is proceeded via the corresponding edge to test the first and last classes of the new list and node. The class is given by the leaf that x_i' eventually reaches.

2.4   SVM Regression Hyperplanes

SVMs for regression must find a function $f(x)$ that has at most ϵ deviation from the actual targets of training samples and, simultaneously, is as flat as possible [12]. In other words, the aim is to estimate the regression coefficients of $f(x)$ with these requirements.

While the former condition is straightforward, errors are allowed as long as they are less than ϵ, the latter needs some further explanation [13]. Resulting values of the regression coefficients may affect the model in the sense that it fits current training data but has low generalization ability, which would contradict the principle of Structural Risk Minimization for SVMs [2]. In order to overcome this limitation, it is required to choose the flattest function in the definition space. Another way to interpret SVMs for regression is that training data are constrained to lie on a hyperplane that allows for some error and, at the same time, has high generalization capacity.

Suppose a linear regression model can fit the training data. Consequently, function f has the form (10).

$$f(x) = \langle w, x \rangle - b .$$ (10)

The conditions for the flattest function (smallest slope) that approximates training data with ϵ precision can be translated into the optimization problem (11).

$$\begin{cases} \min_{w,b} \|w\|^2 \\ \text{subject to} \begin{cases} y_i - \langle w, x_i \rangle + b \le \epsilon \\ \langle w, x_i \rangle - b - y_i \le \epsilon \end{cases} \\ i = 1, 2, ..., m . \end{cases}$$ (11)

It may very well happen that the linear function f is not able to fit all training data and consequently SVMs will again allow for some relaxation, analogously to the corresponding situation for classification.

Therefore, the positive slack variables ξ_i and ξ_i^*, both attached to each sample, are introduced into the condition for approximation of training data and, also as before, the sum of these indicators for errors is minimized. The primal optimization problem in case of regression then translates to (12).

$$\begin{cases} \min_{w,b} \|w\|^2 + C \sum_{i=1}^{m} (\xi_i + \xi_i^*) \\ \text{subject to} \begin{cases} y_i - \langle w, x_i \rangle + b \le \epsilon + \xi_i \\ \langle w, x_i \rangle - b - y_i \le \epsilon + \xi_i^* \\ \xi_i, \xi_i^* \ge 0 \end{cases} \\ i = 1, 2, ..., m . \end{cases}$$ (12)

If a linear function is not at all able to fit training data, a nonlinear function has to be chosen, as well. The procedure follows the same steps as before in SVMs for classification.

When a solution for the optimization problem is reached, the predicted target for a test sample is computed as (13).

$$f(x_i') = \langle w, \Phi(x_i') \rangle - b, i = 1, 2, ..., p .$$ (13)

Also as in the classification problem, the regression coefficients are rarely transparent and the predicted target actually is derived from other computations.

In order to verify the accuracy of the technique, the value of the root mean square error (RMSE) is computed as in (14).

$$RMSE = \sqrt{\frac{1}{p}\sum_{i=1}^{p}(f(x_i') - y_i')^2} \ . \tag{14}$$

2.5 Solving the Optimization Problem within SVMs

The standard algorithm of finding the optimal solution relies on an extension of the Lagrange multipliers technique. Corresponding dual problem may be expressed as (15) for classification.

$$\begin{cases} \min_{\{\alpha_i\}_{i=1,2,\dots,m}} Q(\alpha) = \sum_{i=1}^{m}\alpha_i - \frac{1}{2}\sum_{i=1}^{m}\sum_{j=1}^{m}\alpha_i\alpha_j y_i y_j K(x_i,x_j) \\ \text{subject to} \begin{cases} \sum_{i=1}^{m}\alpha_i y_i = 0 \\ \alpha_i \geq 0 \end{cases} \\ i = 1,2,\dots,m \ . \end{cases} \tag{15}$$

Conversely, in the most general case of nonlinear regression, the dual problem is restated as (16). For reasons of a shorter reference, $\alpha_i(*)$ denotes both α_i and α_i^*, in turn.

$$\begin{cases} \min_{\{\alpha_i(*)\}_{i=1,2,\dots,m}} Q(\alpha(*)) = \sum_{i,j=1}^{m}(\alpha_i - \alpha_i^*)(\alpha_j - \alpha_j^*)K(x_i,x_j) \\ -\epsilon\sum_{i=1}^{m}(\alpha_i + \alpha_i^*) + \sum_{i=1}^{m}y_i(\alpha_i - \alpha_i^*) \\ \text{subject to} \begin{cases} \sum_{i=1}^{m}(\alpha_i - \alpha_i^*) = 0 \\ 0 \leq \alpha_i(*) \leq C \end{cases} \\ i = 1,2,\dots,m \ . \end{cases} \tag{16}$$

The optimum Lagrange multipliers $\alpha_i(*)$s are determined as the solutions of the system by setting the gradient of the objective function to zero. For more mathematical explanation, see [3], [14].

3   Evolutionary Adaptation of the Hyperplane Coefficients to the Training Data

Apart from its emergence as a simple complementary method for solving the SVM derived optimization problem, the EA-powered determination of hyperplane coefficients had to be explored and improved with respect to runtime, prediction accuracy and adaptability.

3.1 Motivation and Aim

Despite the originality and performance of the learning vision of SVMs, the inner training engine is intricate, constrained, rarely transparent and able to converge only for certain particular decision functions. This has brought the motivation to investigate and put forward an alternative training approach that benefits from the flexibility and robustness of EAs.

The technique adopts the learning strategy of the SVMs but aims to simplify and generalize its solving methodology, by offering a transparent substitute to the initial 'black-box'. Contrary to the canonical technique, the evolutionary approach can at all times explicitly acquire the coefficients of the decision function, without any further constraints. Moreover, in order to converge, the evolutionary method does not require the positive (semi-)definition properties for kernels within nonlinear learning. Eventually, the evolutionary approach demonstrates to be an efficient tool for real-world application in vital domains like disease diagnosis and prevention or spam filtering.

There have been numerous attempts to combine SVMs and EAs, however, this method differs from the reported ones: The learning path remains the same, except that the coefficients of the decision function are now evolved with respect to the optimization objectives regarding accuracy and generalization. Several potential structures, enhancements and additions had been proposed, tested and confirmed using available benchmarking test problems.

3.2 Literature Review: Previous EA-SVM Interactions

The chapter focuses on the evolution of the coefficients for decision functions of a learning strategy similar to that of SVMs. However, there are other works known to combine SVMs and EAs. One evolutionary direction tackles model selection, which concerns the adjustment of hyperparameters within SVMs, i.e. the penalty for errors C and parameters of the kernel, and is generally performed through grid search or gradient descent methods. Alternatively, determination of hyperparameters can be achieved through evolution strategies [15]. Another aspect envisages the evolution of the form for the kernel, which can be performed by means of genetic programming [16]. The Lagrange multipliers involved in the expression of the dual problem can be evolved by means of evolution strategies and particle swarm optimization [5]. Inspired by the geometrical SVM learning, [17] also reports the evolution of w and C, while using erroneous learning ratio and lift values as the objective function. Current paper therefore extends the work in the hybridization between EAs and SVMs by filling the gap of a direct evolutionary solving of the primal optimization problem of determining the decision hyperplane, which has never been performed before, to the best of our knowledge.

3.3   Evolving the Coefficients of the Hyperplane

The evolutionary approach for support vector training (ESVM) considers the adaptation of a flexible hyperplane to the given training data through the evolution of the optimal coefficients for its equation. After the necessary revisions in the learning objectives due to a different way of solving the optimization task, the corresponding EA is adopted in a canonical formulation for real-valued problems [18]. For the purpose of a comparative analysis between ESVMs and SVMs, there are solely the classical polynomial and radial kernels that are used in this chapter to shape the decision functions.

Representation

An individual c encodes the coefficients of the hyperplane, w and b (17). Individuals are randomly generated such that $w_{i'} \in [-1, 1], i' = 1, 2, ..., n$, $b \in [-1, 1]$.

$$c = (w_1, ..., w_n, b) . \tag{17}$$

Fitness Assignment

Prior to deciding on a strategy to evaluate individuals, the objective function must be established in terms of the new approach to address the optimization goals.

Since ESVMs depart from the standard mathematical treatment of SVMs, a different general (nonlinear) optimization problem is derived [19] . Accordingly, w is also mapped through Φ into H. As a result, the squared norm that is involved in the generalization condition is now $\|\Phi(w)\|^2$. At the same time, the equation of the hyperplane consequently changes to (18).

$$\langle \Phi(w), \Phi(x_i) \rangle - b = 0. \tag{18}$$

The scalar product is used in the form (19) and, besides, the kernel is additionally employed to address the norm in its simplistic equivalence to a scalar product.

$$\langle u, w \rangle = u^T w . \tag{19}$$

In conclusion, the optimization problem for classification is reformulated as (20).

$$\begin{cases} \min_{w,b} K(w, w) + C \sum_{i=1}^{m} \xi_i, C > 0 \\ \text{subject to } y_i(K(w, x_i) - b) \geq 1 - \xi_i, \\ \xi_i \geq 0, i = 1, 2, ..., m . \end{cases} \tag{20}$$

At the same time, the objectives for regression are transposed to (21).

$$\begin{cases} \min_{w,b} K(w,w) + C\sum_{i=1}^{m}(\xi_i + \xi_i^*) \\ \text{subject to } \begin{cases} y_i - K(w,x_i) + b \le \epsilon + \xi_i \\ K(w,x_i) - b - y_i \le \epsilon + \xi_i^* \\ \xi_i, \xi_i^* \ge 0 \end{cases} \\ i = 1,2,...,m \ . \end{cases} \tag{21}$$

The ESVMs targeting multi-class situations must undergo similar transformation with respect to the expression of the optimization problem [20], [21]. As 1aa is concerned, the aim of every j^{th} ESVM is expressed as to determine the optimal coefficients, w^j and b^j, of the decision hyperplane which best separates the samples with outcome j from all the other samples in the training set, such that (22) takes place.

$$\begin{cases} \min_{w^j,b^j} K(w^j,w^j) + C\sum_{i=1}^{m}\xi_i^j, \\ \text{subject to } y_i(K(w^j,x_i) - b^j) \ge 1 - \xi_i^j, \\ \xi_i^j \ge 0, i = 1,2,...,m, j = 1,2,...,k \ . \end{cases} \tag{22}$$

Within the 1a1 and DDAG approaches, the aim of every ESVM becomes to find the optimal coefficients w^{lj} and b^{lj} of the decision hyperplane which best separates the samples with outcome l from the samples with outcome j, such that (23) holds.

$$\begin{cases} \min_{w^{lj},b^{lj}} K(w^{lj},w^{lj}) + C\sum_{i=1}^{m}\xi_i^{lj}, \\ \text{subject to } y_i(K(w^{lj},x_i) - b^{lj}) \ge 1 - \xi_i^{lj}, \\ \xi_i^{lj} \ge 0, i = 1,2,...,m, l,j = 1,2,...,k, i \ne j \ . \end{cases} \tag{23}$$

The fitness assignment now derives from the objective function of the optimization problem and is minimized. Constraints are handled by penalizing the infeasible individuals through the introduction of a function $t : R \to R$ which returns the value of the argument, if negative, while zero otherwise.

Classification and regression variants simply differ in terms of objectives and constraints. Thus the expression of the fitness function for the former is (24), with the corresponding indices in the multi-class situations [22], [23], [24].

$$f(w,b,\xi) = K(w,w) + C\sum_{i=1}^{m}\xi_i + \sum_{i=1}^{m}[t(y_i(K(w,x_i) - b) - 1 + \xi_i)]^2. \tag{24}$$

As for the latter, the fitness assignment is defined in the form (25) as found in [25], [26], [27].

$$f(w, b, \xi) = K(w, w) + C \sum_{i=1}^{m} (\xi_i + \xi_i^*) + \sum_{i=1}^{m} [t(\epsilon + \xi_i - y_i +$$

$$K(w, x_i) - b)]^2 + \sum_{i=1}^{m} [t(\epsilon + \xi_i^* + y_i - K(w, x_i) + b)]^2. \qquad (25)$$

Selection and Variation Operators

The efficient tournament selection and the common genetic operators for real encoding, i.e. intermediate crossover and mutation with normal perturbation, are applied.

Stop Condition

The EA stops after a predefined number of generations and outputs the optimal coefficients for the equation of the hyperplane. Moreover, ESVM is transparent at all times during the evolutionary cycle, thus w and b may be observed as they adapt throughout the process.

Test Step

Once the coefficients of the hyperplane are found, the class for an unknown data sample can be determined directly following (26) .

$$class(x_i') = sgn(K(w, x_i') - b), i = 1, 2, ..., p. \qquad (26)$$

Conversely for regression, the target of test samples can be obtained through (27) .

$$f(x_i') = K(w, x_i') - b, i = 1, 2, ..., p. \qquad (27)$$

For the multi-class tasks, the label is found by employing the same specific mechanisms, only this time the resulting decision function applied to the current test sample takes the form (28) .

$$f(x_i') = K(w^j, x_i') - b^j, i = 1, 2, ..., p, j = 1, 2, ..., k. \qquad (28)$$

3.4 Preexperimental Planning: The Test Cases

Experimentation had been conducted on five real-world problems, coming from the UCI Repository of Machine Learning Databases[1], i.e. diabetes mellitus diagnosis [28] , spam detection [29] , [30], iris recognition [20] , soybean

---

[1] Available at http://www.ics.uci.edu/~mlearn/MLRepository.html

Table 1 Data set properties

Data	Diabetes	Iris	Soybean	Spam	Boston
No. of samples	768	150	47	4601	506
No. of features	8	4	35	57	13
No. of classes	2	3	4	2	-

disease diagnosis [31] and Boston housing [25], [26] (see Table 1). The motivation for the choice of test cases was manifold. Diabetes and spam are two-class problems, while soybean and iris are multi-class. Differentiating, on the one hand, diabetes diagnosis is a better-known benchmark, but spam filtering is an issue of current major concern; moreover, the latter has a lot more features and samples, which makes a huge difference for classification as well as for optimization. On the other hand, while soybean has a high number of attributes, iris has only four, but a larger number of samples. Finally, Boston housing is a representative regression task. For all reasons mentioned above, the selection of test problems certainly contained all the variety of situations that had been necessary for the objective validation of the ESVM approach. The experimental design was set to employ holdout cross-validation: For each data set, 30 runs of the ESVM were conducted – in every run, approximately 70% random cases were appointed to the training set and the remaining 30% went into test. The necessity for data normalization in diabetes, spam and iris was also observed.

4 Discovering ESVMs

While the target of the EA was straightforward, addressing several inside interactions had been not. Moreover, application of the ESVM to the practical test cases had yielded more research questions yet to be resolved. Finally, further implications of being able to instantly operate on the evolved coefficients had been realized.

4.1 A Naïve Design

As already stated, the coefficients of the hyperplane, w and b, are encoded into the structure of an individual. But, since the conditions for hyperplane optimality additionally refer the indicators for errors, ξ_i, $i = 1, 2, ..., m$, the problem becomes how to comprise them in the evolutionary solving. One simple solution could be to depart from the SVM geometrical strict meaning of a deviation and simply evolve the factors of indicators for errors. Thus, the structure of an individual changes to (29), where $\xi_j \in [0, 1], j = 1, 2, ..., m$.

Table 2 Manually tuned SVM hyperparameter values for the evolutionary and canonical approach

	Diabetes	Iris	1a1/1aa	Soybean	Spam	Boston
ESVMs						
$p(\sigma)$	$p = 2$	$\sigma = 1$		$p = 1$	$p = 1$	$p = 1$
SVMs						
$p(\sigma)$	$p = 1$	$\sigma = 1/m$		$p = 1$	$p = 1$	$\sigma = 1/m$

Table 3 Manually tuned EA parameter values for the naïve construction

	Diabetes	Iris	1a1	Soybean	Spam	Boston
ps	100	100	100	100	200	
ng	250	100	100	250	2000	
cp	0.40	0.30	0.30	0.30	0.50	
mp	0.40	0.50	0.50	0.50	0.50	
emp	0.50	0.50	0.50	0.50	0.50	
ms	0.10	0.10	0.10	0.10	0.10	
ems	0.10	0.10	0.10	0.10	0.10	

$$c = (w_1, ..., w_n, b, \xi_1, ..., \xi_m) . \tag{29}$$

The evolved values of the indicators for errors can now be addressed in the proposed expression for fitness evaluation. Also, mutation of errors is now constrained, preventing the ξ_is from taking negative values.

Once the primary theoretical aspects had been completed, the experimental design had to be inspected and the practical evaluation of the viability of ESVMs had to be conducted. The hyperparameters both approaches share were manually chosen (Table 2). The error penalty C was invariably set to 1. For certain (e.g. radial, polynomial) kernels, the optimization problem is relatively simple, due to Mercer's theorem, and is also implicitly solved by classical SVMs [32]. Note that ESVMs are not restricted to using the traditional kernels, but these had been employed within to enable comparison with classical SVMs. Therefore, as a result of multiple testing, a radial kernel was used for the iris data set, a polynomial one was employed for diabetes, while for spam, soybean and Boston, a linear surface was applied. In the regression case, ϵ was set to 0.

An issue of crucial importance for demonstrating the feasibility of any EA alternative relies on the simplicity to determine appropriate parameters. The EA parameter values were initially manually determined (Table 3).

In order to validate the manually found EA parameter values and to probe the ease in their choice, the tuning method of Sequential Parameter Optimization (SPO) [33] was applied. The SPO builds on a quadratic regression model, supported by a Latin hypercube sampling (LHS) methodology and noise reduction, by incrementally increased repetition of runs. Parameter bounds were set as follows:

Table 4 SPO tuned EA parameter values for the naïve representation

	Diabetes	Iris	Soybean	Spam	Spam +Chunks	Boston
ps	198	46	162	154	90	89
ng	296	220	293	287	286	1755
pc	0.87	0.77	0.04	0.84	0.11	0.36
pm	0.21	0.57	0.39	0.20	0.08	0.5
epm	0.20	0.02	0.09	0.07	0.80	0.47
ms	4.11	4.04	0.16	3.32	0.98	0.51
ems	0.02	3.11	3.80	0.01	0.01	0.12

- Population size (ps) - 5/2000
- Number of generations (ng) - 50/300
- Crossover probability (pc) - 0.01/1
- Mutation probability (pm) - 0.01/1
- Error mutation probability (epm) - 0.01/1
- Mutation strength (ms) - 0.001/5
- Error mutation strength (ems) - 0.001/5

Since the three multi-class techniques behave similarly in the manually tuned multi-class experiments (Table 5), automatic adjustment was run only for the most widely used case of 1a1. The best parameter configurations for all problems as determined by SPO are depicted in Table 4.

Test accuracies/errors obtained by manual tuning are presented in Table 5. Differentiated (spam/non spam for spam filtering and ill/healthy for diabetes) results are also depicted.

Table 5 Accuracy/RMSE of the manually tuned naïve ESVM version on the considered test sets, in percent

	Average	Worst	Best	StD
Diabetes (overall)	**76.30**	71.35	80.73	2.24
Diabetes (ill)	50.81	39.19	60.27	4.53
Diabetes (healthy)	90.54	84.80	96.00	2.71
Iris 1aa (overall)	**95.85**	84.44	100.0	3.72
Iris 1a1 (overall)	**95.18**	91.11	100.0	2.48
Iris DDAG (overall)	**94.96**	88.89	100.0	2.79
Soybean 1aa (overall)	**99.22**	88.24	100	2.55
Soybean 1a1 (overall)	**99.02**	94.11	100.0	2.23
Soybean DDAG (overall)	**98.83**	70.58	100	5.44
Spam (overall)	**87.74**	85.74	89.83	1.06
Spam (spam)	77.48	70.31	82.50	2.77
Spam (non spam)	94.41	92.62	96.30	0.89
Boston	**4.78**	5.95	3.96	0.59

Table 6 Accuracies of the SPO tuned naïve ESVM version on the considered test sets, in percent

	LHS$_{best}$	StD	SPO	StD
Diabetes (overall)	75.82	3.27	**77.31**	2.45
Diabetes (ill)	49.35	7.47	52.64	5.32
Diabetes (healthy)	89.60	2.36	90.21	2.64
Iris (overall)	**95.11**	2.95	**95.11**	2.95
Soybean (overall)	99.61	1.47	**99.80**	1.06
Spam (overall)	89.27	1.37	**91.04**	0.80
Spam (spam)	80.63	3.51	84.72	1.59
Spam (non spam)	94.82	0.94	95.10	0.81
Boston	5.41	0.65	**5.04**	0.52

Table 6 summarizes the performances and standard deviations of the best configuration of an initial LHS sample and of the SPO.

SPO indicates that for all cases, except for the soybean data, crossover probabilities were dramatically increased, while often reducing mutation probabilities, especially for errors. However, the relative quality of SPO's final best configurations against the ones found during the initial LHS phase increases with the problem size. It must be stated that in most cases, results achieved with manually determined parameter values are only improved by SPO, if at all, by more effort, i.e. increasing population size or number of generations.

The computational experiments show that the proposed technique produces equally good results as compared to the canonical SVMs and it is further explained in subsection 4.6. Furthermore, the smaller standard deviations prove the higher stability of ESVMs.

As concerns the difficulty in setting the EA, SPO confirms: Distinguishing the performance of different configurations is difficult even after computing a large number of repeats. Consequently, the "parameter optimization potential" justifies employing a tuning method only for the larger problems, diabetes, spam and Boston. Especially for the small problems, well performing parameter configurations are seemingly easy to find. This brings evidence in support of the (necessary) simplicity in tuning the parameters inside the evolutionary alternative solving.

It must be stated that for the standard kernels, one cannot expect the ESVM to be better than the standard SVM, since the kernel transformation that induces learning is the same. However, the flexibility of the EAs as optimization tools makes ESVMs an attractive choice from the performance perspective, due to their prospective ability to additionally evolve problem-tailored kernels, regardless of whether they are positive (semi-)definite or not, which is impossible under SVMs.

4.2   Chunking within ESVMs

A first problem appears for large data sets, i.e. spam filtering, where the amount of runtime needed for training is very large. This stems from the large genomes employed, as indicators for errors of every sample in the training set are included in the representation. Consequently, this problem was tackled by an adaptation of a chunking procedure [34] inside ESVM.

A chunk of N training samples is repeatedly considered. Within each chunking cycle, the EA, with a population of half random individuals and half previously best evolved individuals, runs and determines the coefficients of the hyperplane. All training samples are tested against the obtained decision function and a new chunk is constructed based on $N/2$ randomly and equally distributed incorrectly placed samples and half randomly samples from the current chunk. The chunking cycle stops when a predefined number of iterations with no improvement in training accuracy passes (Algorithm 1).

Algorithm 1. ESVM with Chunking

Require: The training samples
Ensure: Best obtained coefficients and corresponding accuracy
  begin
  Randomly choose N training samples, equally distributed, to make a chunk;
  while a predefined number of iterations passes with no improvement **do**
    if first chunk **then**
      Randomly initialize population of a new EA;
    else
      Use best evolved hyperplane coefficients and random indicators for errors
      to fill half of the population of a new EA and randomly initialize the other
      half;
    end if
    Apply EA and find coefficients of the hyperplane;
    Compute side of all samples in the training set with evolved hyperplane coefficients;
    From incorrectly placed, randomly choose (if exist) N/2 samples, equally distributed;
    Randomly choose the rest up to N from the current chunk and add all to a new one;
    if obtained training accuracy if higher than the best one obtained so far **then**
      Update best accuracy and hyperplane coefficients; set improvement to true;
    end if
  end while
  Apply best obtained coefficients on the test set and compute accuracy
  return   accuracy
  end

ESVM with chunking was applied to the spam data set. Manually tuned parameters had the same values as before, except the number of generations

Table 7 Accuracy/RMSE of the manually tuned ESVM with chunking version on the considered test sets, in percent

	Average	Worst	Best	StD
Spam (overall)	**87.30**	83.13	90.00	1.77
Spam (spam)	83.47	75.54	86.81	2.78
Spam (non spam)	89.78	84.22	92.52	2.11

Table 8 Accuracies of the SPO tuned ESVM with chunking version on the considered test sets, in percent

	LHS$_{best}$	StD	SPO	StD
Spam (overall)	87.52	1.31	**88.37**	1.15
Spam (spam)	86.26	2.66	86.35	2.70
Spam (non spam)	88.33	2.48	89.68	2.06

for each run of the EA which is now set to 100. The chunk size, N, was chosen as 200 and the number of iterations with no improvement, i.e. repeats of the chunking cycle, was designated to be 5. Values derived from the SPO tuning are presented in the chunking column from Table 4.

Results of manual and SPO tuning are shown in Tables 7 and 8. The novel approach of ESVM with chunking produced good results in a much smaller runtime; it runs 8 times faster than the previous one, at a cost of a small loss in accuracy. Besides solving the EA genome length problem, proposed mechanism additionally reduces the large number of computations that derives from the reference to the many training samples in the expression of the fitness function.

4.3 A Pruned Variant

Although already a good alternative approach, the ESVM may still be improved concerning simplicity. The current optimization problem requires to treat the error values, which in the present EA variant are included in the representation. These severely complicate the problem by increasing the genome length (variable count) by the number of training samples. Moreover, such a methodology strongly departs from the canonical SVM concept. Therefore, it had been investigated whether the indicators for errors could be computed instead of evolved.

Since ESVMs directly and interactively provide hyperplane coefficients at all times, the generic EA representation (17) can be kept and the indicators can result from geometrical calculations. In case of classification, the procedure follows [1]. The current individual, which is the current separating hyperplane, is considered and supporting hyperplanes are determined through the mechanism below. One first computes (30).

$$\begin{cases} m_1 = min\{K(w, x_i)|y_i = +1\} \\ m_2 = max\{K(w, x_i)|y_i = -1\} \, . \end{cases} \tag{30}$$

Then (31) proceeds.

$$\begin{cases} p = |m_1 - m_2| \\ w' = \frac{2}{p}w \\ b' = \frac{1}{p}(m_1 + m_2) \, . \end{cases} \tag{31}$$

For every training sample x_i, the deviation to its corresponding supporting hyperplane (32) is obtained.

$$\delta(x_i) = \begin{cases} K(w', x_i) - b' - 1, \text{ if } y_i = +1 \\ K(w', x_i) - b' + 1, \text{ if } y_i = -1 \\ i = 1, 2, ..., m \, . \end{cases} \tag{32}$$

If sign of deviation equals class, corresponding $\xi_i = 0$; else, the (normalized) absolute deviation is returned as the indicator for error. Experiments showed the need for normalization of the computed deviations in the cases of diabetes, spam and iris, while, on the contrary, soybean requires no normalization. The different behavior can be explained by the fact that the first three data sets have a larger number of training samples. The sum of deviations is subsequently added to the expression of the fitness function. As a consequence, in the early generations, when the generated coefficients lead to high deviations, their sum, considered from 1 to the number of training samples, takes over the whole fitness value and the evolutionary process is driven off the course to the optimum. The form of the fitness function remains as before (24), obviously without taking the ξ_is as arguments.

The proposed method for acquiring the errors for the regression situation is as follows. For every training sample, one firstly calculates the difference between the actual target and the predicted value that is obtained with the coefficients of the current individual (regression hyperplane), as in (33).

$$\delta_i = |K(w, x_i) - b - y_i|, i = 1, 2, ..., m \, . \tag{33}$$

Secondly, one tests the difference against the ϵ threshold, following (34).

$$\begin{cases} \text{if } \delta_i < \epsilon \qquad \text{then } \xi_i = 0 \\ \text{else} \qquad \qquad \xi_i = \delta_i - \epsilon \\ i = 1, 2, ..., m \, . \end{cases} \tag{34}$$

The newly obtained indicators for errors can now be employed in the fitness evaluation of the corresponding individual, which changes from (25) to (35):

Table 9 Manually tuned parameter values for the pruned approach

	Diabetes	Iris	Soybean	Spam	Boston
ps	100	100	100	150	200
ng	250	100	100	300	2000
pc	0.4	0.30	0.30	0.80	0.50
pm	0.4	0.50	0.50	0.50	0.50
ms	0.1	4	0.1	3.5	0.1

Table 10 SPO tuned parameter values for the pruned representation

	Diabetes	Iris	Soybean	Spam	Boston
ps	190	17	86	11	100
ng	238	190	118	254	1454
pc	0.13	0.99	0.26	0.06	0.88
pm	0.58	0.89	0.97	0.03	0.39
ms	0.15	3.97	0.08	2.58	1.36

$$f(w, b) = K(w, w) + C \sum_{i=1}^{m} \xi_i . \tag{35}$$

The function to be fitted to the data is thus still required to be as flat as possible and to minimize the errors of regression that are higher than the permitted ϵ. Experiments on the Boston housing problem demonstrated that the specific method for computing the deviations does not require any additional normalization.

The problem related settings and SVM hyperparameters were kept the same as for naïve approach, except ϵ which was set to 5 for the regression problem, which reveals that the pruned representation apparently needs a more generous deviation allowance within training.

The EA first proceeded with the manual values for parameters from Table 9. Subsequent parameter values derived from SPO on the pruned variant are shown in Table 10.

Results obtained after manual and SPO tuning are depicted in Tables 11 and 12. The automated performance values were generated by 30 validation runs for the best found configurations after the initial design and SPO, respectively.

Results of automated tuning are similar to those of the manual regulation which once again demonstrates the easy adjustability of the ESVM. Additionally, the performance spectra of LHS was plotted in order to compare the hardness of finding good parameters for our two representations on the spam and soybean problems (Figs. 3 and 4). The Y axis represents the fractions of all tried configurations; therefore the Y value corresponding to each bar denotes the percentage of configurations that reached the accuracy of the X axis where the bar is positioned.

Table 11 Accuracy/RMSE of the manually tuned pruned ESVM version on the considered test sets, in percent

	Average	Worst	Best	StD
Diabetes (overall)	**74.60**	70.31	82.81	2.98
Diabetes(ill)	45.38	26.87	58.57	6.75
Diabetes (healthy)	89.99	86.89	96.75	2.66
Iris 1aa (overall)	**93.33**	86.67	100	3.83
Iris 1a1 (overall)	**95.11**	73.33	100	4.83
Iris DDAG (overall)	**95.11**	88.89	100	3.22
Soybean 1aa (overall)	**99.22**	88.24	100	2.98
Soybean 1a1 (overall)	**99.60**	94.12	100	1.49
Soybean DDAG (overall)	**99.60**	94.12	100	1.49
Spam (overall)	**85.68**	82	88.26	1.72
Spam (spam)	70.54	62.50	77.80	4.55
Spam (non spam)	95.39	92.66	97.44	1.09
Boston	**5.07**	6.28	3.95	0.59

Table 12 Accuracies of the SPO tuned pruned ESVM version on the considered test sets, in percent

	LHS_{best}	StD	SPO	StD
Diabetes (overall)	72.50	2.64	**73.39**	2.82
Diabetes(ill)	35.50	10.14	43.20	6.53
Diabetes (healthy)	92.11	4.15	89.94	3.79
Iris (overall)	**95.41**	2.36	**95.41**	2.43
Soybean (overall)	**99.61**	1.47	99.02	4.32
Spam (overall)	89.20	1.16	**89.51**	1.17
Spam (spam)	79.19	3.13	82.02	3.85
Spam (non spam)	95.64	0.90	94.44	1.42
Boston	4.99	0.66	**4.83**	0.45

The diagrams illustrate the fact that naïve ESVM is harder to parameterize than the pruned approach: When SPO finds a configuration for the latter, it is already a promising one, as it can be concluded from the higher corresponding bars.

It is interesting to remark that the pruned representation is not that much faster. Although the genome length is drastically reduced, the runtime consequently gained is however partly lost again when computing the values for the slack variables. This draws from the extra number of scalar products that must be calculated due to (30), (32) and (33). As run length itself is a parameter in present studies, an upper bound of the necessary effort is rather obtained. Closer investigation may lead to a better understanding of suitable run lengths, e.g. in terms of fitness evaluations. However, the pruned representation has its advantages. Besides featuring smaller genomes, less parameters are needed, because the slack variables are not evolved and thus

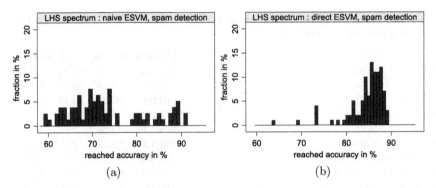

Fig. 3 Comparison of EA parameter spectra, LHS with size 100, 4 repeats, (a) for the naïve (7 parameters) and (b) the pruned (5 parameters) representation on the spam problem

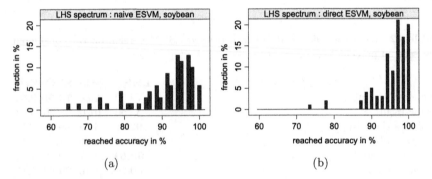

Fig. 4 Comparison of EA parameter spectra, LHS with size 100, 4 repeats, (a) for the naïve (7 parameters) and (b) the pruned (5 parameters) representation on the soybean problem

two parameters are eliminated. As a consequence, it can be observed that this representation is easier to tune.

The best configurations for the pruned representation perform slightly worse as compared to the results recorded for the naïve representation. The independent evolution of the slack variables seems to result in a better adjustment of the hyperplane as opposed to their strict computation. Parameter tuning beyond a large initial design appears to be infeasible, as performance is not significantly improved in most cases. If at all, it is successful for the larger problems of diabetes, spam and Boston. This indicates once more that parameter setting for the ESVM is rather easy, because there is a large set of good performing configurations. Nevertheless, there seems to be a slight tendency towards fewer good configurations (harder tuning) for the large problems.

Table 13 SPO tuned parameter values for the pruned representation with crowding

	Diabetes	Iris	Spam
ps	92	189	17
ng	258	52	252
pc	0.64	0.09	0.42
pm	0.71	0.71	0.02
ms	0.20	0.20	4.05

4.4   A Crowding Variant

In addition to the direct pruned representation, a crowding [35] variant of the EA had also been tested. Within crowding, test for replacement is done against the most similar parent of the current population. Crowding based EAs are known to provide good global search capabilities. This is of limited value for the kernel types employed in this study, but it is important for nonstandard kernels. It is desirable, however, to investigate whether the employment of a crowding-based EA on the pruned representation would maintain the performance of the technique or not. All the other elements of the EA remained the same and the values for parameters as determined by SPO are shown in Table 13. The crowding experiment was chosen to be run only on the representative tasks for many samples (diabetes and spam), features (spam) and classes (iris).

Note that only automated tuning was performed for the pruned crowding ESVM and results can be found in Table 14.

Table 14 Accuracies of the SPO tuned pruned version with crowding on the considered test sets, in percent

	LHD$_{best}$	StD	SPO	StD
Diabetes (overall)	74.34	2.30	**74.44**	2.98
Diabetes(ill)	43.68	6.64	45.32	7.04
Diabetes (healthy)	90.13	3.56	90.17	3.06
Iris (overall)	**95.63**	2.36	94.37	2.80
Spam (overall)	88.72	1.49	**89.45**	0.97
Spam (spam)	80.14	5.48	80.79	3.51
Spam (non spam)	94.25	1.66	95.07	1.20

SPO revealed that for the crowding variant, some parameter interactions dominate the best performing configurations: For larger population sizes, smaller mutation steps and larger crossover probabilities are better suited, and with greater run lengths, performance increases with larger mutation step sizes. For the original pruned variant, no such clear interactions could be attained. However, in both cases, many good configurations were detected.

4.5   Integration of SVM Hyperparameters

For practical considerations, a procedure for a dynamic choice of model hyperparameters was further included within the pruned construction. Having judged from performed experiments, the parameter expressing the penalty for errors C seemed of no significance within the ESVM technique; it was consequently dropped from the parameters pool. Further on, by simply inserting one more variable to the genome, the kernel parameter (p or $sigma$) could also be evolved. In this way, benefiting from the evolutionary solving of the primal problem, model selection was actually performed at the very same time.

The idea was tested through an immediate manual tuning of the EA parameters; the values are depicted in Table 15. For reasons of generality with respect to the new genomic variable, an extra mutation probability (hpm) and mutation strength (hms) respectively, were additionally set. The corresponding gene also had a continuous encoding, the hyperparameter being rounded at kernel application. The soybean task was not considered for this experiment anymore, as very good results had already been achieved.

Table 15 Manually tuned parameter values for the all-in-one pruned representation

	Diabetes	Iris	Boston	Spam
ps	100	50	100	5
ng	250	280	2000	480
pc	0.4	0.9	0.5	0.1
pm	0.4	0.9	0.5	0.1
hpm	0.4	0.9	0.9	0.1
ms	0.1	1	0.1	3.5
hms	0.5	0.1	0.1	0.1

The resulting values for the SVM hyperparameters were identical to our previous manual choice (Table 2), with one exception in the diabetes task, where sometimes a linear kernel is obtained.

Results of the all-inclusive technique (Table 16), similar in accuracy or regression error to the prior ones and obtained at no additional cost, sustain the inclusion of model selection and point to the next extension, the coevolution of nonstandard kernels.

4.6   ESVMs Versus SVMs

In order to validate the aim of this work, that is to offer a simpler, yet equally performing alternative to SVM training, this section compares the ESVM results with those of canonical SVMs run in R on the same data sets. The reasons for this choice of a contrast were twofold: The R software already

Table 16 Accuracy/RMSE of the manually tuned all-in-one pruned ESVM version on the considered test sets, in percent

	Average	Worst	Best	StD
Diabetes (overall)	**74.20**	66.66	80.21	3.28
Diabetes(ill)	46.47	36.99	63.08	6.92
Diabetes (healthy)	89.23	81.40	94.62	3.46
Iris (overall)	**96.45**	93.33	100	1.71
Spam (overall)	**88.92**	85.39	91.48	1.5
Spam (spam)	79.98	68.72	94.67	5.47
Spam (non spam)	94.79	84.73	96.91	2.22
Boston	**5.06**	6.19	3.97	0.5

contains a standard package for a SVM implementation and objectivity is achieved only in similar experimental setup and test cases. However, search for the outcome of the application of other related approaches (as described in subsection 3.2) on the same data sets revealed only results on the diabetes task: A classification accuracy of 74.48% and a standard deviation of 4.30% came out of a 20-fold cross-validation within evolution of Lagrange multipliers in [32] and an accuracy of 76.88% and a standard deviation of 0.25% averaged over 20 trials was obtained through the evolution of the SVM hyperparameters in [15].

The results, obtained after 30 runs of holdout cross-validation, are illustrated in Table 17. After having performed manual tuning for the SVM hyperparameters, the best results were obtained as in the corresponding row of Table 2. It is worthy to note a couple of differences between our ESVM and the SVM implementation: In the Boston housing case, despite the employment of a linear kernel in the former, the latter produces better results for a radial function, while, in the diabetes task, the ESVMs employ a degree two polynomial and SVMs target it linearly.

The results for each problem were compared via a Wilcoxon rank-sum test. The p-values (see Table 17) suggest to detect significant differences only in the cases of Soybean and Boston data sets. However, the absolute difference is not large for Boston housing, rendering SVM a slight advantage. It may be more relevant for the Soybean task, where ESVM is better.

Table 17 Accuracy/RMSE of canonical SVMs on the considered test sets, in percent, as compared to those obtained by ESVM and p-values from a Wilcoxon rank-sum test

	SVM	StD	ESVM	StD	p-value
Diabetes	**76.82**	1.84	**77.31**	2.45	0.36
Iris	**95.33**	3.16	**96.45**	2.36	0.84
Spam	**92.67**	0.64	**91.04**	0.80	0.09
Soybean	**92.22**	9.60	**99.80**	1.06	3.98×10^{-5}
Boston	**3.82**	0.7	**4.78**	0.59	1.86×10^{-6}

Although, in terms of accuracy, the ESVM approach had not achieved better results for some of the test problems, it has many advantages: The decision surface is always transparent even when working with kernels whose underlying transformation to the feature space cannot be determined. The simplicity of the EA makes the solving process easily explained, understood, implemented and tuned for practical usage. Most importantly, any function can be used as a kernel and no additional constraints or verifications are necessary.

From the opposite perspective, the training is relatively slower than that of SVM, as the evaluation always relates to the training data. However, in practice (often, but not always), it is the test reaction that is more important. Nevertheless, by observing the relationship between each result and the corresponding size of the training data, it is clear that SVM performs better than ESVM for larger problems; this is probably due to the fact that, in these cases, much more evolutionary effort would be necessary. The problem of handling large data sets is thus worth investigating deeper in future work.

5   Conclusions and Outlook

The evolutionary learning technique proposed in this chapter resembles the vision upon learning of SVMs but solves the inherent optimization problem by means of an EA. An easier and more flexible alternative to SVMs is put forward and undergoes several enhancements in order to provide a viable alternative to the classical paradigm. These developments are summarized below:

- Two possible representations for the EA (one simpler, and a little faster, and one more complicated, but also more accurate) that determines the coefficients are imagined.
- In order to boost the suitability of the new technique for any issue, a novel chunking mechanism for reducing size in large problems is also proposed; obtained results support its employment.
- The use of a crowding-based EA is inspected in relation to the preservation of performance. Crowding would be highly necessary in the immediate coevolution of nonstandard kernels.
- Finally, an all-inclusive ESVM construction for the practical perspective is developed and validated.
- On a different level, an additional aim was to address and solve real-work tasks of high importance.

Several conclusions can be eventually drawn and the potential of the technique can be further strengthened through the application of two enhancements:

- As opposed to SVMs, ESVMs are much easier to understand and use.
- ESVMs do not impose any kind of constraints or requirements.

- Moreover, the evolutionary solving of the optimization problem enables the acquirement of function coefficients directly and at all times within a run.
- SVMs, on the other hand, are somewhat faster, as the kernel matrix is computed only once.
- Performances are comparable, for different test cases ESVMs and SVMs take the lead, in turn.

Although already a suitable alternative, the novel ESVM can still be enhanced in several ways:

- The requirement for an optimal decision function actually involves two criteria: the surface must fit to training data but simultaneously generalize well. So far, these two objectives are combined in a single fitness expression. As a better choice for handling these conditions, a multicriterial approach could be tried instead.
- Additionally, the simultaneous evolution of the hyperplane and of nonstandard kernels will be achieved. This approach is highly difficult by means of SVM standard methods for hyperplane determination, whereas it is straightforward for ESVMs. A possible combination can be achieved through a cooperative coevolution between the population of hyperplanes and that of GP-evolved kernels.

6 Appendix - Definition of Employed Concepts

There are a series of notions that appear throughout the chapter. Their meaning, use and reference are explained in what follows:

Evolutionary algorithm (EA) [18] - a metaheuristic optimization algorithm in the field of artificial intelligence that finds its inspiration in what governs nature: A population of initial individuals (genomes) goes through a process of adaptation through selection, recombination and mutation; these phenomena encourage the appearance of fitter solutions. The best performing individuals are selected in a probabilistic manner as parents of a new generation and gradually the system evolves to the optimum. The fittest individual(s) obtained after a certain number of iterations is (are) the solution to the problem.

Support vector machine (SVM) [2] - a supervised learning method for classification and regression: Given a set of samples, the method aims for the optimal decision hyperplane to model the data and establish an equilibrium between a good training accuracy and a high generalization ability; according to [36], a possible definition of an SVM could be "a system for efficiently training linear learning machines in kernel-induced feature spaces, while respecting the insights of generalization theory and exploiting optimization theory".

Classification / regression hyperplane - the decision hyperplane whose defining coefficients must be determined; within classification, it must differentiate between samples of different classes, while, with respect to regression, it represents the surface on which the data are restrained to be positioned.

multi (k)-class SVM [10] - SVMs are implicitly built for binary classification tasks; for problems with k outcomes, $k > 2$, the technique considers the labels and corresponding samples two by two and uses common approaches like one-against-one and one-against-all to combine the obtained classifiers.

Primal problem - the direct form of the optimization task within SVMs of the determination of the decision hyperplane, while balancing between accuracy and generalization capacity. It is **dualized** in the standard SVM solving by Lagrange multipliers.

Kernel - a function of two variables that defines the scalar product between them; within SVMs, it is employed for the "kernel trick" - a technique to write a nonlinear operator as a linear one in a space of higher dimension as a result of Mercer's theorem.

Crowding-based EA [35] - a technique that was introduced as a method of maintaining diversity: new obtained individuals replace only similar individuals in the population: A percentage G (*generation gap*) of the individuals is chosen via fitness proportional selection in order to create an equal number of offspring; for each of these offspring, CF (a parameter called *crowding factor*) individuals from the current population are randomly selected – the offspring then replaces the most similar individual from these.

References

1. Bosch, R.A., Smith, J.A.: Separating hyperplanes and the authorship of the disputed federalist papers. American Mathematical Monthly 105, 601–608 (1998)
2. Vapnik, V.: The Nature of Statistical Learning Theory. Springer, Heidelberg (1995)
3. Haykin, S.: Neural Networks: A Comprehensive Foundation. Prentice-Hall, New Jersey (1999)
4. Cortes, C., Vapnik, V.: Support vector networks. Machine Learning 1, 273–297 (1995)
5. Mierswa, I.: Evolutionary learning with kernels: A generic solution for large margin problems. In: Proc. of the Genetic and Evolutionary Computation Conference, vol. 1, pp. 1553–1560 (2006)
6. Burges, C.J.C.: A tutorial on support vector machines for pattern recognition. Data Mining and Knowledge Discovery 2, 121–167 (1998)
7. Boser, B.E., Guyon, I.M., Vapnik, V.: A training algorithm for optimal margin classifiers. In: Proceedings of the 5th Annual ACM Workshop on Computational Learning Theory, vol. 1, pp. 11–152 (1992)

8. Courant, R., Hilbert, D.: Methods of Mathematical Physics. Wiley Interscience, Hoboken (1970)
9. Mercer, J.: Functions of positive and negative type and their connection with the theory of integral equations. Transactions of the London Philosophical Society (A) 209, 415–446 (1908)
10. Hsu, C.-W., Lin, C.-J.: A comparison of methods for multi-class support vector machines. IEEE Transactions on Neural Networks 13, 415–425 (2004)
11. Platt, J.C., Cristianini, N., Shawe-Taylor, J.: Large margin dags for multiclass classification. In: Proc. of Neural Information Processing Systems, vol. 1, pp. 547–553 (2000)
12. Smola, A.J., Scholkopf, B.: A tutorial on support vector regression. Technical Report NC2-TR-1998-030. NeuroCOLT2 Technical Report Series (1998)
13. Rosipal, R.: Kernel-based Regression and Objective Nonlinear Measures to Access Brain Functioning. PhD thesis Applied Computational Intelligence Research Unit School of Information and Communications Technology University of Paisley, Scotland (2001)
14. Stoean, R.: Support vector machines. An evolutionary resembling approach. Universitaria Publishing House Craiova (2008)
15. Friedrichs, F., Igel, C.: Evolutionary tuning of multiple svm parameters. In: Proc. 12th European Symposium on Artificial Neural Networks, vol. 1, pp. 519–524 (2004)
16. Howley, T., Madden, M.G.: The genetic evolution of kernels for support vector machine classifiers. In: Proc. of 15th Irish Conference on Artificial Intelligence and Cognitive Science, vol. 1 (2004),
http://www.it.nuigalway.ie/m_madden/profile/pubs.html
17. Jun, S.H., Oh, K.W.: An evolutionary statistical learning theory. International Journal of Computational Intelligence 3, 249–256 (2006)
18. Eiben, A.E., Smith, J.E.: Introduction to Evolutionary Computing. Springer, Heidelberg (2003)
19. Stoean, R., Preuss, M., Stoean, C., Dumitrescu, D.: Concerning the potential of evolutionary support vector machines. In: Proc. of the IEEE Congress on Evolutionary Computation, vol. 1, pp. 1436–1443 (2007)
20. Stoean, R., Dumitrescu, D., Preuss, M., Stoean, C.: Different techniques of multi-class evolutionary support vector machines. In: Proc. of Bio-Inspired Computing: Theory and Applications, vol. 1, pp. 299–306 (2006)
21. Stoean, R., Stoean, C., Preuss, M., Dumitrescu, D.: Evolutionary multi-class support vector machines for classification. In: Proceedings of International Conference on Computers and Communications - ICCC 2006, Baile Felix Spa - Oradea, Romania, vol. 1, pp. 423–428 (2006)
22. Stoean, R., Dumitrescu, D., Stoean, C.: Nonlinear evolutionary support vector machines. application to classification. Studia Babes-Bolyai, Seria Informatica LI, 3–12 (2006)
23. Stoean, R., Dumitrescu, D.: Evolutionary linear separating hyperplanes within support vector machines. In: Scientific Bulletin, University of Pitesti. Mathematics and Computer Science Series, vol. 11, pp. 75–84 (2005)
24. Stoean, R., Dumitrescu, D.: Linear evolutionary support vector machines for separable training data. In: Annals of the University of Craiova. Mathematics and Computer Science Series, vol. 33, pp. 141–146 (2006)
25. Stoean, R., Preuss, M., Dumitrescu, D., Stoean, C.: ϵ - evolutionary support vector regression. In: Symbolic and Numeric Algorithms for Scientific Computing, SYNASC 2006, vol. 1, pp. 21–27 (2006)

26. Stoean, R., Preuss, M., Dumitrescu, D., Stoean, C.: Evolutionary support vector regression machines. In: IEEE Postproc. of the 8th International Symposium on Symbolic and Numeric Algorithms for Scientific Computing, vol. 1, pp. 330–335 (2006)
27. Stoean, R.: An evolutionary support vector machines approach to regression. In: Proc. of 6th International Conference on Artificial Intelligence and Digital Communications, vol. 1, pp. 54–61 (2006)
28. Stoean, R., Stoean, C., Preuss, M., El-Darzi, E., Dumitrescu, D.: Evolutionary support vector machines for diabetes mellitus diagnosis. In: Proceedings of IEEE Intelligent Systems 2006, London, UK, vol. 1, pp. 182–187 (2006)
29. Stoean, R., Stoean, C., Preuss, M., Dumitrescu, D.: Evolutionary support vector machines for spam filtering. In: Proc. of RoEduNet IEEE International Conference, vol. 1, pp. 261–266 (2006)
30. Stoean, R., Stoean, C., Preuss, M., Dumitrescu, D.: Evolutionary detection of separating hyperplanes in e-mail classification. Acta Cibiniensis LV, 41–46 (2007)
31. Stoean, R., Stoean, C., Preuss, M., Dumitrescu, D.: Forecasting soybean diseases from symptoms by means of evolutionary support vector machines. Phytologia Balcanica 12 (2006)
32. Mierswa, I.: Making indefinite kernel learning practical, technical report. Technical report. Artificial Intelligence Unit, Department of Computer Science, University of Dortmund (2006)
33. Bartz-Beielstein, T.: Experimental research in evolutionary computation - the new experimentalism. Natural Computing Series. Springer, Heidelberg (2006)
34. Perez-Cruz, F., Figueiras-Vidal, A.R., Artes-Rodriguez, A.: Double chunking for solving svms for very large datasets. In: Proceedings of Learning 2004, Elche, Spain, vol. 1 (2004),
    eprints.pascal-network.org/archive/00001184/01/learn04.pdf
35. DeJong, K.A.: An Analysis of the Behavior of a Class of Genetic Adaptive Systems. PhD thesis University of Michigan, Ann Arbor (1975)
36. Cristianini, N., Shawe-Taylor, J.: An Introduction to Support Vector Machines. Cambridge University Press, Cambridge (2000)

Part II
Connectionist Learning

Meta-learning and Neurocomputing – A New Perspective for Computational Intelligence

Ciro Castiello

Abstract. In this chapter an analysis of computational mechanisms of induction is brought forward, in order to assess the potentiality of meta-learning methods versus the common base-learning practices. To this aim, firstly a formal investigation of inductive mechanisms is accomplished, sketching a distinction between fixed and dynamical bias learning. Then a survey is presented with suggestions and examples which have been proposed in literature to increase the efficiency of common learning algorithms. The peculiar laboratory for this kind of investigation is represented by the field of connectionist learning. To explore the meta-learning possibilities of neural network systems, knowledge-based neurocomputing techniques are considered. Among them, some kind of hybridisation strategies are particularly analysed and addressed as peculiar illustrations of a new perspective of Computational Intelligence.

1 Introduction

The key for knowledge enlargement is induction. Humans make use of induction to unveil the natural mechanisms of physical phenomena and to lay the foundations of concepts and theories. Inductive processes for knowledge acquisition have constituted the subject of intellectual speculations and methodological investigations for the historical research concerning human perception. That is due to the inherent riskiness connected with induction: differently from the reliable conclusions of a correct deduction, no guarantee comes with the generalisation from observed cases. However, the information gain produced by induction lets us review non-deductive inferences as the main drive of knowledge acquisition, which is often referred to as *learning.*

Far from intervening in the intriguing epistemological debate about inductive discovery (some contributions of ours can be found in [15, 16]), in

Ciro Castiello
Dipartimento di Informatica, Università degli Studi di Bari
via Orabona, 4, 70125 – Bari, Italy
e-mail: `castiello@di.uniba.it`

A.-E. Hassanien et al. (Eds.): Foundations of Comput. Intel. Vol. 1, SCI 201, pp. 117–142.
springerlink.com   © Springer-Verlag Berlin Heidelberg 2009

this chapter we undertake a theoretical study in the realm of Computational Intelligence (CI), which stands as a laboratory where formal inductive investigations and computational methods are combined. In a typical scenario of base-learning, applying a learner over some data produces an hypothesis which depends on the fixed bias embedded in the learner. As opposite to base-learning, *meta-learning* offers the opportunity to increase the potentiality of the inductive process by adapting its mechanisms to the domain of tasks under study, thus converting the search for the best suitable bias into a dynamical strategy [3, 20, 23, 33, 46, 62, 63]. The formal analysis of the inductive methods performed in this chapter is devoted to assess the foundation of generalisation processes and to evaluate their limitations. By referring to the literature, we intend to provide also a review of computational methods for automatic learning, which have been proposed to increase the efficiency of learning algorithms.

A number of issues are connected with such analysis, involving the characterisation of learning problems in terms of their different complexity; the exploitation of past experience in novel learning contexts (which is known in literature as *lifelong learning* [56]); the problem of knowledge collecting and transferring. We are interested in discussing those questions with specific reference to the field of connectionist learning, reviewing the meta-learning practice as a new perspective for the CI methodologies. In fact, a number of hybridisation strategies can be adopted to overcome the lack of transparency in connectionist systems, which obstacles any kind of guided search different from purely data investigation. In this context, knowledge-based neurocomputing methods [21] may represent the ultimate direction of the meta-learning paradigm: our review includes peculiar methodologies designed on the basis of neuro-fuzzy and neuro-evolutionary hybridisations [1, 14, 17].

On the basis of these assumptions, the chapter is organised as follows. A discussion about the intrinsic limitations of fixed-bias learning methods is carried out in the following section. In section 3, a detailed analysis of the peculiar nature of meta-learning approaches is presented, together with the illustration of different implementation strategies. In section 4 the meta-learning issues are closely related to the realm of connectionist systems, producing motivations for the resorting to knowledge-based methods. Hybridisation strategies are proposed in section 5 as tools for building up meta-learning frameworks based on neural network modelling. Section 6 closes the chapter with a conclusive discussion, pointing out the overall ensemble of critical topics debated.

2 Learning through Fixed Biases

In this section we carry on an analysis of artificial learning methods useful for generating inductive hypotheses from data. Particularly, the *classification mapping* constitutes the main object of investigation in our study. When dealing with classification tasks, a set of input values is related to a set of target output values, referred to as *classes* or *categories*. If the classification task admits only two output values, then it is commonly referred to as a *concept*, halving an instance domain into subsets of positive and negative examples. For the sake of simplicity, the formalisation we are going to propose in the following refers to inductive concept learning (the generalisation to any kind of target function learning is straightforward).

Let $\mathcal{H}$ be an hypothesis space, $\mathcal{C}$ be a concept space and $\mathcal{X}$ be an instance space.

Definition 1. *Let* $\mathcal{T} = \{T \in \mathfrak{P}(\mathcal{X}) \mid |T| \text{ is finite}\}$ *be the training space. A learner* $\mathcal{L}$ *is defined as a mapping:*

$$\mathcal{L} : \mathcal{T} \to \mathcal{H} \tag{1}$$

between the training space $\mathcal{T}$ *and the hypothesis space* $\mathcal{H}$.

Remark 1. The mapping performed by $\mathcal{L}$ is aimed to identify an hypothesis $h \in \mathcal{H}$, in order to approximate a target concept $c \in \mathcal{C}$.

Remark 2. The search for the hypothesis h is performed moving from the analysis of a training set $T \in \mathcal{T}$, whose elements $\mathbf{x}_k$ $(k = 1, \ldots, K)$ are m-dimensional feature vectors, drawn from the instance space $\mathcal{X}$.

Remark 3. In case of induction by examples (supervised learning), the training set T has to include the target output related to each vector $\mathbf{x}_k \in \mathcal{X}$, so that the learner $\mathcal{L}$ could be able to recognise positive and negative examples. Therefore, to characterise the supervised learning regime, we should refer to the following definition of the training space:

$$\mathcal{T} = \{T \in \mathfrak{P}(\mathcal{X} \times \{0,1\}) \mid |T| \text{ is finite}\}, \tag{2}$$

which represents the proper training space for supervised learning tasks, such as the case schematised in figure 1.

The obtained performance of a learner is typically evaluated in terms of generalisation results, that is referring to the learner behaviour when facing a test set, composed of instances drawn from the $\mathcal{X}$ space. In this way, it is possible to discriminate among different inductive hypotheses.

Definition 2. *Given* $h \in \mathcal{H}, c \in \mathcal{C}$, *such that* $h, c : \mathcal{X} \to \{0,1\}$, *let* $\Psi[\cdot, \cdot]$ *be a loss function related to the learner* $\mathcal{L}$. *A successful induction is achieved when the learner* $\mathcal{L}$ *is able to detect the hypothesis* h *minimising the value* $\Psi[h(\mathbf{x}), c(\mathbf{x})]$, *that is the distance between* $h(\mathbf{x})$ *and* $c(\mathbf{x})$, *for all* $\mathbf{x} \in \mathcal{X}$.

To succeed in its task of generalisation, the learner $\mathcal{L}$ cannot neglect a set of assumptions that are collectively embedded into the concept of *bias*. Here we report a definition of bias, in the way it has been expressed in [58]:

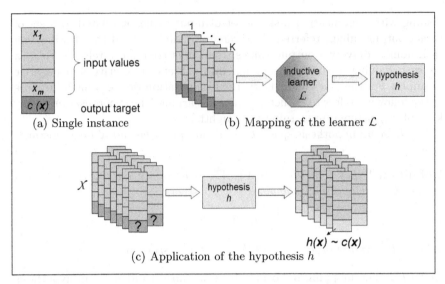

(a) Single instance   (b) Mapping of the learner $\mathcal{L}$

(c) Application of the hypothesis h

Fig. 1 The process of inductive learning by examples: (a) definition of the sample instance $\mathbf{x}_k \in \mathcal{X}$, with $\mathbf{x}_k = (x_{k1}, \ldots, x_{km})$, coupled with the target output $c(\mathbf{x}_k)$; (b) the search of the hypothesis h performed by the learner $\mathcal{L}$; (c) application of the obtained hypothesis h to new instances drawn from $\mathcal{X}$

Definition 3. *The* bias *is constituted by all factors that influence hypothesis selection, except for the presented examples of the concept being learnt. These factors include the criteria adopted to perform the search inside the space of hypotheses and its specific configuration.*

Therefore bias assumptions concretely concern the influence which intervenes over the learning process. The formalisation of the mapping task (1) is commonly referred to as base-learning paradigm [61, 62], with the employment of a fixed form of bias. This, in turn, yields for the learner $\mathcal{L}$ a fixed hypothesis space $\mathcal{H}_{\mathcal{L}}$. Learning is managed at base level since the quality of the hypotheses normally improves with an increasing number of example data. Nevertheless, a bias should prove its appropriateness to produce good performances. Biases are tools of generalisation and an induction may fail when the generalisation is scarce (data overfitting); the generalisation is excessive (overgeneralisation); the generalisation occurs over some wrong properties of training examples (bad generalisation). In some cases the blame for overfitting and overgeneralisation could be laid upon the training set, which may not be representative of the problem at hand. However, our study is aimed at investigating the general laws underlying the learning process, bringing forward the following basic question:

a. Is it possible to evaluate a proper number of training examples to achieve successful learning?

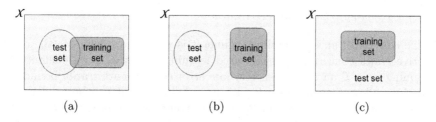

(a)   (b)   (c)

Fig. 2 Three kinds of test set sampling regimes: (a) identically and independently drawn; (b) off-training set; (c) entire instance space

b. There exists an excelling learning strategy that we could apply across the board?
c. Can we recognise classes of problems which are inherently easy (or difficult) to solve?

An attempt to answer these questions can be made by weakening our requests, in order to say more about the learning potential of a fixed-bias learning procedure. In particular, we should content to deal with some looser results, defining specific learning scenarios, as we are going to relate in the following section.

2.1 Formal Results on the Possibilities of Inductive Learning

The optimum performance a learner $\mathcal{L}$ may achieve is expressed by definition 2 in terms of a loss function. However, the performance of $\mathcal{L}$ is directly observable only on instances included into the particular training set T derived from $\mathcal{T}$ to instruct the learner; nothing can be said about the generalisation over the entire instance space $\mathcal{X}$. Generally, the learner behaviour is evaluated over a test set, collecting instances from $\mathcal{X}$ by means of different kinds of sampling regimes. A discriminating factor in this process is represented by determining whether the learner evaluation must be conducted by allowing the employment of already seen instances (that are instances included into the training set). Instances can be identically and independently drawn (iid) from $\mathcal{X}$ to generate a test set. Alternatively, the off-training set regime (ots) prevents test set instances from overlapping with the training set instances. Even the entire instance space may be regarded as a test set: this is a special case of non-ots sampling regime. In figure 2 the differences among the described sampling regimes are graphically illustrated.

In the following, we are going to review the above mentioned learning conditions, presenting some of the results known in literature.

The iid sampling regime and PAC learning

In the context of an iid sampling regime, it is possible to delineate the distinctive traits concerning a supervised learning scenario. If we content of obtaining from $\mathcal{L}$ an hypothesis h whose error could be arbitrarily bounded and if we allow $\mathcal{L}$ to produce h with a probability that can be arbitrarily bounded, then we are asking for a *probably* generation of an hypothesis that is a good *approximation* of a target concept. The probably approximately correct (PAC) learning [59] represents the prevalent paradigm in computational learning theory: assuming a stationary probability distribution $\mathcal{D}$ to generate training instances drawn from the instance space $\mathcal{X}$, the following definition can be stated:

Definition 4. *A concept space $\mathcal{C}$ is PAC-learnable by a learner $\mathcal{L}$ if, for all $c \in \mathcal{C}$, $\delta < 1$, $\epsilon < 1$, there is an m such that, after inspecting m instances, $\mathcal{L}$ is able to produce with probability at least $(1 - \delta)$ an hypothesis h such that, for all $\mathbf{x} \in \mathcal{X}$, $h(\mathbf{x}) \neq c(\mathbf{x})$ with probability smaller than ϵ.*

Remark 4. If a (fixed bias) PAC-learner is employed to produce an hypothesis h, a low generalisation error would be observed with high probability, however generated a test set by means of the probability distribution $\mathcal{D}$ (adopted to draw training instances), together with the iid sampling regime.

Within the PAC-learning scenario formalised in the definition 4, a first relevant result can be assessed, regarding the number m of instances to be inspected in order to achieve satisfactory generalisation outcomes [44].

Theorem 1. *For a finite hypothesis space $\mathcal{H}$, the number m of instances needed to PAC-learn a concept space $\mathcal{C}$ must satisfy the following lower bound:*

$$m \geq \frac{1}{\epsilon} \left(\ln |\mathcal{H}| + \ln(1/\delta) \right). \tag{3}$$

The result expressed by (3) involves the logarithm of the size of the hypothesis space as a measure of the complexity of $\mathcal{H}$, thus implying a difficulty when dealing with an infinite space. To remedy this drawback, a different measure can be considered, based on the relationship existing between the hypothesis space $\mathcal{H}$ and the instance space $\mathcal{X}$. If we consider an hypothesis $h \in \mathcal{H}$ and a subset of instances $S \subseteq \mathcal{X}$, then it is possible to distinguish the instances in S satisfying h from those falsifying the hypothesis values. (In other words, every h imposes some dichotomies on S.)

Definition 5. *The hypothesis space $\mathcal{H}$ is said to* shatter *the set of instances S if it contains enough hypotheses to represent every possible dichotomy of S.*

A measure of the complexity of $\mathcal{H}$ can be expressed in terms of the *Vapnik-Chervonenkis dimension* [60]:

Definition 6. *The Vapnik-Chervonenkis dimension $VC(\mathcal{H})$ defined over $\mathcal{X}$ represents the maximum number of instances that can be shattered by $\mathcal{H}$ (i.e. the size of the largest finite subset of $\mathcal{X}$ shattered by $\mathcal{H}$).*

Beside defining a new measure for expressing the power of an hypothesis set, the VC dimension can be employed also to assess the strength of the bias related to the learner $\mathcal{L}$ performing the mapping (1).

Definition 7. *The* bias strength *entailed by $\mathcal{H}$ can be defined by:*

$$\frac{1}{VC(\mathcal{H})},$$

Remark 5. If $VC(\mathcal{H}) = \infty$ (that is, any subset of $\mathcal{X}$ can be shattered by $\mathcal{H}$, possessing highest representation power), the bias strength is zero (weakest bias).

As an alternative formulation of (3), the following result can be proved [25, 29].

Theorem 2. *The minimum value m for an hypothesis space $\mathcal{H}$ is of the order of:*

$$\frac{1}{\epsilon}\left[VC(\mathcal{H}) + \ln\frac{1}{\delta}\right]. \tag{4}$$

By means of (4) it is possible to observe how PAC-learning depends on the adopted bias, and the number of instances required to PAC-learn a task depends on the hypothesis space considered. Particularly, PAC-learning with $m < \infty$ is only possible if the hypothesis space $\mathcal{H}$ is such that $VC(\mathcal{H}) < \infty$. Learning through a finite number of instances is impossible without a hard bias. The above considerations provide an answer to the first question posed in section 2 when starting our formal analysis.

The ots sampling regime and the no free lunch theorems.

The results expressed by (3) and (4) originates from the assumption of a fixed probability distribution $\mathcal{D}$ and the iid test sampling regime. When we consider generalisation performance only, adopting an off-training set (ots) regime to sample the test set, we could achieve more stringent results about how far a learner can go.

Definition 8. *Let $c \in \mathcal{C}$ be a target concept and $h \in \mathcal{H}$ be its approximating hypothesis derived by a learner $\mathcal{L}$. It is possible to associate to $\mathcal{L}$ the error function err:*

$$err(h, c) = \sum_{\mathbf{x} \in \mathcal{X}} p_{\mathcal{D}}(\mathbf{x})[1 - \delta(h(\mathbf{x}), c(\mathbf{x}))], \tag{5}$$

defined in terms of the probability density function $p_{\mathcal{D}}(\mathbf{x})$ of the stationary distribution of probability $\mathcal{D}$ (being $\delta(\cdot, \cdot)$ the Kronecker delta function).

Remark 6. The reported form of the function *err* refers to a generic iid sampling regime; when dealing with the ots sampling regime, equation (5) should be related also to the training set T.

Definition 9. *The* ots-error function *associated to $\mathcal{L}$ is defined as:*

$$err_{ots}(h, c, T) = \frac{\sum\limits_{\mathbf{x} \notin T} p_{\mathcal{D}}(\mathbf{x})[1 - \delta(h(\mathbf{x}), c(\mathbf{x}))]}{\sum\limits_{\mathbf{x} \notin T} p_{\mathcal{D}}(\mathbf{x})}. \tag{6}$$

Remark 7. By means of (6) we are able to express that the error is evaluated over a test set whose elements do not overlap with the training instances.

Indicating by E the random variable such that

$$E = err(\mathcal{H}, \mathcal{C}, \mathcal{T}),$$

we are going to express the "risk measure", related to the employment of a learner $\mathcal{L}$, by the expected value $\mathbb{E}_{\mathcal{L}}(E|c, T)$. Under the condition of the off-training set error, the conservation law for generalisation performance [50] and the no free lunch (nfl) theorems [67, 68] state that there is no way to find a learning method which can outperform all the others with respect to all possible tasks. In fact, when averaged over all targets, the generalisation performance of any two algorithms is identical. A version of the nfl theorems can be stated as follows [66].

Theorem 3. *For any two learners $\mathcal{L}_A$, $\mathcal{L}_B$ and uniformly averaged over all c, the following result can be proved:*

$$\mathbb{E}_{\mathcal{L}_A}(E|c, T) - \mathbb{E}_{\mathcal{L}_B}(E|c, T) = 0. \tag{7}$$

The question b. raised in section 2, concerning the existence of a preferable learning strategy to be applied in most circumstances, can be addressed at the end of the reported analysis, which ultimately denies any superiority among learners. In particular, if algorithm A outperforms algorithm B in some circumstances, there are different situations where B outperforms A. Moreover, there exists only a tiny fraction of the possible targets where a significant difference can be detected between the performances of A and B [65].

In bias terms, the results expressed by (7) state that a fixed bias cannot tackle every kind of learning problems, but will be limited to a subset that we shall refer to as *domain* or *area of expertise* (there is no holy grail bias). This assertion gives rise to several issues, involving on the one hand the well-known theoretical limitations of classical inductive learning [24], on the other hand the possibilities of novel research directions aimed at developing more effective learning paradigms. In the following sections, we are going to investigate the relationships between learning problems and hypothesis spaces, addressing the problem of determining what type of prior knowledge and information sources can be exploited by a learner.

Fig. 3 Different learning
algorithms covering dif-
ferent regions inside the
universe of tasks

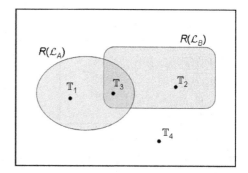

3 Learning through Dynamical Biases

In definition 1 the inductive learning process has been formalised as a search
over an hypothesis space $\mathcal{H}$ and we have pointed out how, in a base-learning
context, a learner $\mathcal{L}$ is compelled to move inside a limited domain of expertise. Figure 3 depicts the universe of all tasks as a rectangular area and the
regions $R(\mathcal{L}_A)$ and $R(\mathcal{L}_B)$ represent the subsets of targets which can be efficiently learnt by the learners $\mathcal{L}_A$ and $\mathcal{L}_B$, respectively. The respective bias
configurations allow the learner $\mathcal{L}_A$ to better learn task $\mathbb{T}_1$ and the learner
$\mathcal{L}_B$ to fit target $\mathbb{T}_2$, while task $\mathbb{T}_3$ can be efficiently tackled by both learners. Task $\mathbb{T}_4$, not being included inside the regions $R(\mathcal{L}_A)$ and $R(\mathcal{L}_B)$, will
be poorly learnt by the two learners in the example. The urge toward the
meta-learning approach is intended to understand what helps a learner $\mathcal{L}$
to dominate inside a region $R(\mathcal{L})$ (by analysing the properties of $\mathcal{L}$ and the
properties of the tasks in $R(\mathcal{L})$), and how it could be possible to shift the
region $R(\mathcal{L})$ according to the tasks at hand (by modifying the bias of $\mathcal{L}$).

Actually, various meanings can be found in literature coupled with the
meta-learning catchword [19, 27, 38, 52]. The wholeness of references, however, agrees to conceive the meta-learning approach as a research apparatus
for discovering learning algorithms that improve their efficiency through experience. The aim is to extend the traditional base-learning mechanism, to
provide a meta-learning approach for moving from a fixed bias scenario to a
dynamical bias discovering.

In the following sections, we are going to take account of various mechanisms designed to let an artificial intelligent system develop a more articulated
learning strategy than those based on fixed biases.

3.1 Bias Combination, Bias Selection, Transfer of Learning

To overcome the limitations of base-learning, an approach consists in simply
grouping a number of fixed bias techniques to let the single biases compensate each other. This kind of strategy is often referred to as *ensemble methods*

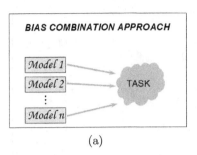
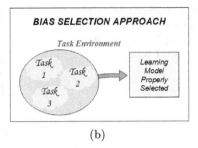

(a)   (b)

Fig. 4 Different strategies of meta-learning: (a) the bias combination and (b) the bias selection approaches

or *model combination* [26]. It may take place by employing one model only: different hypotheses are drawn using different training sets with the same learning procedure. In this context, several procedures may be developed by simply differentiating the sampling mechanism for the example data. This is the case of *bagging* [9], a classification technique which produces several training sets by randomly sampling with replacement from the original training set, and *boosting* [51] where training sets are generated adopting different distributions of probability in the sampling process.

More appropriate bias combination approaches are based on the employment of several learners, whose arrangement in a single ensemble can be accomplished in a simple way just by grouping the various hypotheses deriving from a number of learners applied on the same task. Multistrategy hypothesis boosting [18] combines different learners to improve prediction accuracy. In *stacked generalisation* [64] bias combination takes place by redefining the input representation for the task: the training set is replaced by the vector of classifications generated by the involved learners and the final hypothesis emerges from the "stacked" levels of task redescriptions.

In the bias combination approaches no dynamic selection of bias occurs and the dominant region remains ultimately fixed. Moreover, the trade-off for an accuracy enhancement is usually paid in terms of interpretability, since ensembles provide no comprehensible explanation for their judgements.

As an alternative to combination, bias selection consists in devising suitable methods for choosing appropriate learning algorithms. In general, this approach relies on characterisation of tasks, so that it could be possible to produce a mapping between the task characteristics and some kind of learning strategies. Practically, it is possible to employ different ways of characterisation, the most popular of them being based on the extraction of a number of measures from the data related to the tasks to be tackled. In particular, it has been variously studied how statistical and information-theoretic measures can be fruitfully exploited to discriminate the effectiveness of learning algorithms in solving distinct problems [13, 39, 43]. Moreover, it could be possible to produce a ranking of learning algorithms on the basis of their performances [7, 8, 47]. Different characterisations have been also proposed, where features other than those derived from data are adopted for discrimination. In landmarking [45] the performance of several models is related to the performance obtained by simpler and faster designed algorithms, called landmarkers. Other approaches are based on the employment of decision-tree induced from data to characterise the tasks [5].

The selection process is commonly focused more on task characterisation than learning model design. In this way, bias selection strategy differs from model combination in the sense that, instead of grouping a set of learners for a single target, it aims at gathering a group of tasks in terms of their bias requirements, in order to define a proper learning method for all of them (in figure 4 the bias combination and bias selection approaches are graphically represented).

Defining a task environment is a somewhat troublesome problem, related to the last basic question raised in section 2. In general, we can refer to [32], where a task environment is defined as a group of functions or tasks sharing a class of properties. In [4] the task environment is defined by associating to the set of all possible learning problems a probability distribution, determining the frequency of each problem[1]. An example of environment is a collection of real-world inductive problems, with an appropriate distribution of biases associated, that a successful inductive strategy aims to approximate. In any case, the group of related tasks sharing the same environment exhibits valuable properties, since a learning system applied to the entire environment is able to improve its performance, with respect to the results obtained if the single tasks are independently tackled.

The difficulties related to the bias selection approach can be relaxed if we consider a learning system which does not discover the relatedness of the tasks presented, but assumes it. Such a mechanism of bias tuning, performed by examining a group of tasks within an environment, is referred to as *transfer of learning*. If the tasks are indeed related, transfer can successfully ease learning. To some extent, transfer of learning process steps aside from bias selection mechanisms, since a bias is selected for a group of tasks, but there is no general information about connections between learners and tasks; no posterior bias selection for new problems is therefore favoured.

Transfer of learning has been chiefly explored in the context of connectionist learning: within this learning scenario, biases can be expressed in terms of network architecture and parameters, and these features are shared by networks engaged on related tasks. Transfer of learning has been implemented in different ways. Transfer in sequence has been exemplified by the consolidation system [53, 54]. This mechanism preserves the network parameters for the already learnt tasks and uses them for initiating the investigation of new, related tasks. In [48] a framework is introduced for the communication

---

[1] In this way, the task environment is defined as the pair $(\mathcal{C}, Q)$, being $\mathcal{C}$ the universe of tasks (assuming concept learning problems) and Q a probability distribution on $\mathcal{C}$, controlling which learning problem the learner is likely to observe.

between agents which exchange each other the knowledge deriving from neural networks. This kind of network transfer is useful to speed up the learning process of the networks, improve their accuracy and reduce the number of training samples for future learning.

Transfer of learning paves the way to a broader debate concerning the relevance of knowledge representation and reuse for learning models involved in real-world contexts, with particular reference to connectionist approaches. In the following section, we are going to deepen our analysis of inductive learning mechanisms by focusing the investigation on the realm of artificial neural systems, observing how knowledge-based methods could be helpful for developing peculiar forms of meta-learning.

4   Knowledge-Based Methods in Neurocomputing

In our account for the definition of the proper bias for a learning model, here we refer to the particular domain of neurocomputing methods. As known, artificial neural networks (ANNs) provide a general method for learning functions from examples, thus representing one of the chief tool of induction, and find application in many practical situations [6, 34, 35, 49]. Our intent consists in analysing ANNs as a particular type of base-learning models, enclosing their study in the framework of the meta-learning strategies.

We start our investigation by examining the bias/variance dilemma and its implications into the knowledge-data trade-off. In particular, we are going to pinpoint the role of knowledge in mitigating the dilemma, dealing with the particular learning scenario related to connectionist models. In this way, it is possible to assess the relevance of knowledge-based mechanisms in enhancing the learning capabilities of neural networks. Additionally, we shall deepen the analysis of the so-called multitask learning model.

4.1   The Bias/Variance Dilemma and the Prior Knowledge Exploitation

In the previous sections it has already been asserted how bias-free inference is doomed to produce poor generalisation performances when attempting to learn complex problems. In the field of statistical research, model-free estimation (also known as *tabula rasa* learning) has been investigated with reference to the nonparametric statistical inference. The limitations of nonparametric methods have been notably understood in what is known as *bias/variance dilemma*:

> *"The essence of the dilemma lies in the fact that estimation error* [of a learning model] *can be decomposed into two components, known as bias and variance; whereas incorrect models lead to high bias, truly model-free inference suffers from high variance. Thus, model-free (tabula rasa) approaches to complex*

inference tasks are slow to "converge", in the sense that large training samples are required to achieve acceptable performance." [31]

Again, the bias is summoned to play a role in evaluating the effectiveness of a learning model: bias-based mechanisms should be employed in order to limit the side-effects due to the variance contribution. Formally, the bias/variance dilemma is obtainable from a statistical analysis of the generalisation error, performed when we want to derive an hypothesis h approximating a target concept c. Admitting an uncertainty degree in determining $c(\mathbf{x})$ (being $\mathbf{x} \in \mathcal{X}$), we shall consider the expectation $\mathbb{E}[c(\mathbf{x})|\mathbf{x}]$.

Theorem 4. *Let $\mathcal{T}_K$ be the space of all the training sets of size K and $T = \{\mathbf{x}_1, \ldots, \mathbf{x}_K\}$ a training set in $\mathcal{T}_K$. Let h_T denote an hypothesis generated from the analysis of the data included in $T \in \mathcal{T}_K$. The estimation error of h_T in approximating c can be decomposed as:*

$$\mathbb{E}_T[(h(\mathbf{x}, T) - \mathbb{E}[c(\mathbf{x})|\mathbf{x}])^2] = \underbrace{(\mathbb{E}_T[h(\mathbf{x}, T)] - \mathbb{E}[c(\mathbf{x})|\mathbf{x}])^2}_{(bias)^2} +$$

$$+ \underbrace{\mathbb{E}_T[(h(\mathbf{x}, T) - \mathbb{E}_T[h(\mathbf{x}, T)])^2]}_{(variance)} \qquad (8)$$

where $\mathbb{E}_T$ denotes the expectation with respect to the average over each training set $T \in \mathcal{T}_K$.

Statistical bias can be related to the notion of bias that we defined in the artificial learning context. When we focus on neural network analysis, the bias measures the extent to which the average (over all training sets) of the network function differs from the desired function $\mathbb{E}[c(\mathbf{x})|\mathbf{x}]$. Conversely, the variance measures the extent to which the network function is sensitive to the particular choice of the dataset [6]. In this way, it is straightforward to spotlight how bias-free models relapse to a kind of futile blind search, where the large amount of variance can be compensated only by a virtually limitless training set, while learning models characterised by restricted domains of expertise are endowed with an excessively hard bias, which makes them inflexible and unable to generalise adequately.

The bias/variance dilemma expressed by (8) is such that reducing bias often increases variance and vice versa. Commonly in connectionist systems the employment of additional hidden units is suggested to increase the complexity of the model, thus reducing the bias component. Nevertheless, this kind of approach leads also to a boost in variance amount, in contrast with the requirement of minimising the total generalisation error. To some extent, the bias/variance dilemma can be brought back to the knowledge-data trade-off which reflects the bivalent nature of learning. Two ideal conditions, in fact, can be identified, corresponding on the one hand to a complete and correct base of knowledge (yielding a form of analytical learning that has no need of training instances), on the other to a purely data-driven learning (where the total lack of prior knowledge should be compensated by a virtually limitless data availability). Obviously, in real case situations these asymptotic conditions attenuate into a quite different context where originates from some knowledge domain – which often turns out to be imprecise and lacunose – and a certain finite amount of data.

The powerful processing capabilities of neural networks contrast with their intrinsic impermeability to human understanding, since the numeric representation embedded into a neural model is hardly comprehensible for users (neural networks regarded as black boxes). Prior knowledge can be exploited in several ways as a form of hint to assist neural learning: the paradigm of *Knowledge Based Neurocomputing* (KBN) promotes the use of explicit representation of knowledge inside a neurocomputing system [21]. The key point for KBN approaches is that knowledge has to be injected inside or obtained from a connectionist model in such a format that can be easily understood by humans (that is, in a symbolic or well-structured form). In this way, generalisation performance can be enhanced by exploiting the prior knowledge related to the problem to be learnt: this mechanism contributes in specifying a preference for a particular hypothesis (or class of hypotheses), thus acting as a form of bias for the mapping task (1) performed by a neural network. This attempt to look inside the black box delineates a meta-learning approach in connectionist contexts.

In the next section, we are going to review a peculiar form of hint deriving from a knowledge sharing process which is represented by *multitask learning*, an approach aimed at learning a bias by identifying the shared features in a set of related problems.

4.2 Multitask Learning

Training samples give an account about the particular problem at hand, while the information supplied by prior knowledge is usually more general, encompassing also the domain where the problem evolves. Usually, training data are useful when a specific task has to be learnt, and prior information may be employed to generalise over any problem inside a larger environment. Conversely, multitask learning represents an attempt to perform a kind of transfer of learning (see section 3.1), aiming at exploiting training samples concerning a number of related tasks, to extract more general domain-related knowledge. The search in the hypothesis space is devoted to the identification of a class of solutions which could be helpful both in identifying the proper hypothesis for a particular task and in easing the application of the learning model to different problems. By doing so, multitask learning defines a way for learning a bias, based on the common properties of a set of tasks.

In order to better focalise the multitask learning methodology, we will inspect its basic principles from a theoretical point of view, referring to the mathematical formalisation proposed in [32]. The keystone of the methodology consists in representing a family of hypotheses by a *manifold* (or a mixture of manifolds) inside the hypothesis space. Moving from the analysis of different training samples, collected from a number of related tasks, it could be possible to argue the position of the manifold on which all the hypotheses belonging to the family are lying. The solution for a particular task can be identified by performing a search over the manifold to locate a point on it (instead of considering the entire hypothesis space).

Definition 10. *Let $\{c_n\} = (c_1, \ldots, c_N)$ an ensemble of N related concept tasks, to be studied in a supervised learning regime, sharing the same task domain. Let $T_n = \{(\mathbf{x}_{kn}, c_n(\mathbf{x}_{kn}))\}_{k=1}^{K_n}$ be the training set associated to the concept task c_n, composed of K_n training examples. The hypothesis h_n approximating the concept task c_n can be expressed as:*

$$h_n = g(\varphi, \alpha_n), \tag{9}$$

where g defines the particular form of the manifold, φ indicates the position of the surface with respect to the hypothesis space and α_n represents the exact position of the task-specific hypothesis h_n on the manifold, among all the hypotheses located on the surface.

Remark 8. By means of the definition 10, the output of an hypothesis h_n, when applied on the instances of a particular training set T, can be expressed as:

$$h_n(\mathbf{x}) = g(\varphi, \alpha_n)(\mathbf{x}), \tag{10}$$

with $\mathbf{x} \in T$.

Remark 9. In the single-task learning, the hypothesis is expressed in terms of a point on the manifold:

$$h_n = g^\star(\alpha_n),$$

being $g^\star$ a particular form of the function g degenerated into a single point.

Different task families correspond to different manifold surfaces, whose topology depends on the particular domain. In figure 5 two examples of manifolds are reported, with F_1 and F_2 illustrating the shapes of different task families.

Definition 11. *The empirical risk function associated to the multitask learning of $\{c_n\}$ is defined as:*

$$ER_{ml}(\alpha_1, \ldots, \alpha_N, \varphi) = \frac{1}{N} \sum_{n=1}^{N} \frac{1}{K_n} \sum_{k=1}^{K_n} \Psi[g(\varphi, \alpha_n)(\mathbf{x}_{kn}), c_n(\mathbf{x}_{kn})], \tag{11}$$

where $\Psi[\cdot, \cdot]$ is the loss function that evaluates the effectiveness of each learnt hypothesis.

Remark 10. A learning model involved in the search for the suitable hypotheses for $\{T_n\}$, has to minimise the empirical risk in definition 11.

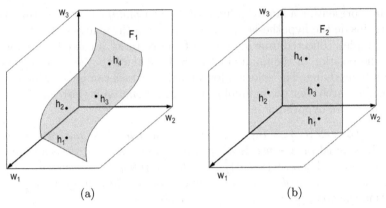

(a) (b)

Fig. 5 Two examples of manifolds: (a) generic shape manifold; (b) affine, axis-aligned manifold. The points h_i $(i = 1, \ldots, 4)$ represent the hypotheses derived for four distinct tasks sharing the same environment; the hypotheses belong to the classes of solutions F_1 in case (a) and F_2 in case (b). It can be noted how the affine axis-aligned manifold depicted in (b) can be referred to an application of the shared representation method [11]: four neural networks, each composed by three layers, are trained sharing one of the layers. The weights of the three layers assume different values along the axes w_1, w_2, w_3 and, in this particular example, the shared layer corresponds to the one moving on w_1

Remark 11. The empirical risk function associated to the single-task learning is defined as:

$$ER_{sl}(\alpha_n) = \frac{1}{K_n} \sum_{k=1}^{K_n} \Psi[g^\star(\alpha_n)(\mathbf{x}_{kn}), c_n(\mathbf{x}_{kn})]. \tag{12}$$

A comparison between the equations (11) and (12) helps us to highlight the distinctive characteristics of multitask learning: differently from the case (12), all the tasks are simultaneously involved and the training samples are employed to identify the suitable parameters $(\alpha_1, \ldots, \alpha_N, \varphi)$ at once. On the other hand, single task learning does not provide for a transfer of learning, since the information is not retained and exploited while tackling different related tasks.

A particular implementation of the multitask learning methodology can be found in [11], where the training samples of a group of related tasks are employed to improve the generalisation performance of neural networks. The central tenet of this approach consists in sharing what is learnt by different networks while tasks are tackled in parallel. To achieve this aim, instead of simultaneously training similar neural networks on a number of tasks, a single connectionist system is realised, which exhibits the same topology of the original networks, but joins their outputs in a single configuration. In other words, the neural networks are trained in such a way that they share their first layers and differentiate in the last layers, thus learning the multiple tasks in parallel while using a shared representation.

5 Hybrid Methods for Meta-learning

In our excursus concerning different forms of meta-learning strategies, we dedicate this section to some kinds of *hybrid approaches*, i.e. particular techniques which realise meta-learning processes by properly combining several methods from a number of distinct research fields. To some extent, the hybrid approaches can be brought back to the topic of knowledge exploitation, discussed in the previous section. In a recent line of enquiry, related to philosophy of science, the mechanism of hybridisation has been appraised as a more correct attitude for developing consistent research in the field of Artificial Intelligence (AI) [22]. Hybridisation, in fact, appears to be a much more effective practice to produce successful models, in place of the abused appeal to "paradigms" (that disorderly evolve in a quite exaggerated number in AI contexts, with respect to what happens in more consolidated sciences, such as physics).

Our intent consists in analysing some integration methodologies which represent plausible approaches in neurocomputing, emphasising a new perspective of Computational Intelligence related to the meta-learning possibilities offered by the hybridisation of neural networks with different models. In particular, here we examine the integration of neural models with two kinds of approaches that have proved their effectiveness in combining with connectionist systems in a plethora of applications: *evolutionary programming* and *fuzzy logic*. The employment of evolutionary approaches for implementing meta-learning strategies represents the basis of a recent work for the adaptive optimisation of neural networks [1]. Besides, a particular neuro-fuzzy learning scheme constitutes the core of an AI framework proposed as an illustrative model of meta-learning mechanism which includes the possibility of lifelong learning [14].

5.1 Supporting the Bias Search: The Employment of Evolutionary Techniques

Meta-Learning Evolutionary Artificial Neural Networks (MLEANN) [1] have been proposed as an automatic computational framework, intended to adaptively define the architecture, connection weights and learning algorithm parameters of a neural network, according to the problem at hand. Moving from sceptical considerations about the possibility of defining once and for all particular network configurations yielding superior learning capabilities in different tasks (in accordance with the no free lunch theorems), the recourse to evolutionary algorithms is stimulated to escape the task of manually designing a neural system which could be smaller, faster and capable of better generalisation performances.

An evolutionary program is an iterative probabilistic algorithm, used in optimisation problems, which maintains a population of individuals for each iteration [30, 36, 42]. Each individual, representing a potential solution for

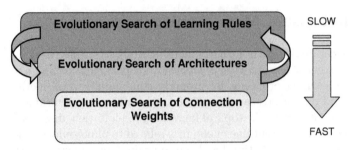

Fig. 6 The general scheme of the MLEANN framework

the problem at hand and implemented as some data structure (genotype encoding), is evaluated in terms of a particular function, which is specifically designed for the problem, in order to express the fitness of the individuals. In this way, at each iteration a new population can be generated on the basis of the fittest elements: they are selected (according to some prefixed rule) to produce an offspring by means of "genetic" operators which are usually represented by unary transformations (mutation), and higher order transformations (crossover). The successive generations of several populations converge to a final offspring (hopefully) encoding an optimal solution. The tedious trial and error process, performed to find an optimal network, may be eliminated by employing the evolutionary search procedures, with the advantage to adapt the neural models to a dynamic environment [69, 70, 71].

Particularly, the general framework for MLEANNs proposed in [1] is structured in such a way to articulate the evolution at various levels. At the lowest level, network weights are evolved to replace weight training; at the next level, evolution is introduced into network architecture; at the highest level, the learning mechanisms become matter of evolution. Evolutionary search of connection weights is formulated as a global search on the basis of a prefixed architecture: the weight configurations are represented by binary strings and proper genetic operators are applied to generate new individuals. This kind of algorithm appears to be less dependant on the initial weight settings when compared with gradient based techniques. A similar procedure has been designed to introduce evolution at the architecture level, provided a prefixed learning algorithm and a suitable coding scheme to translate a network topology into a genotype representation. Finally, the evolutionary search for learning rule parameters is carried on by encoding the genotypes as real-valued coefficients. The general scheme of the MLEANN framework is depicted in figure 6.

The fitness values of the randomly generated networks at lowest levels are evaluated at the learning rule level in a parallel fashion employing different learning algorithms, and then the evolutionary steps (parents selection, offspring reproduction) are applied as usual. In this way, evolutionary training acts as a form of attraction basin search procedure (to locate a good region in the space), which is improved by incorporating into the evolution the local search implemented by the learning algorithms. The right arrow in figure 6 indicate that, progressing from higher to lower levels, evolution takes place at a faster time scale. The figure shows also that, according to the prior knowledge available, a reasonable approach consists in exchanging the learning rules and architecture levels (for instance, when information about network topology is applicable or a particular class of architectures is pursued). A number of experimental results are reported in [1] to demonstrate how the MLEANN approach can be helpful in automatically designing neural models, characterised by simpler structure and better generalisation performance.

5.2 Expressing Knowledge by Rules: the Neuro-fuzzy Integration

In the context of CI, one of the most productive approach is represented by the fruitful marriage between the neural learning models and the fuzzy logic. There has been a good deal of studies devoted to understand the special role of fuzzy logic among classical knowledge-based approaches. Actually, the peculiarity of fuzzy reasoning relies in its capacity of adapting to real-world situations: by admitting multivalued membership degrees, fuzzy sets propose a more corresponding modelling of phenomena than binary dichotomies of classical logic [72]. Moreover, even if fuzzy logic renounces the rigid correctness of traditional mathematical methods, still preserves a formal apparatus that enables a rigorous treatment of reasoning processes. The correspondence with typical human cognitive patterns is further enhanced by the natural inclination of fuzzy approaches toward natural language, allowing a form of "computing with words" [73].

Although connectionist and fuzzy approaches basically refer to different computational strategies (being the first ones related to inductive methods to learn from examples, the others to logical reasoning for knowledge manipulation), they share a similar way to process inexact information. Additionally, when we turn to consider fuzzy systems, which store information as banks of fuzzy rules, we are concerned with methods to estimate functions or models with partial descriptions of the overall system behaviour. Experts may provide this heuristic knowledge, or neural networks may adaptively infer it from sample data. In this way, the neuro-fuzzy integration allows the fuzzy inference systems to adjust linguistic information using numerical data on the one hand, enabling neural networks, on the other hand, to manage interpretable knowledge [37, 40, 41].

Moving from the above considerations, we are going to present the MIND-FUL (Meta-INDuctive neuro-FUzzy Learning) framework, embodying a meta-learning strategy based on the neuro-fuzzy integration: it represents the subject of our current line of research and it has been thoroughly described in [14]. MINDFUL represents a peculiar approach which differs to some extent from the majority of meta-learning schemes examined in our review: most of them end up assimilating the concept of bias with the specific learner dynamically chosen for the task at hand. Model combination and model selection approaches, for instance, may involve different learning strategies which constitute a pool of bias-specific candidates. In MINDFUL, a single learning scheme is employed, endowed with the capability of improving its performance: a neuro-fuzzy system plays the twofold role of base-learner (to tackle ordinary predictive learning tasks) and meta-learner (to produce some form of meta-knowledge). By doing so, the proposed strategy is characterised on a number of key points. Firstly, the idea of meta-learning is translated to a more qualified level, since it is not intended as simply picking a learning procedure among a pool of candidates, but it focuses on a deeper analysis of the learning model behaviour, in order to understand and possibly to improve it. Moreover, the choice for a single learning model should be able to preserve the uniformity of the whole system and reduce its complexity, even in terms of comprehensibility. Finally, the neuro-fuzzy strategy, applied both at base and meta-level, endows also the meta-learning procedure with the benefits deriving from the integration of the connectionist paradigm with fuzzy logic.

Descending into details, the framework is structured on the basis of two predictive models employing the same neuro-fuzzy learning scheme. In particular, a learner $\mathcal{L}_{base}$, working on a training set T_{base} available for a particular problem, is responsible of the inductive process for realising the mapping (1). Starting from the analysis of the input data in T_{base}, the neuro-fuzzy model produces a fuzzy rule base which codifies the processed information in a linguistically comprehensible fashion. Each fuzzy rule is expressed in the form:

$$\text{IF } x_1 \text{ is } A_1 \text{ AND} \ldots \text{AND } x_n \text{ is } A_n$$
$$\text{THEN } y_1 \text{ is } b_1 \text{ AND} \ldots \text{AND } y_m \text{ is } b_m, \tag{13}$$

where A_i are fuzzy sets defined over the input components x_i, $i = 1, \ldots, n$; b_j are fuzzy singletons defined over the output components y_j, $j = 1, \ldots, m$. Gaussian membership functions are employed to describe the fuzzy sets A_i, and the fuzzy inference system evaluates the fulfilment degree of each rule using the product operator as the particular T-norm interpreting the AND connective. The inferred output is computed by evaluating the average of the activation strengths corresponding to each rule in the fuzzy rule base, weighted by the fuzzy singletons. The described fuzzy inference model is correlated with a particular neural network designed on purpose, the neuro-fuzzy network [12], reflecting in its topology the structure of the fuzzy inference system. In fact, the free parameters of the neuro-fuzzy network are associated to the parameters of the Gaussian functions and fuzzy singletons. The learning scheme of the network is articulated in two successive steps, intended to firstly initialise a knowledge structure and then to refine the obtained fuzzy rule base. An unsupervised learning of the network is performed to cluster the input data and initialise the fuzzy rule base. The knowledge base is successively refined by a

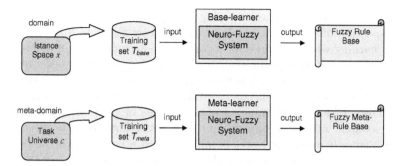

Fig. 7 The working scheme proposed for the base and the meta-learner

supervised learning of the network, thus enhancing the accuracy of the fuzzy inductive inference.

Based on the same neuro-fuzzy working engine, a meta-learner $\mathcal{L}_{meta}$ attempts to learn how to modify the learning bias when different kinds of tasks are faced, thus acting as a supporting unity for $\mathcal{L}_{base}$. To achieve this aim, a meta-training set T_{meta} has to be inspected, which establishes a relationship between a set of domain characteristics (meta-features), pertaining to the base-level predictive tasks, and the base-learner parameter settings, which stand as a form of learning bias for $\mathcal{L}_{base}$. Particularly, the meta-features representing task properties can be derived from the analysis of each base-level training set T_{base} inspected by $\mathcal{L}_{base}$. A number of descriptors are extracted, ranging from general meta-features (number of input components, number of output components), to statistical and information content meta-features (correlation coefficient, class entropy, noise-signal ratio, etc.). The definition of the bias involves the parameter settings engaged in the previously described neuro-fuzzy modelling strategy pertaining to $\mathcal{L}_{base}$, particularly the bias configurations which proved their usefulness in constructing effective predictive models during the activity of $\mathcal{L}_{base}$. The collection of the meta-features and the parameter settings inside the meta-training set T_{meta} is a prelude for the meta-learning activity of $\mathcal{L}_{meta}$. The neuro-fuzzy working scheme is replied to express in form of fuzzy rule base a kind of meta-knowledge encompassing the cross-task level information, which can be employed to improve the effectiveness of subsequent base-learning experiences. In this way, the overall working scheme of the MINDFUL framework (graphically represented in figure 7) configures as a parameter selection method of meta-learning, which can be related to similar strategies proposed in literature [28, 55]. A massive session of experiments (performed on synthetic and real-world data) demonstrates the effectiveness of the MINDFUL framework in dealing with established task domains, even when compared with different approaches [14].

Additionally, the MINDFUL framework paves the way for a more comprehensive meta-learning experience. Actually, it should be noted that, while T_{base} depends solely on the availability of training samples for the task at hand, the construction of the meta-training set T_{meta} may take place only after the inspection of predictive problems executed by $\mathcal{L}_{base}$, in a form of incremental learning which provides for a continuous accumulation of meta-knowledge. This kind of strategy presents some analogies with the transfer of learning (see section 3.1) and the lifelong learning framework. In fact, moving from the observation that the continuous evaluation of different tasks is one of the key points of human learning success, the idea of studying learning in a lifelong context offers the opportunity to transfer knowledge between learning problems. If a learner could be able to face a collection of tasks over its entire lifetime, it might employ the information accumulated in the previous $n - 1$ tasks to solve the n-th problem, thus implicitly performing a dynamical search of the right bias, in a form of "learning to learn" approach [57]. In this way, on the one hand it could be possible to achieve better accuracy performances, on the other hand good generalisation results may be pursued even when scarce data are available.

6   Summary and Discussion

The applied research in the field of artificial intelligent systems often deals with empirical evaluations of learning algorithms to illustrate the selective superiority of a particular model. This kind of approach, with multiple models evaluated on multiple datasets, is characterised by a "case study" formulation (which has been notably addressed in literature [2, 10]). The selective superiority demonstrated by a learner in a case study application reflects the inherent nature of the so-called base-learning strategies, where data-based models exhibit generalisation capabilities when tackling a particular task. Precisely, base-learning approaches are characterised by the employment of a fixed bias, that is the ensemble of all the assumptions, restrictions and preferences presiding over the learner behaviour. This means a restricted domain of expertise for each learner, and a reduction in its overall scope of application. The limitations of base-learning strategies can be stated even by referring to some theoretically established results: the no free lunch theorems express the fundamental performance equality of any chosen couple of learners (when averaged on every task), and deny the superiority of specific learning models outside the case study dimension.

In this context, the chapter proposes an organic perspective in the realm of meta-learning, presenting a survey with the aim of critically inspecting (even if not exhaustively) the dynamical bias discovering approach. We started examining some kind of basic techniques, such as the bias combination processes, where the simple ensemble of multiple models could lead to an improvement of generalisation performances. Successively, we progressed toward the analysis of more comprehensive approaches, involving the concept of transfer of learning and considerations related to the similarity assessment between tasks. The attention has been mainly focused on a particular type of inductive learning model, represented by neural networks, and we referred to the knowledge-based approach as a research line which attempts to fill the gap between knowledge-driven and data-driven learning, thus profiting of prior knowledge to attune the bias in neural generalisation. The field of connectionist learning has been further investigated by illustrating different approaches aimed at realising meta-learning capabilities in neural network systems. On the one hand, the multitask learning methodology has been introduced as a tool for bias learning, based on the common properties of a set of related tasks. On the other hand, it was argued how the integration of neural networks with other computational strategies may be beneficial to define hybrid methods with dynamical bias discovering capabilities. Particularly, we referred to the MLEANN and the MINDFUL frameworks, which involve the adoption of evolutionary strategies and neuro-fuzzy combination, respectively. In this way, a novel perspective can be established for Computational Intelligence, consisting in the augmented possibilities provided by the hybridisation strategies on the way of meta-learning attainment.

References

1. Abraham, A.: Meta-learning evolutionary artificial neural networks. Neurocomputing Journal 56, 1–38 (2004)
2. Aha, D.W.: Generalizing from case studies: a case study. In: Proceedings of the Ninth International Conference on Machine Learning (MLC 1992) (1992)
3. Anderson, M.L., Oates, T.: A review of recent research in reasoning and metareasoning. AI Magazine 28(1), 7–16 (2007)
4. Baxter, J.: A model of inductive bias learning. Journal of artificial intelligence research 12, 149–198 (2000)
5. Bensusan, H., Giraud-Carrier, C., Kennedy, C.: A higher-order approach to meta-learning. In: Proc. of the ECML 2000 workshop on MetaLearning: Building Automatic Advice Strategies for Model Selection and Method Combination, pp. 109–117 (2000)
6. Bishop, C.M.: Neural Networks for Pattern Recognition. Oxford University Press, Oxford (1995)
7. Brazdil, P., Soares, C., Costa, J.: Ranking learning algorithms: using IBL and meta-learning on accuracy and time results. Machine Learning 50(3), 251–277 (2003)
8. Brazdil, P.B., Soares, C.: Ranking classification algorithms based on relevant performance information. In: Proc. of the ECML 2000 Workshop on Meta-Learning: Building Automatic Advice Strategies for Model Selection and Method Combination, Barcelona, Spain (2000)
9. Breiman, L.: Bagging Predictors. Machine Learning 24, 123–140 (1996)
10. Brodley, C.: Addressing the selective superiority problem: automatic algorithm/model class selection. In: Proceedings of the Tenth International Conference on Machine Learning (MLC 1993) (1993)
11. Caruana, R.A.: Multitask learning. Machine Learning 28, 7–40 (1997)

12. Castellano, G., Castiello, C., Fanelli, A.M.: KERNEL: a System for Knowledge Extraction and Refinement by NEural Learning. In: Damiani, E., Howlett, R.J., Jain, L.C., Ichalkaranje, N. (eds.) Knowledge-Based Intelligent Information Engineering Systems & Allied Technologies, pp. 443–447. IOS Press, Amsterdam (2002)
13. Castiello, C., Castellano, G., Fanelli, A.M.: Meta-data: Characterization of input features for meta-learning. In: Torra, V., Narukawa, Y., Miyamoto, S. (eds.) MDAI 2005. LNCS (LNAI), vol. 3558, pp. 457–468. Springer, Heidelberg (2005)
14. Castiello, C., Castellano, G., Fanelli, A.M.: MINDFUL: a framework for Meta-INDuctive neuro-FUzzy Learning. Information Sciences 178, 3253–3274 (2008)
15. Castiello, C., Fanelli, A.M.: Hybrid strategies and meta-learning: an inquiry into the epistemology of artificial learning. Research on Computing Science 16, 153–162 (2005)
16. Castiello, C., Fanelli, A.M.: Hybrid Systems for Meta-Learning (Part I): Epistemological Concerns. In: Proceedings of the International Conference on Computational Intelligence for Modelling Control and Automation (CIMCA), pp. 180–185 (2005)
17. Castiello, C., Fanelli, A.M., Torsello, M.A.: The Mindful system: a meta-learning strategy into action. In: Proceedings of Information Processing and Management of Uncertainty in Knowledge-Based Systems (IPMU 2006), pp. 548–555 (2006)
18. Chan, P.K., Stolfo, S.J.: Experiments on multistrategy learning by meta-learning. In: Proc. Second International Conference Information and Knowledge Management, pp. 314–323 (1993)
19. Chan, P.K., Stolfo, S.J.: Toward parallel and distributed learning by meta-learning. In: Working Notes AAAI Work. Know. Disc. Databases, pp. 227–240 (1993)
20. Chan, P.K., Stolfo, S.J.: On the accuracy of meta-learning for scalable data mining. J. Intelligent Information Systems 9, 5–28 (1997)
21. Cloete, I., Zurada, J.M. (eds.): Knowledge-Based Neurocomputing. The MIT Press, Cambridge (2000)
22. Cordeschi, R., Tamburrini, G.: Intelligenza artificiale: la storia e le idee. In: Burattini, E., Cordeschi, R. (eds.) Intelligenza Artificiale. Manuale per le discipline della comunicazione. Carocci (2001)
23. Desjardins, M., Gordon, D.: Evaluation and selection of bias in machine learning. Machine Learning 20, 5–22 (1995)
24. Dietterich, T.G.: Limitations of Inductive Learning. In: Proceedings of the sixth international workshop on Machine Learning (IWML 1989), pp. 124–128 (1989)
25. Dietterich, T.G.: Machine Learning. Annual Review of computer science 4, 255–306 (1990)
26. Dietterich, T.G.: Machine Learning Research: Four Current Directions. AI Magazine 18(4), 97–136 (1997)
27. Domingos, P.: Knowledge Discovery Via Multiple Models. Intelligent Data Analysis 2, 187–202 (1998)
28. Duch, W., Grudzinski, K.: Meta-learning: searching in the model space. In: Proc. of the Int. Conf. on Neural Information Processing (ICONIP), pp. 235–240 (2001)
29. Ehrenfeucht, A., Haussler, D., Kearns, M., Valiant, L.: A general lower bound on the number of examples needed for learning. In: Proceedings of the 1988 workshop on computational learning theory, pp. 110–120. Morgan Kaufmann, San Mateo (1988)

30. Fogel, D.: Evolutionary Computation: Towards a New Philosophy of Machine Intelligence, 2nd edn. IEEE Press, Los Alamitos (1999)
31. Geman, S., Bienenstock, E., Doursat, R.: Neural network and the bias/variance dilemma. Neural Computation 4(1), 1–58 (1992)
32. Ghosn, J., Bengio, Y.: Bias learning, knowledge sharing. IEEE Trans. Neural Network 14, 748–765 (2003)
33. Giraud-Carrier, C., Vilalta, R., Brazdil, P.: Introduction to the special issue on meta-learning. Machine Learning 54, 187–193 (2004)
34. Haykin, S.: Neural Networks - A Comprehensive Foundation, 2nd edn. Prentice Hall, Englewood Cliffs (1999)
35. Hertz, J., Krogh, A., Palmer, R.G.: Introduction to the theory of neural computation. Addison Wesley, Reading (1991)
36. Holland, J.: Adaptation in Natural and Artificial Systems. University of Michigan Press (1975)
37. Jang, J.S.R., Sun, C.T.: Neuro-Fuzzy Modeling and Control. Proceedings of the IEEE 83, 378–406 (1995)
38. Kalousis, A., Hilario, M.: Model Selection Via Meta-Learning: a Comparative Study. In: Proceedings of the 12th International IEEE Conference on Tools with AI. IEEE Press, Los Alamitos (2000)
39. Kopf, C., Taylor, C., Keller, J.: Meta-analysis: from data characterisation for meta-learning to meta-regression. In: Proceedings of PKDD, workshop on data mining decision support, meta-learning and ILP (2000)
40. Kosko, B.: Neural Networks and Fuzzy Systems: a Dynamical Systems Approach to machine intelligence. Prentice Hall, Englewood Cliffs (1991)
41. Lin, C.T., Lee, C.S.G.: Neural Fuzzy System: a Neural-Fuzzy Synergism to Intelligent Systems. Prentice-Hall, Englewood Cliffs (1996)
42. Michalewicz, Z.: Genetic Algorithms + Data Structure = Evolution Programs. Springer, Heidelberg (1992)
43. Michie, D., Spiegelhalter, D.J., Taylor, C.C.: Machine learning, neural and statistical classification. Ellis Horwood Series in Artificial Intelligence (1994)
44. Mitchell, T.: Machine Learning. McGraw-Hill, New York (1997)
45. Pfahringer, B., Bensusan, H., Giraud-Carrier, C.: Meta-learning by landmarking various learning algorithms. In: Proc. of the Seventeenth International Conference on Machine Learning, pp. 743–750 (2000)
46. Prodromidis, A.L., Chan, P.K., Stolfo, S.J.: Meta-learning in distributed data mining systems: Issues and approaches. In: Kargupta, H., Chan, P. (eds.) Advances of Distributed Data Mining. AAAI Press, Menlo Park (2000)
47. Prudêncio, R.B.C., Ludermir, T.B.: Meta-learning approaches to selecting time series models. Neurocomputing 61, 121–137 (2004)
48. Quirolgico, S.: Communicating Neural Networks in a Multi-Agent System. Ph.D. thesis, University of Maryland, Baltimora (2002)
49. Rumelhart, D.E., Widrow, B., Lehr, M.: The basic ideas in neural networks. Communications of the ACM 37(3), 87–92 (1994)
50. Schaffer, C.: A conservation law for generalization performance. In: Proceedings of the eleventh International Conference on Machine Learning (ICML 1994), pp. 259–265 (1994)
51. Schapire, R.E.: A Brief Introduction to Boosting. In: Proceedings of the sixteenth International Joint Conference on Artificial Intelligence (1999)
52. Schweighofer, N., Doya, K.: Meta-Learning in Reinforcement Learning. Neural Networks 16, 5–9 (2003)

53. Silver, D.L.: Selective Transfer of Neural Network task knowledge. Ph.D. thesis, University of Western Ontario, London, Ontario (2000)
54. Silver, D.L., Mercer, R.E.: Toward a Model of Consolidation: The Retention and Transfer of Neural Net task knowledge. In: Proceedings of INNS World Congress on Neural Networks, pp. 164–169 (1995)
55. Soares, C., Brazdil, P.B., Kuba, P.: A Meta-Learning Method to Select the Kernel Width in Support Vector Regression. Machine Learning 54, 195–209 (2004)
56. Thrun, S.: Lifelong learning algorithms. In: Thrun, S., Pratt, L. (eds.) Learning to learn, pp. 181–209. Kluwer Academic Publishers, Dordrecht (1998)
57. Thrun, S., Pratt, L. (eds.): Learning to Learn. Kluwer Academic Publishers, Dordrecht (1998)
58. Utgoff, P.: Shift of bias for inductive concept learning. In: Machine learning: an artificial intelligence approach. Morgan Kauffman, San Mateo (1986)
59. Valiant, L.: A theory of the learnable. Comm. of ACM 27, 1134–1142 (1984)
60. Vapnik, V.N., Chervonenkis, A.Y.: On the uniform convergence of relative frequencies of events to their probabilities. Theory of probability and its applications 16, 264–280 (1971)
61. Vilalta, R., Drissi, Y.: Research directions in Meta-Learning. In: Proceedings of the International Conference on Artificial Intelligence (ICAI 2001), Las Vegas, Nevada (2001)
62. Vilalta, R., Drissi, Y.: A perspective view and survey of Meta-Learning. Artificial Intelligence Review 18, 77–95 (2002)
63. Vilalta, R., Giraud-Carrier, C., Brazdil, P.: Meta-Learning: Concepts and Techniques. In: Maimon, O., Rokach, L. (eds.) Data Mining and Knowledge Discovery Handbook: A Complete Guide for Practitioners and Researchers, ch. 1. Springer, Heidelberg (2005)
64. Wolpert, D.H.: Stacked Generalisation. Neural Networks 5, 241–259 (1992)
65. Wolpert, D.H.: A rigorous investigation of evidence and occam factor in Bayesian reasoning. Technical Report, The Santa Fe Institute (2000)
66. Wolpert, D.H.: The Supervised Learning No-Free-Lunch Theorems. In: Roy, R., Koppen, M., Ovaska, S., Furuhashi, T., Hoffmann, F. (eds.) Proceedings of the Sixth Online World Conference on Soft Computing in Industrial Applications, pp. 25–42 (2001)
67. Wolpert, D.H., Macready, W.G.: No Free Lunch Theorems for Search. Technical Report, Santa Fe Institute (1995)
68. Wolpert, D.H., Macready, W.G.: No Free Lunch Theorems for Optimization. IEEE Transactions on Evolutionary Computation 1(1), 67–82 (1997)
69. Yao, X.: Evolving Artificial Neural Networks. Proceedings of the IEEE 87(9), 1423–1447 (1999)
70. Yao, X., Liu, Y.: Making Use of Population Information in Evolutionary Artificial Neural Networks. IEEE Transactions on Systems, Man, and Cybernetics - Part B 28(3), 417–425 (1998)
71. Yao, X., Liu, Y.: Towards Designing Artificial Neural Networks by Evolution. Applied Mathematics and Computation 91, 83–90 (1998)
72. Zadeh, L.A.: Fuzzy Sets. Information and Control 8, 338–353 (1965)
73. Zadeh, L.A., Kacprzyk, J. (eds.): Computing with Words in Information. Physica-Verlag (1999)

Three-Term Fuzzy Back-Propagation

M. Hadi Mashinchi and Siti Mariyam H.J. Shamsuddin

The disadvantages of the fuzzy BP learning are its low speed of error convergence and the high possibility of trapping into local minima. In this paper, a fuzzy proportional factor is added to the fuzzy BP's iteration scheme to enhance the convergence speed. The added factor makes the proposed method more dependant on the distance of actual outputs and desired ones. Thus in contrast with the conventional fuzzy BP, when the slop of error function is very close to zero, the algorithm does not necessarily return almost the same weights for the next iteration. According to the simulation's results, the proposed method is superior to the fuzzy BP in terms of generated error.

1 Introduction

To simulate the human's inexact interference system, researchers have come out with multi-valued logics. Fuzzy logic is among the most renowned ones which has been applied successfully in many applications, particularly since the 80s. On the other hand, artificial neural networks (ANN) have been inspired from the humans' neural networks and the way they can learn new things and adapt themselves to new environments. Albeit, fuzzy set theory and ANNs are being used solely in many applications by theoreticians and practitioners, but each portrays advantages unique to itself. Fuzzy neural networks (FNN) were proposed to have all of their positive points in an encapsu-

M. Hadi Mashinchi
Faculty of Computer Science and Information System, Skudai, Johor, 81310, Malaysia
e-mail: h_mashinchi@yahoo.com

Siti Mariyam H.J. Shamsuddin
Faculty of Computer Science and Information System, Skudai, Johor, 81310, Malaysia
e-mail: mariyam@utm.my

A.-E. Hassanien et al. (Eds.): Foundations of Comput. Intel. Vol. 1, SCI 201, pp. 143–158.
springerlink.com   © Springer-Verlag Berlin Heidelberg 2009

lated unit. Furthermore, FNNs are general model of ANNs. While they can be applied as conventional neural networks, they are capable of dealing with uncertainties. Their applicability in inexact environments has made them a very useful technique where a system is not provided with exact information. Particularly in any environment which adaptation is required and either the provider or addresser or both are human, FNNs should be subsituted by ANNs. The FNNs are applicable in three cases as is shown Fig. 1. Note that generally if both sides, provider and addresser, are sensors and machines then the inputs and outputs are naturally crisp values, so no need of applying FNNs. This occures since the machines needs a crisp value to activate their effectors at final stages. Also sensors produce signals which are not fuzzy most of the time for the system. Thus as Fig. 1 shows, FNNs are mostly required when at least one side og the system is a human. In this paper we consider the more generalize case which is Case 3. In Case 3 the FNN is trained according to some linguistic expresssions and then it acts in real environment where it receives some new linguistic expression and comes up with linguistic expression as result. As an example if an automatic consultant wants to give some advices to a costumer regarding stock market, it must have the ability to be trained based on linguistic expressions. Basically nature of stock market is on inexactness, as an example suppose stock A has *vast increase* on its value while the values of stockes B and C *fairly decreased* or *slightly increased* respectively. And with assumption that stock D gets influence from them we want to know how its value goes. The output of the system is the suggestion which is given to the costumer and it is mostly a linguistic expression as well. So such system receives some historical input-output data from experts and learn them so in case new input is given to the system, it can guess the output according the previous learnt data. The FNNs have been applied for many applications such as image processing [13], market prediction [16], polynomial fuzzy functions [14, 15] and agriculture [17].

Since FNNs have been proposed, most of the attentions have been diverted towards enhancing the learning ability. Direct fuzzification is among the first learning attempts for FNNs [2]. This menthod works based on the classical BP for ANNs. So that initially the inputs, outputs, weights and biases are considered crisp values. Then the BP iteration scheme is constructed. Having the iteration scheme the fuzzy values are subsituted to the BP formula. The main problem of direct fuzzification is that the itration scheme is derived based on the assumption of crisp values, while if the real nature of values are considered then th eiteration shceme would be different. According to the given reason this method has been rejected from the theoritical point of view [22]. Another attempt was a learning method for FNNs with triangular weights. This learning method was a generalized form of crisp BP and could only deal with symetric triangular fuzzy values [4]. To be able to deal with all kinds of convex fuzzy values a genetic-based learing approach was proposed [1]. Basically genetic-based learning methods are good global opptimizers but they are slow to find very close solutions to the exact ones. Therefore the reduction of the error value gets slow after some stages. The most recent learning approach are the fuzzy BP [6, 22] and the genetic-fuzzy BP [7]. In the recent fuzzy BP a learning iteration scheme is derived based on the derivation of min-max function so in contrast with the direct fuzzification, the iterative formula is constructed based on fuzzy values. This method has the ability to deal with all kinds of convex fuzzy values. But the major problem of BP based learning approaches are their lack of ability to find the global optimum point. They return a point which has a very small derivation, which it doesn't mean it is the global optimum point. So they are called local optimizer methods. In another study conducted by authors the ability of global optimization of genetic-based learning methods are merged with fuzzy BP, as a local optimizer [7].

In spite of all existing learning methods, there still is a great demand in enhancing the speed of learning processes. In this study, a proportional factor, which was firstly introduced by Zweiri et al. [10] for conventional ANNs, is added to fuzzy BP and called three-term fuzzy BP. The proposed method has the ability to search for global optimum set of weights since the it does not return the same weights in iteration scheme if the current state is a local optimum solution. So if the algorithm find a local optimum sets of weights which has the derivation of zero on the error function, since the actual outputs and desired outputs are not the same, the new weights are assigned differently. This additional factor has two major benefits; first the it increases the speed of convergence and secondly it reduces the possibility of trapping into local minimma.

The organization of the rest of the paper is as following. Section 2 of this paper introduces FNNs which is followed by Section 3, outlining the three-term BP idea for ANNs. The development of the three-term fuzzy BP is discussed in Section 4. In Section 5, a simulation for two datasets is conducted to show the superiority of the proposed method over the fuzzy BP. The conclusions and future works are discussed in the last section.

2 Artificial Fuzzy Neural Networks

In contrast with conventional neural networks, FNNs are able to receive fuzzy data, make a fuzzy inference and consequently come out with fuzzy decisions. Moreover, they can deal with crisp information which makes them a generalized model of conventional neural networks. Different FNNs can be made by considering fuzzy or crisp values of weights, inputs and outputs. Table 1 shows these variation of FNNs [4]. The first case of FNNs can be applied for classification of fuzzy numbers. The second and fourth cases are proper to learn from fuzzy if-then rules. The fourth case is more generalized of second one and ofcourse have more difficulties for training. The third case can be applied for fuzzification of real numbers. In the case 5, the output should be a crisp value so basically this case has not exists in the reality. The cases 6 and 7, the fuzzification of weights is not necessary because the outputs are crisp values.

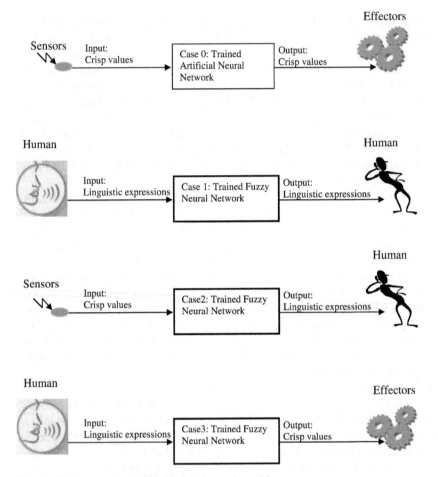

Fig. 1 Different cases of FNNs in the real world

In an FNN, the input-output relation is computed based on Zadeh's extension principle and fuzzy arithmetic [20]. Fig. 2 shows a three-layer FNN where all the inputs, outputs, weights and biases are fuzzy values. In Fig. 2, $\tilde{x}$, $\tilde{O}$, $\tilde{w}$, $\tilde{v}$ and $\tilde{b}$ are the fuzzy inputs, outputs, second and third layer weights, and biases, respectively.

The idea of FNNs was first initiated by Lee et al. in 1974 [5]. But it did not receive any attention until 1987 when Kosko proposed the fuzzy associated memories [21]. Since then, many researches have been conducted in the field of FNNs. Having the qualitative knowledge in neural networks was proposed by Narazaki and Ralescu in 1991 [8]. Buckley and Hayashi conducted the direct fuzzification of neural networks in 1993 [2] which was followed by the work done by Ishibuchi et al. with triangular weights [4]. Recently, FNNs with polygonal fuzzy weights based on fuzzy BP has been proposed [6, 22]. The

Table 1 FNNs' variations [4]

Type	weights	inputs	outputs
Case 0 of ANNs:	crisp value	crisp value	crisp value
Case 1 of FNNs:	crisp value	fuzzy	crisp value
Case 2 of FNNs:	crisp value	fuzzy	fuzzy
Case 3 of FNNs:	fuzzy	crisp value	fuzzy
Case 4 of FNNs:	fuzzy	fuzzy	fuzzy
Case 5 of FNNs:	crisp value	crisp value	fuzzy
Case 6 of FNNs:	fuzzy	crisp value	crisp value
Case 7 of FNNs:	fuzzy	fuzzy	crisp value

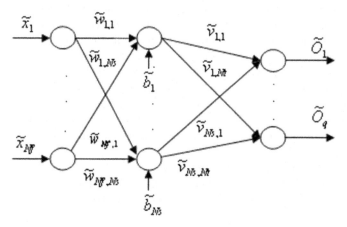

Fig. 2 A general three-layer fuzzy neural network

input-output relationship in a three-layer, $n \times h \times q$, FNN can be computed as following 1.

$$\tilde{O}_q = \sum_{ns=1}^{h} \tilde{Z}_{ns} \cdot \tilde{v}_{ns,q} \quad q = 1, \cdots, m \tag{1}$$

where in (1), $\tilde{Z}_h$ is the output of the second layer which is computed by (2) and $f(*)$ demonstrates the sigmoid activation function (3).

$$\tilde{Z}_{ns} = f(\sum_{nf=1}^{n} \tilde{x}_{nf} \cdot \tilde{w}_{nf,ns} + \tilde{b}_{ns}) \quad ns = 1, \cdots, h \tag{2}$$

$$f(\tilde{x}) = \frac{1}{1 + e^{-2\tilde{x}}} \tag{3}$$

To find the proper fuzzy weights and biases that have the ability to map the fuzzy inputs to corresponding outputs the fuzzy values should be broken to their level sets. If a very curve fuzzy weights are required, more level sets should be considered. Breaking the fuzzy values makes the learning process very time consuming. Each level sets of inputs, which are intervals themselfs, should be maped to corresponding output intervals. Thus for them the optimized intervals of weights should be defined. Furthermore the intervales should be defined in which they are nested from level set zero to level set one.

3   Three-Term BP

BP algorithms is the most common learning method of ANNs. It backpropagates the error based on the derivations. Training process is usually done by weights updating iteratively using a mean-square error function. The common BP algorithm applies two parameters; learning rate α and momentum factor β for weight adjustment. Both parameters have direct influence on the performance of training. Large value of α speed up the training process while if it is set to a very large value the process becomes unstable and it can not find the optimized weights. On the other hand, the smaller value of α the slower rate of convergence [9, 23]. The application of the other parameter, β, can allow us to use the greater value of α without unstability porblem. The advantagouse of utilizing β are summaried as follows [9, 23]:

1. Might smooth out oscillations occurs in learning,
2. Larger α value can be used if β is added in weight adaptation,
3. Encourages movement in the same direction of successive steps.

The proposal of the new term in three-term BP was due to the slow rate of error convergence. The three-term BP was firstly proposed by Zweiri et al. in 2000 [10] and comprehensively studied in [11], and recently a study on the stability analysis of three-term BP has been undertaken [12]. This third term, γ, is proportional to the distance between the actual output/s and the desired output/s. The learning iteration scheme in three-term BP is given as following (4).

$$w(t+1) = w(t) - \alpha \cdot \frac{\partial E(t)}{\partial w(t)} + \beta \cdot \Delta w(t-1) + \gamma \cdot D(W(t)) \qquad (4)$$

Where in 4, $w(t+1)$ is the new weight and α, β and γ are learning rate, momentum value and proportional factor, respectively. The function $D(*)$, returns the distance between the desired output/s and actual output/s. The actual output/s is computed according the current state of weights stored in matrix $W(t)$.

In addition of enhancing the speed of convergence, the third term is useful to escape from local minimums [9, 23]. The charactristics of α, β and γ in three-term BP are summarized in Tables 2, 3 and 4, respectively.

Table 2 FNNs' variations [9, 23]

Learning rate α
Propotional to derivation of error
Added to increase the convergence speed
Too large value will make the training process unstable

Table 3 FNNs' variations [9, 23]

Momentum rate β
Propotional to the previous value of the incremental change of the weights
Added to smooth out the oscillation and increase the convergence speed
Enable the network to use larger value of α

Table 4 FNNs' variations [9, 23]

Proportional factor γ
Propotional to the difference between the output and the target at each iteration
Added to increase the convergence speed and escape from local minima
Does not increase the complexity of algorithm and running time

In conventional neural networks if a small learning rate is used, the convergence speed would be very slow and there is a great possibility for the system to trap into local minimum. Conversely, if a very large learning rate is applied, the learning convergence is become unstable. The proposal of the third term has been aimed at solving these problems [9, 11, 23]. The outperforming performance of the three-term BP has been shown in classification [9] and service selection [3].

4 Three-Term Fuzzy BP

Many learning methods have been proposed for FNNs; triangular fuzzy weight learning [4], genetic-based approach [1], Fuzzy BP [6, 22], and the genetic fuzzy BP [7] are among the recent ones. Although fuzzy BP is the strongest model regarding its stringent mathematical development but increasing the convergence rate is still demandable for many applications. Fuzzy BP algorithm stops the iteration whenever the derivation of error gets very close to zero. This point is the local optimum solution but while it has the slope of zero no more new weight assignment is presued. To avoid local optimum problem and enhance the speed of the fuzzy BP algorithm the third

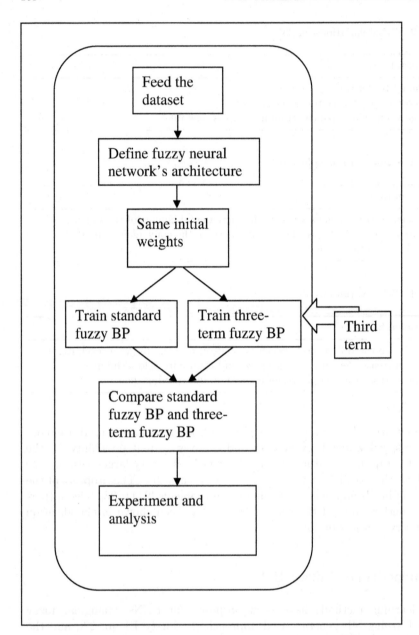

Fig. 3 The framework of study

parameter is added. The framwork of the study is given in Fig. 3. The algorithm of the proposed method is given in Algorithm 1.

The step 2.2 in the Algorithm 1 does not need to undergo a new process. The distance between the actual outputs and desired ones are computed in

---

Algorithm 1. Three-term fuzzy BP

1: Initialize the weights
2: **while** (stoping criteria met)
    2.1: standard = Compute_standard_Fuzzy_BP(current_weights,pre_weights)
    2.2: third_term = Compute_distance(actual_outputs,desired_outputs)
    2.3: next_weights = current_weights - standard + γ·third_term
  endwhile
3: **return** (next_weights)

---

the standard fuzzy BP, so simply the algorithm reuses the computed value from step 2.1. This means that the three-term fuzzy BP has same complexity as fuzzy BP and the running time for both learning methods are almost the same.

The following gives a more mathematical details of the learning iteration scheme of step 2 of the Algorithm 1. The main three-term formula for a 3-layer FNN shown in the algorithm, is given in (5). The error value for a FNN is defined in (6).

$$\widetilde{w}(t+1) = \widetilde{w}(t) - \alpha \cdot \frac{\partial \widetilde{E}(t)}{\partial \widetilde{w}(t)} + \beta \cdot \Delta\widetilde{w} + \gamma \cdot D(\widetilde{W}(t)) \tag{5}$$

$$E = \tfrac{1}{2}([D_{(k)}^1(l) - \textstyle\sum_{j=1}^{Nt} R_{j(k)}^1(l)]^2 + [D_{(k)}^2(l) - \sum_{j=1}^{Nt} R_{j(k)}^2(l)]^2)$$
$$l = 1, \cdots p \tag{6}$$

where the value of $\frac{\widetilde{E}(t)}{\partial w(t)}$ is computed according to (7).

$$\frac{\partial E(k)}{\partial z(k)} = (\sum_{j=1}^{Nt} R_{j(k)}^1 - D_{(k)}^1) \cdot \frac{\partial R_{j(k)}^1}{\partial z} + (\sum_{j=1}^{Nt} R_{j(k)}^2 - D_{(k)}^2) \cdot \frac{\partial R_{j(k)}^2}{\partial z} \tag{7}$$

In (7) $D_{(k)}^1(l)$ and $D_{(k)}^2(l)$ are left and right boundaries of the desired intervals of level set for element of the training set, respectively. $R_{(k)}^1(l)$ and $R_{(k)}^2(l)$ are left and right boundaries of actual intervals of level set k for l^{th} element of the training set, and given in (8)(9) respectively.

$$R_{j(k)}^1 = min([v_j^1(k) \cdot f(Agg_j^1(k))], [v_j^1(k) \cdot f(Agg_j^1(k))]) \tag{8}$$

$$R_{j(k)}^1 = min([v_j^2(k) \cdot f(Agg_j^1(k))], [v_j^2(k) \cdot f(Agg_j^2(k))]) \tag{9}$$

In (8) and (9) , Agg_j is the output of neuron j in the second layer and is computed by (10) and (11).

$$Agg_j^1(k) = b_j^1(k)+$$
$$\textstyle\sum_{i=1}^{Nf} min(x_i^1(k) \cdot w_{ij}^1(k), x_i^1(k) \cdot w_{ij}^2(k), x_i^2(k) \cdot w_{ij}(k)^1, x_i^2(k) \cdot w_{ij}^2(k)) \tag{10}$$

$$Agg_j^2(k) = b_j^2(k) +$$
$$\sum_{i=1}^{Nf} max(x_i^2(k) \cdot w_{ij}^1(k), x_i^1(k) \cdot w_{ij}^2(k), x_i^2(k) \cdot w_{ij}(k)^1, x_i^2(k) \cdot w_{ij}^2(k))$$
$$(11)$$

In order to conduct the fuzzy interval arithmetic, all the fuzzy values are considered as the following level set representation; $\widetilde{x}_i(k) = [x_i^1(k), x_i^2(k)]$, $\widetilde{b}_i(k) = [b_i^1(k), b_i^2(k)]$, $\widetilde{w}_i j(k) = [w_{ij}^1(k), w_{ij}^2(k)]$, and $\widetilde{w}_i j(k) = [w_{ij}^1(k), w_{ij}^2(k)]$.

The distance function $D(\widetilde{W})(t)$ in (5) is computed by (12).

$$D(\widetilde{W})(t) =$$
$$\frac{1}{L} \sum_{l=1}^{L} ([D_{(k)}^1(l) - 2 + \sum_{j=1}^{Nt} R_{j(k)}^1(l)]^2 + [D_{(k)}^2(l) - 2 + \sum_{j=1}^{Nt} R_{j(k)}^2(l)]^2$$
$$l = 1, \cdots p$$
$$(12)$$

For each kth level set, the training formula (5) should be executed. The third term leads the algorithms not necessarily to return the same weights if the value of $\frac{\partial \widetilde{E}(t)}{\partial \widetilde{w}(t)}$ is very close to zero.

5   Simulations and Results

Simulation and comparison of the proposed learning method and fuzzy BP is conducted with two datasets. The first dataset illustrated in Table 5 and Fig. 4 has been used by Liu et al. [6, 22]. The second dataset depicted in Table 6 has been applied by Aliev et al. [1] for verification of their proposed method. FNNs are very useful tools for fuzzy regression [1, 18, 19, 24]. The given dataset by Aliev has been desined for fuzzy regression with application in evaluation in the manufacturing process of aluminum heaters. We use the three-term fuzzy BP learning method for intermediate quality evaluation in the manufacturing process of aluminum heaters. Since the qualities are determined by human, they are expressed by fuzzy numbers. Having nine fuzzy input-output data we applied FNN to construct a mapping function with the least error [1].

For comparison same fully connected 3-layer network with identical initial state is considered. The stopping criterion is set to 400 iterations for each of 1000 trials. Without lose of any generality and only for sake of simplicity, two level sets, the supports and cores, are considered for both datasets.

Table 5 Liu's dataset [6, 22]

No.	Rule
1	IF (x is small) and (y is small) THEN (z is small)
2	IF (x is large) and (y is large) THEN (z is large)
3	IF (x is small) and (y is large) THEN (z is medium)
4	IF (x is large) and (y is small) THEN (z is medium)

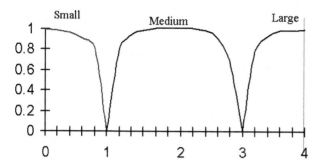

Fig. 4 Fuzzy values of small, medium and large in Liu's fuzzy rules [6, 22]

Table 6 Aliev's dataset [1]

No.	I1	I2	O
1	(0,0.1,0.2)	(0,0.1,0.2)	(-0.083,0.037,0.166)
2	(0.4,0.6,0.8)	(0.3,0.5,0.6)	(-0.558,0.912,2.284)
3	(0.8,0.9,1.2)	(0.2,0.5,0.9)	(-1.665,0.729,4.068)
4	(0.6,0.7,0.8)	(0.9,0.95,1.0)	(2.08,3.405,5.155)
5	(0.4,0.5,0.6)	(0.2,0.6,0.9)	(-0.945,1.360,3.428)
6	(0.0,0.1,0.2)	(0.3,0.5,0.6)	(-0.022, 0.861,1.403)
7	(0.6,0.7,0.8)	(0.2,0.6,0.9)	(-1.247, 1.321,3.825)
8	(0.4,0.5,0.6)	(0.2,0.5,0.9)	(-0.649,0.937,3.131)
9	(0.0,0.1,0.2)	(0.7,0.8,0.9)	(1.182,2.149,3.073)

Table 7 A comparison between average generated between fuzzy BP and the proposed three-term fuzzy BP for Liu's dataset [6, 22]

ID	α	β	γ	Fuzzy BP's error	3-term fuzzy BP's error
D1	0.04	0.02	0.001	1.063189	0.727247
D2	0.05	0.025	0.002	0.723729	0.292399
D3	0.06	0.03	0.003	0.376728	0.11259

Considering α, β and γ, different modes of three-term fuzzy BP and fuzzy BP are investigated as it has been done in [11, 23]. The final average generated errors for Liu's and Aliev's datasets are captured and shown in Tables 7 and 8, respectively. The given errors are computed by root mean squared (RMS) error function. The graphs depicted in Fig. 5 and 6 show the average error convergence for fuzzy BP and proposed three-term fuzzy BP for 1000 trials.

According to Tables 7 and 8, all modes of the three term fuzzy BP generate less error. It is observed that a very large values of α, β and γ and make the

Table 8 A comparison between average generated between fuzzy BP and the proposed three-term fuzzy BP for Aliev's dataset [1]

ID	α	β	γ	Fuzzy BP's error	3-term Fuzzy BP's error
D1	0.04	0.02	0.001	2.251948	2.135802
D2	0.05	0.025	0.002	2.219752	1.979147
D3	0.06	0.03	0.003	2.200649	1.8592892

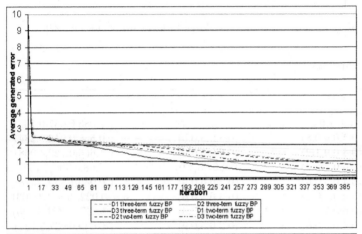

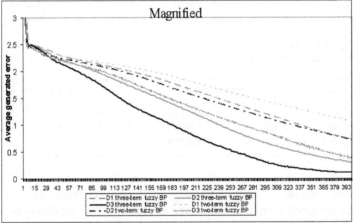

Fig. 5 Convergence rate of proposed method and standard fuzzy BP for Liu's [6, 22]

learning process unstable. As an example, mode C3 of the three-term fuzzy BP has many fluctuations in final iterations.

To simulate the behaviour of the trained FNN, as an example the trained FNN with Liu's dataset, some unseen fuzzy inputs are given to the FNN as

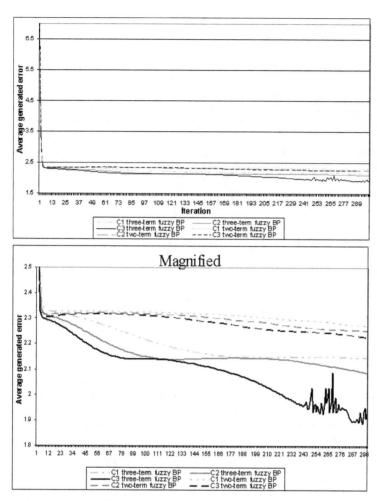

Fig. 6 Convergence rate of proposed method and standard fuzzy BP for Aliev's dataset [1]

given in Table 9. The given outputs in Table 9 are the linguistic terms that we guess that should be naturally generated based on the training set. But we do not know what is their actual fuzzy membership function. To find the actual fuzzy values associate with these pre-guessed linguitic expressions, the trained FNN is fed with the inputs and the actual outputs are observed.

This can be viewed from another angle; the trained FNN generates the numeric meaning for these pre-guessed linguistic expressions. So for example if the FNN receives medium and small as inputs then the output should be something in between of them (based on the training set), we guess. So we call this pre-guessed output, *small-medium*. The numeric output of the trained FNN which is a fuzzy membership gives a numerical meaning of expression

Table 9 Unseen fuzzy inputs and their pre-guessed fuzzy outputs

No.	Rule
1	IF (x is small) and (y is medium) THEN (z is small-medium)
2	IF (x is medium) and (y is large) THEN (z is large-medium)
3	IF (x is medium) and (y is medium) THEN (z is medium)

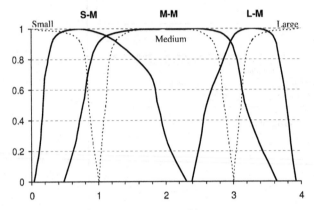

Fig. 7 Fuzzy outputs for unseen fuzzy inputs

Table 10 Bahaviour of the trained FNN with Liu's dataset for unseen fuzzy inputs

Level set	small-medium	medium-medium	medium-large
level set 0:	[0.033,2.313]	[0.475,3.639]	[2.387,3.920]
level set 0.2:	[0.121,2.122]	[0.651,3.491]	[2.455,3.868]
level set 0.4:	[0.162,1.892]	[0.785,3.235]	[2.467,3.802]
level set 0.6:	[0.212,1.783]	[0.853,3.152]	[2.728,3.744]
level set 0.8:	[0.229,1.638]	[0.946,3.034]	[2.807,3.651]
level set 1:	[0.433,0.973]	[1.632,2.717]	[3.137,3.452]

small-medium. The achived fuzzy outputs are shown in Fig. 7 and the detailed
level sets boundaries are given in table 10. According to Fig. 7 clearly the
meaning of the pre-guessed is very close the ones observed from the trained
FNN which shows the system works properly.

6   Discussions and Future Research

According to the results, it appears that the three-term fuzzy BP is superior to the fuzzy BP in terms of generating error. While both methods
sharply dipped the error at the first iterations, the standard fuzzy BP slightly

decreases error for the rest of iterations. On the other hand, the three-term fuzzy BP is faster in reaching the local optimized state in the rest of the iterations. Also according to the experiments, the greater values of α, β and γ give higher speed of convergence. However, if these coefficients are set as large values, the convergence may become unstable. The processing time is also almost the same for both proposed method and fuzzy BP. It may seem that the third term requires some computations, so the three-term fuzzy BP's process is more time consuming than fuzzy BP. In fact the third term is an essential part and should be computed in each iteration of the fuzzy BP itself. Therefore the three term BP algorithm can reuse the stored value from the fuzzy BP's process.

The stability analysis is an essential research which should be carried out for three-term fuzzy BP. The value of the proportional factor has direct effect on the convergence rate, therefore as a future research one may conduct a study on three-term fuzzy BP with dynamic proportional coefficient.

Furthermore, we show how FNNs are capable of generating numerical meanings for linguistic expressions. For the control systems, linguistic expressions are meaningless, so FNNs are proper tools to correspond human's expressions with numerical signals.

Acknowledgements. This research is supported by grant number 78183, Ministry of Higher Education (MOHE), Malaysia and Research Management Centre (RMC), Universiti Teknologi Malaysia (UTM).

References

1. Aliev, R.A., Fazlollahi, B., Vahidov, R.M.: Genetic algorithm-based learning of fuzzy neural network, Part 1: feed-forward fuzzy neural networks. Fuzzy Sets and Systems 118(2), 351–358 (2001)
2. Buckley, J.J., Hayashi, Y.: Direct fuzzification of neural networks. In: 1st Asian Fuzzy System Symposium, pp. 560–567 (1993)
3. Cai, H., Sun, D., Cao, Q., Pu, F.: A novel BP algorithm based on three-term and applica-tion in service selection of ubiquitous. In: 9th International Conference on Computing, Control, Automation, Robotics and Vision, pp. 1–6 (2006)
4. Ishibuchi, H., Kwon, K., Tanaka, H.A.: Learning of fuzzy neural networks with triangular fuzzy weights. Fuzzy Sets and Systems 71(3), 277–293 (1995)
5. Lee, S.C., Lee, E.T.: Fuzzy sets and neural networks. Journal of Cybernetics 4, 83–103 (1974)
6. Liu, P., Hongxing, L.: Efficient learning algorithms for three-layer feedforward regular fuzzy neural networks. IEEE Trans. On Neural Networks 15(2), 545–558 (2004)
7. Mashichi, M.H., Mashinchi, M.R., Shamsuddin, S.M., Pedrycz, W.: Genetically tuned fuzzy BP learning method based on derivation of min-max function for fuzzy neural networks. In: International Conference on Genetic and Evolutionary Methods Nevada, pp. 213–219 (2007)

8. Narazaki, H., Ralescu, L.A.: A method to implement qualitative knowledge in multi-layered neural networks. Fuzzy Logic and Fuzzy Control, 82–95 (1991)
9. Shamsuddin, S.M., Saman, F.I., Darus, M.: Three backpropagatiopn algorithm for classification problem. Neural Network World 17(4), 363–376 (2007)
10. Zweiri, Y.H., Seneviratne, L.D.: Optimization and stability of a three-term BP algorithm Neural networks. In: CI 2000, pp. 1514–1540 (2000)
11. Zweiri, H.Y., Whidborne, J.F., Althoefer, K., Seneviratne, L.D.: A three-term backpropagation algorithm. Neurocomputing 50, 305–318 (2003)
12. Zweiri, H.Y., Seneviratne, L.D., Althoefer, K.: Stability analysis of a three-term back-propagation algorithm. Neural Networks 18, 1341–1347 (2005)
13. Liu, P.: Representation of digital image by fuzzy neural network. Fuzzy Sets and Systems 130(1), 109–123 (2002)
14. Abbasbandy, S., Otadi, M.: Numerical solution of fuzzy polynomials by fuzzy neural networks. Applied Mathematics and Computations 181(2), 1084–1089 (2006)
15. Abbasbandy, S., Neirto, J.J., Amirfagharian, M.: Best approximation of fuzzy functions. Nonlinear Studies 14(1), 87–102 (2007)
16. Kuo, R.J., Wu, P., Wang, C.P.: An intelligent sales forcasting system through integration of artificial intelligence and fuzzy neural networks with fuzzy weight elimination. Neural networks 15(7), 909–925 (2002)
17. Fukuda, S., Hiramatsu, K., Mori, M.: Fuzzy neural network model for habitat prediction and HEP for habitat quality estimation focusing on Japanese medaka (Oryais latipes) in agricultural canals. Paddy Water Environ. 4(3), 119–124 (2006)
18. Ishibuchi, H., Tanaka, H.: Regression analysis with internal model by neural net. In: Proc. Internat. Joint Conf. on Neural Networks, pp. 1594–1599 (1991)
19. Miyazaki, A., Kwan, K., Ishibuchi, H., Tanaka, H.: Fuzzy regression analysis by fuzzy neural networks and its application. IEEE Transactions on Fuzzy Systems (1995)
20. Klir, G.J., Yuan, B.: Fuzzy sets and fuzzy logic: theory and applications. Prentice Hall PTR, New York (1995)
21. Kosko, B.: Fuzzy associated memories. In: Kandel, A. (ed.) Fuzzy expert systems, Addison-Weley, Reading (1987)
22. Liu, P., Li, H.X.: Fuzzy neural network theory and application. World Scientific, River Edge (2004)
23. Saman, F.I.: Three-term backpropagation algorithm for classification problem. Master Thesis Universiti Teknologi Malaysia, Malaysia (2006)
24. Buckley, J., Hayashi, Y.: Fuzzy neural networks. In: Yager, R., Zadeh, L. (eds.) Fuzzy Sets, Neural Networks, and Soft Computing. Van Nostrand Reinhold, New York (1994)

Entropy Guided Transformation Learning

Cícero Nogueira dos Santos and Ruy Luiz Milidiú

This work presents Entropy Guided Transformation Learning (ETL), a new machine learning algorithm for classification tasks. ETL generalizes Transformation Based Learning (TBL) by automatically solving the TBL bottleneck: the construction of good template sets. ETL uses the information gain in order to select the feature combinations that provide good template sets.

We describe the application of ETL to two language independent Text Mining preprocessing tasks: part-of-speech tagging and phrase chunking. We also report our findings on one language independent Information Extraction task: named entity recognition. Overall, we successfully apply it to six different languages: Dutch, English, German, Hindi, Portuguese and Spanish.

For each one of the tasks, the ETL modeling phase is quick and simple. ETL only requires the training set and no handcrafted templates. Furthermore, our extensive experimental results demonstrate that ETL is an effective way to learn accurate transformation rules. We believe that by avoiding the use of handcrafted templates, ETL enables the use of transformation rules to a greater range of Text Mining applications.

1 Introduction

Since the last decade, Machine Learning (ML) has proven to be a very powerful tool to help in the construction of Text Mining systems, which would otherwise require an unfeasible amount of time and human resources. Transformation Based Learning (TBL) is a ML algorithm introduced by Eric Brill [6] to solve Natural Language Processing (NLP) tasks. TBL is a corpus-based,

Cícero Nogueira dos Santos and Ruy Luiz Milidiú
Departamento de Informática
PUC-Rio
Rio de Janeiro, Brazil
e-mail: nogueira@inf.puc-rio.br,milidiu@inf.puc-rio.br

A.-E. Hassanien et al. (Eds.): Foundations of Comput. Intel. Vol. 1, SCI 201, pp. 159–184.
springerlink.com   © Springer-Verlag Berlin Heidelberg 2009

error-driven approach that learns a set of ordered transformation rules which correct mistakes of a baseline classifier. It has been used for several important NLP tasks, such as part-of-speech tagging [6, 15, 13], phrase chunking [39, 30, 12], named entity recognition [16, 32] and semantic role labeling [23].

TBL rules must follow patterns, called templates, that are meant to capture the relevant feature combinations. TBL templates are handcrafted by problem experts. Its quality strongly depends on the problem expert skills to build them. Even when a template set is available for a given task, it may not be effective when we change from a language to another. When the number of features to be considered is large, the effort to manually create templates is extremely increased, becoming sometimes infeasible. Hence, the human driven construction of good template sets is a bottleneck on the effective use of the TBL approach.

In this chapter, we present Entropy Guided Transformation Learning (ETL), a new machine learning algorithm for classification tasks. ETL generalizes Transformation Based Learning by automaticaly solving the TBL bottleneck: the construction of good template sets. ETL uses the information gain in order to select the feature combinations that provide good template sets. We also show that ETL can use the *template evolution* strategy [10] to accelerate transformation learning.

We describe the application of ETL to two language independent Text Mining preprocessing tasks: Part-of-Speech (POS) Tagging and Phrase Chunking (PC). We also report our findings on one language independent Information Extraction task: Named Entity Recognition (NER). Overall, we apply ETL to eleven different corpora in six different languages: Dutch, English, German, Hindi, Portuguese and Spanish. Our goal in these experiments is to assess the robustness and predictive power of the ETL strategy. In Table 1, we enumerate the eleven corpora used throughout this chapter. For each corpus, we indicate its corresponding language, task and the state-of-the-art system.

In all tasks, the ETL modeling phase is quick and simple. ETL only requires the training set and no handcrafted templates. ETL also simplifies the incorporation of new input features, such as capitalization information. For each one of the tasks, ETL shows better results than TBL with handcrafted templates. The last column of Table 1 shows the ETL effectiveness. In this Table, the best observed results are in bold. Using the ETL approach, we obtain state-of-the-art competitive performance results in nine out of the eleven corpus-driven tasks.

The remainder of this chapter is organized as follows. In section 2, we detail the ETL algorithm. In section 3, we report our findings on the application of ETL to POS tagging. In section 4, we detail the application of ETL to the PC task. In section 5, we report our findings on the application of ETL to the NER task. Finally, in section 6, we present our concluding remarks.

Table 1 System performances

Task	Corpus	Language	State-of-the-art		ETL
			Approach	Performance	
POS	Mac-Morpho	Portuguese	TBL	96.60	**96.75**
	Tycho Brahe	Portuguese	TBL	96.63	**96.64**
	Brown	English	TBL	96.67	**96.69**
	TIGER	German	TBL	96.53	**96.57**
PC	SNR-CLIC	Portuguese	TBL	87.71	**88.85**
	Ramshaw & Marcus	English	SVM	**94.22**	92.59
	CoNLL-2000	English	SVM	**94.12**	92.28
	SPSAL-2007	Hindi	HMM + CRF	**80.97**	78.53
NER	LearnNEC06	Portuguese	SVM	**88.11**	87.71
	SPA CoNLL-2002	Spanish	AdaBoost	**79.29**	76.22
	DUT CoNLL-2002	Dutch	AdaBoost	**77.05**	71.97

2 Entropy Guided Transformation Learning

Entropy Guided Transformation Learning (ETL) is a new machine learning algorithm for classification tasks. The ETL approach generalizes Transformation Based Learning by automatically solving the TBL bottleneck: the construction of good template sets. ETL has been sucessfully applied to part-of-speech tagging [13] and phrase chunking [31]. In this section, we first provide an overview of Transformation Based Learning. Next, we detail the ETL learning strategy.

2.1 *Transformation Based Learning*

Transformation Based Learning is a machine learning algorithm for classification tasks introduced by Eric Brill [6]. TBL is a corpus-based, error-driven approach that learns a set of ordered transformation rules which correct mistakes of a baseline classifier. It has been used for several Natural Language Processing tasks, such as part-of-speech tagging [6, 15, 13], noun-phrase and text chunking [39, 30, 12], spelling correction [28], appositive extraction [20], named entity recognition [16, 32] and semantic role labeling [23].

The following three rules illustrate the kind of transformation rules used throughout this chapter.

$$pos[0] = ART \quad pos[1] = ART \rightarrow pos[0] = PREP$$
$$pos[0] = ART \quad pos[1] = V \quad word[0] = a \quad \rightarrow pos[0] = PREP$$
$$pos[0] = N \quad pos[-1] = N \quad pos[-2] = ART \rightarrow pos[0] = ADJ$$

These rules were learned for Portuguese POS tagging. They check the following features: pos[0], the current token part-of-speech tag; pos[1], the one to the right token part-of-speech tag; pos[-1], the one to the left token part-of-speech tag; `pos[-2]`; the two to the left token part-of-speech tag; and `word[0]`, the current token word. The first rule should be read as

> **"IF** the POS tag of the current word is an *article*
> **AND** the POS tag of the next word is also an *article*
> **THEN** change the POS tag of the current word to *preposition* "

TBL rules are composed of two parts: the left hand side and the right hand side. The *left hand side* is a conjunction of feature=value tests, whereas the *right hand side* indicates a value assignment to a target feature. TBL rules must follow patterns, called *rule templates*, that specify which feature combinations should appear in the rule left-hand side. The template set defines the candidate rules space to be searched. Briefly, a template is an uninstantiated rule. The following three templates were used to create the previously shown rules.

```
pos[0]   pos[1]
pos[0]   pos[1]   word[0]
pos[0]   pos[-1]  pos[-2]
```

TBL requires three inputs:

(i) a correctly labeled training set;
(ii) an initial classifier, the *Baseline System* (BLS), which provides a initial labeling for the training examples. Usually, the BLS is based on simple statistics of the correctly labeled training set, such as to apply the most frequent class;
(iii) a set of rule templates.

The TBL algorithm is illustrated in Figure 1. The central idea in the TBL learning process is to greedily learn rules that incrementally reduces the number of classification mistakes produced by the Initial Classifier. At

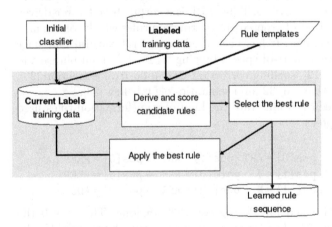

Fig. 1 Transformation Based Learning

Algorithm 1. Transformation Based Learning Pseudo-Code

---
input *LabeledTrainingSet*; *TemplateSet*; *InitialClassifier*
1: *LearnedRules* ← {}
2: *CurrentTrainingSet* ← apply(*InitialClassifier*, *LabeledTrainingSet*)
3: **repeat**
4:     *CandidateRules* ← {}
5:     **for all** *example* ∈ *CurrentTrainingSet* **do**
6:         **if** isWronglyClassified(*example*) **then**
7:             **for all** *template* ∈ *TemplateSet* **do**
8:                 *rule* ← instantiateRule(*template*, *example*)
9:                 *CandidateRules* ← *CandidateRules* + *rule*
10:             **end for**
11:         **end if**
12:     **end for**
13:     *bestScore* ← 0
14:     *bestRule* ← Null
15:     **for all** *rule* ∈ *CandidateRules* **do**
16:         *good* ← countCorrections(*rule*, *CurrentTrainingSet*)
17:         *bad* ← countErrors(*rule*, *CurrentTrainingSet*)
18:         *score* ← *good* − *bad*
19:         **if** *score* > *bestScore* **then**
20:             *bestScore* ← *score*
21:             *bestRule* ← *rule*
22:         **end if**
23:     **end for**
24:     **if** *bestScore* > 0 **then**
25:         *CurrentTrainingSet* ← apply(*bestRule*, *CurrentTrainingSet*)
26:         *LearnedRules* ← *LearnedRules* + *bestRule*
27:     **end if**
28: **until** *bestScore* > 0
output *LearnedRules*

---

each iteration, the algorithm learns the rule that has the highest *score*. The score of a rule r is the difference between the number of errors that r repairs and the number errors that r creates. A pseudo-code of TBL is presented in Algorithm 1. In this pseudo-code, the *apply* function classifies the given training set examples using the given initial classifier or transformation rule. The *isWronglyClassified* function checks whether the example is misclassified or not. This checking is done by comparing the current example class to the correct class. The *instantiateRule* function creates a new rule by instantiating the given template with the given example context values. The *countCorrections* function returns the number of corrections that a given rule would produce in the current training set. Similarly, the *countErrors* function returns the number of misclassifications that a given rule would produce in the current training set. There are also several variants of the TBL algorithm. FastTBL [17] is the most successful, since its reported run time is 13 to 139 faster than the original TBL.

When using a TBL rule set to classify new data, we first apply the Initial Classifier to the new data. Then we apply the learned rule sequence. The rules must be applied following the same order they were learned.

2.2 TBL Bottleneck

TBL templates are meant to capture relevant feature combinations. Templates are handcrafted by problem experts. Therefore, TBL templates are task specific and their quality strongly depends on the problem expert skills to build them. For instance, Ramshaw & Marcus [39] propose a set of 100 templates for the phrase chunking task. Radu Florian [16] proposes a set of 133 templates for the named entity recognition task. Higgins [23] handcrafted 130 templates to solve the semantic role labeling task. Elming [14] handcrafted 70 templates when appling TBL for machine translation. Santos & Oliveira [12] extended the Ramshaw & Marcus template set by adding six templates specifically designed for Portuguese noun phrase chunking.

Even when a template set is available for a given task, it may not be effective when we change from a language to another. When the number of features to be considered is large, the effort to manually create templates is extremely increased, becoming sometimes infeasible. Hence, the human driven construction of good template sets is a bottleneck on the effective use of the TBL approach.

2.3 Entropy Guided Template Generation

Entropy Guided Transformation Learning uses Information Gain in order to select the feature combinations that provide good template sets. The ETL algorithm is illustrated in the Figure 2.

Information Gain, which is based on the data Entropy, is a key strategy for feature selection. The most popular Decision Tree (DT) learning algorithms [38, 42] implement this strategy. Hence, they provide a quick way to obtain entropy guided feature selection. In the ETL strategy, we use DT induction algorithms to automatically generate template sets.

The remainder of this section is organized as follows. First, we review the information gain measure and the DT learning method. Next, we demonstrate how to automatically generate templates from decision trees. Finally, we present a *template evolution scheme* that speeds up the TBL step.

Information Gain

Information Gain (IG) is a statistical measure commonly used to assess feature relevance [11, 18]. IG is based on the Entropy concept, which characterizes the impurity of an arbitrary collection of examples. The entropy of a training set S whose examples assume classes from the set L is defined as

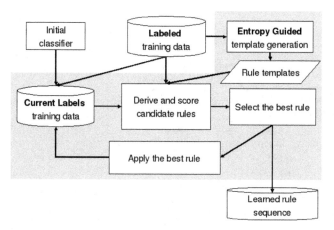

Fig. 2 Entropy Guided Transformation Learning

$$H(S) = -\sum_{l \in L} p(l)\log_2 p(l) \tag{1}$$

where l is a class label from L and $p(l)$ is the probability of an example in S being classified as l.

In feature selection, information gain can be thought as the expected reduction in entropy $H(S)$ caused by using a given feature A to partition the training examples in S. The information gain $IG(S, A)$ of a feature A, relative to an example set S is defined as

$$IG(S, A) = H(S) - \sum_{v \in Values(A)} \frac{|S_v|}{|S|} H(S_v) \tag{2}$$

where $Values(A)$ is the set of all possible values for feature A, and S_v is the subset of S for which feature A has value v [34]. When using information gain for feature selection, a feature A is preferred to feature B if the information gain from A is greater than that from B.

Decision Trees

Decision Tree induction is a widely used Machine Learning algorithm [37]. Quinlan's C4.5 [38] system is the most popular DT induction implementation. It recursively partitions the training set using the feature providing the largest Information Gain. This results into a tree structure, where the nodes correspond to the selected features and the arc labels to the selected feature values. After the tree is grown, a pruning step is carried out in order to avoid overfitting.

In Figure 3, we illustrate the DT induction process for Portuguese POS tagging. Here, the five selected features are: pos[0], the current token

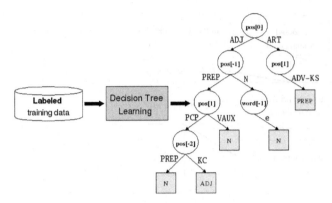

Fig. 3 Decision Tree Learning

part-of-speech; pos[-1], the one to the left token part-of-speech; pos[1], the one to the right token part-of-speech; pos[-2], the two to the left token part-of-speech; and word[-1], the one to the left token word. The feature values are shown in the figure as arc labels.

We use C4.5 system to obtain the required entropy guided selected features. We use pruned trees in all experiments shown here.

Template Extraction

In a DT, the more informative features appear closer to the root. Since we just want to generate the most promising templates, we combine first the more informative features. Hence, as we transverse the DT from the root to a leaf, we collect the features in this path. This feature combination provides an information gain driven template. Additionally, paths from the root to internal nodes also provide good templates.

It is very simple to obtain these templates from C4.5's output. From the given DT, we eliminate the leaves and the arc labels. We keep only the tree structure and the node labels. Next, we execute a depth-first traversal of the DT. For each visited tree node, we create a template that combines the features in the path from the root to this node. Figure 4 illustrates the template extraction process. In this figure, the template in bold is extracted from the tree path in bold.

2.4 Template Evolution

TBL training time is highly sensitive to the number and complexity of the applied templates. Curran & Wong [10] argued that we can better tune the *training time* vs. *templates complexity* trade-off by using an evolutionary template approach. The main idea is to apply only a small number of templates that

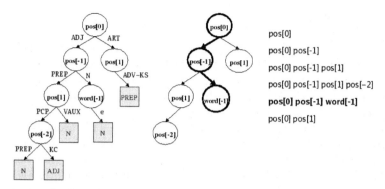

pos[0]

pos[0] pos[-1]

pos[0] pos[-1] pos[1]

pos[0] pos[-1] pos[1] pos[-2]

pos[0] pos[-1] word[-1]

pos[0] pos[1]

Fig. 4 Decision Tree template extraction

evolve throughout the training. When training starts, templates are short, consisting of few feature combinations. As training proceeds, templates evolve to more complex ones that contain more feature combinations. In this way, only a few templates are considered at any point in time. Nevertheless, the descriptive power is not significantly reduced.

ETL provides an easy scheme to implement the template evolution strategy. First, we partition the learned template set by template size. Let T_k be the template set containing all templates of size k, where $k = 1, ..., K$ and K equals to the largest template size. Next, we split the TBL step into K consecutive phases. In phase k, TBL learns rules using only templates from T_k. For instance, using the tree shown in Figure 4, we have four TBL training phases. In Table 2, we show the template sets used in the four TBL phases when the tree shown in Figure 4 is used.

Using the template evolution strategy, the training time is decreased by a factor of five for the English text chunking task. This is a remarkable reduction, since we use an implementation of the *fastTBL* algorithm [35] that is already a very fast TBL version. Training time is a very important issue when modeling a system with a corpus-based approach. A fast ML strategy enables the testing of different modeling options, such as different feature sets. The efficacy of the rules generated by ETL template evolution is quite similar to the one obtained by training with all the templates at the same time.

Table 2 ETL Template Evolution

Phase	Template set
1	pos[0]
2	pos[0] pos[-1]
	pos[0] pos[1]
3	pos[0] pos[-1] pos[1]
	pos[0] pos[-1] word[-1]
4	pos[0] pos[-1] pos[1] pos[-2]

2.5   Related Work

Corston-Oliver & Gamon [9] present a combination of DTs and TBL. They derive candidate rules from the DT, and use TBL to select and apply them. Their work is restricted to binary features only. ETL strategy extracts more general knowledge from the DT, since it builds rule templates. Furthermore, ETL is applied to any kind of discrete features.

Carberry et al. [7] introduce a randomized version of the TBL framework. For each error, they try just a few randomly chosen templates from the given template set. This strategy speeds up the TBL training process, enabling the use of large template sets. However, they use handcrafted templates and variations of them, what implies that a template designer is still necessary.

Hwang et al [24] use DT decomposition to extract complex feature combinations for the Weighted Probabilistic Sum Model (WPSM). They also extract feature combinations that use only some nodes in a tree path. This is required to improve the effectiveness of the WPSM learning. Their work is similar to ours, since they use DT's to feature extraction. Nevertheless, the ETL learned template set is simpler then theirs, and are enough for effective TBL learning.

An evolutionary scheme based on Genetic Algorithms (GA) to automatically generate TBL templates is presented by Milidiú et al. [33]. Using a simple genetic coding, the generated template sets show an efficacy near to the handcrafted templates. The main drawback of this strategy is that the GA step is computationally expensive. If we need to consider a large context window or a large number of features, it becomes infeasible.

3   Part-of-Speech Tagging

Part-of-Speech (POS) tagging is the process of assigning a POS or another lexical class marker to each word in a text [26]. POS tags classify words into categories, based on the role they play in the context in which they appear. The POS tag is a key input feature for NLP tasks like phrase chunking and named entity recognition.

The POS tagging task is modeled as a token classification problem. For each token, its context is given by the features of its adjacent tokens. The number of adjacent tokens defines the size of what is called the *context window*. For instance, a size three context window consists of the current token, the previous and the next ones. This modeling approach is also used in the other two tasks discussed in this chapter.

This section presents the application of the ETL approach to language independent POS tagging. We consider three languages: Portuguese, German and English. We generate ETL models for two Portuguese corpora, Mac-Morpho [1] and Tycho Brahe [25], a German corpus, TIGER [3], and an English corpus, Brown [19]. In Table 3, we show some characteristics of these

Table 3 Part-of-Speech Tagging Corpora

Corpus	Language	Tagset size	Training Data		Test Data	
			Sent.	Tokens	Sent.	Tokens
Mac-Morpho	Portuguese	22	44233	1007671	9141	213794
Tycho Brahe	Portuguese	383	30698	775601	10234	259991
TIGER	German	54	41954	742189	8520	146389
Brown	English	182	47027	950975	10313	210217

corpora. The Mac-Morpho Corpus is tagged with 22 POS tags, while the Tycho Brahe Corpus is tagged with 383 POS tags. The TIGER Corpus is tagged with 54 POS tags, and the Brown Corpus is tagged with 182 POS tags. Both the Tycho Brahe Corpus and the Brown Corpus use more POS tags because these tags also identify morphological aspects such as word number and gender. Each corpus is divided into training and test sets. For the Portuguese corpora, these training and test set splits are the same as reported by Milidiú et al. [31].

3.1 POS Tagging Modeling

A word that appears in the training set is called a *known word*. Otherwise, it is called an *unknown word*. Our POS modeling approach follows the two stages strategy proposed by Brill [6]. First, morphological rules are applied to classify the unknown words. Next, contextual rules are applied to classify known and unknown words.

The morphological rules are based on the following token features:

- up to c characters long word prefixes and suffixes;
- specific character occurrence in a word;
- adding (or subtracting) a c characters long prefix (or suffix) results in a known word;
- occurrence of the word before (or after) a specific word W in a given long list of word bigrams. For instance, if the word appears after "to", then it is likely to be a verb in the infinitive form.

In our experiments, we set the parameter c equal to 5.

With a very simple template set [6], one can effectively perform the morphological stage. For this stage, it is enough to use one feature or two feature templates. The one feature templates use one of the current token features. The two feature templates use one of the current token features and the current token POS.

The contextual rules use the context window features word and POS. We use the ETL strategy for learning contextual rules only.

3.2 ML Modeling

The following ML model configurations provide our best results.

BLS The baseline system assigns to each word the POS tag that is most frequently associated with that word in the training set. If capitalized, an unknown word is tagged as a proper noun, otherwise it is tagged as a common noun.

TBL The results for the TBL approach refer to the contextual stage trained using the lexicalized template set proposed in [6]. This template set uses combinations of words and POS tags in a context window of size 7.

ETL In the ETL learning, we use the features *word* and *POS* in a context window of size 7. In order to overcome the sparsity problem, we only use the 200 most frequent words to induce the DT. In the DT learning step, the POS tag of the word is the one applied by the initial classifier (BLS). On the other hand, the POS tag of the neighbor words are the true ones. We report results for ETL trained with all the templates at the same time and also using template evolution.

3.3 Mac-Morpho Corpus

Santos et al. [13] present a TBL system with state-of-the-art performance for the Mac-Morpho Corpus. Therefore, for the Mac-Morpho Corpus, we only report the performance of ETL, TBL and BLS systems.

In Table 4, we summarize the performance results of the three systems. Both ETL and TBL systems reduce the BLS error in at least 64%. The ETL accuracy is similar to the one of TBL. Its value, 96.75%, is equivalent to the best one reported so far for the Mac-Morpho Corpus.

Using the template evolution strategy, training time is reduced by nearly 73% and there is no performance loss. This is a remarkable reduction, since we use an implementation of the *fastTBL* algorithm [35] that is already a very fast TBL version.

3.4 Tycho Brahe Corpus

Santos et al. [13] present a TBL system with state-of-the-art performance for the Tycho Brahe Corpus. Therefore, for the Tycho Brahe Corpus, we only report the performance of ETL, TBL and BLS systems.

In Table 5, we summarize the performance results of the three systems. Both ETL and TBL systems reduce the BLS error in 62%. ETL and TBL systems achieved similar performance. Therefore, ETL has state-of-the-art performance for the Tycho Brahe Corpus.

Using the template evolution strategy, training time is reduced by nearly 63% and there is no performance loss.

Table 4 System performances for the Mac-Morpho Corpus

System	Accuracy (%)	# Templates
ETL	**96.75**	72
TBL	96.60	26
BLS	90.71	–

Table 5 System performances for the Tycho Brahe Corpus

System	Accuracy (%)	# Templates
ETL	**96.64**	43
TBL	96.63	26
BLS	91.12	–

3.5 TIGER Corpus

We have not found any work reporting the state-of-the-art performance for the TIGER Corpus. Therefore, for the TIGER Corpus, we report the performance of ETL, TBL and BLS systems.

Table 6 System performances for the TIGER Corpus

System	Accuracy (%)	# Templates
ETL	**96.57**	66
TBL	96.53	26
BLS	90.53	–

In Table 6, we summarize the performance results of the three systems. Both ETL and TBL systems reduce the BLS error by at least 63%. ETL and TBL systems achieved similar performance.

The TnT tagger [4], which is based on Hidden Markov Models, is a state-of-the-art system for German. It has been trained and tested on the NEGRA Corpus [41], which is the predecessor of TIGER Corpus and uses the same tagset. For the NEGRA Corpus, Brants [4] reports an overall accuracy of 96.7%. Since both ETL and TBL systems performances for the TIGER Corpus are very close to the TnT performance for the NEGRA Corpus, we believe that both ETL and TBL systems achieve state-of-the-art performance for German POS tagging.

Using the template evolution strategy, training time is reduced by nearly 61% and there is no performance loss.

3.6 Brown Corpus

Eric Brill [6] presents a TBL system with state-of-the-art performance for the Brown Corpus. Therefore, for the Brown Corpus, we only report the performance of ETL, TBL and BLS systems.

Table 7 System performances for the Brown Corpus

System	Accuracy (%)	# Templates
ETL	**96.69**	63
TBL	96.67	26
BLS	90.91	–

In Table 5, we summarize the performance results of the three systems. Both ETL and TBL systems reduce the BLS error in 63%. ETL and TBL systems achieved similar performance. Therefore, ETL has state-of-the-art performance for the Brown Corpus.

Using the template evolution strategy, training time is reduced by nearly 68% and there is no performance loss.

4 Phrase Chunking

Phrase Chunking (PC) consists in dividing a text into non-overlapping phrases [40]. It provides a key feature that helps on more elaborated NLP tasks such as NER and semantic role labeling. In the example that follows, we use brackets to indicate the eight phrase chunks in the sentence. In this example, there are four Noun Phrases (NP), two Verb Phrases (VP) and two Prepositional Phrases (PP).

> [NP He] [VP reckons] [NP the current account
> deficit] [VP will narrow] [PP to] [NP only
> # 1.8 billion] [PP in] [NP September]

This section presents the application of the ETL approach to language independent PC. We apply ETL to four different corpora in three different languages. The selected corpora are: SNR-CLIC, a Portuguese Noun Phrase chunking corpus [21]; Ramshaw & Marcus (R&M), an English Base Noun Phrase chunking corpus [39]; CoNLL-2000, an English Phrase chunking corpus [40]; and SPSAL-2007, a Hindi Phrase chunking corpus [2]. All the corpora are tagged with both POS and PC tags. Table 8 shows some characteristics of these corpora. The NP chunking task consists in recognizing non-overlapping text segments that contain NPs only.

Phrase Chunking Modeling

We approach PC as a token classification problem, in the same way as in the CONLL-2000 shared task [40]. For both corpora SNR-CLIC and Ramshaw & Marcus, we use the $IOB1$ tagging style, where: O, means that the word is not a NP; I, means that the word is part of a NP and B is used for the leftmost word of a NP beginning immediately after another NP. This tagging style is shown in the following example.

Table 8 Phrase Chunking Corpora

Corpus	Language	Phrases	Training Data		Test Data	
			Sentenc.	Tokens	Sentenc.	Tokens
SNR-CLIC	Portuguese	NP	3514	83346	878	20798
R&M	English	NP	8936	211727	2012	47377
CoNLL-2000	English	All	8936	211727	2012	47377
SPSAL-2007	Hindi	All	924	20000	210	5000

```
He/I  reckons/O  the/I  current/I  account/I  deficit/I  will/O
narrow/O  to/O  only/I  #/I  1.8/I  billion/I  in/O  September/I
```

For both corpora CoNLL-2000 and SPSAL-2007, we use the $IOB2$ tagging style, where: O, means that the word is not a phrase; $B - XX$, means that the word is the first one of a phrase type XX and $I - XX$, means that the word is inside of a phrase type XX. This tagging style is shown in the following example.

```
He/B-NP  reckons/B-VP  the/B-NP  current/I-NP  account/I-NP
deficit/I-NP will/B-VP narrow/I-VP  to/B-PP  only/B-NP  #/I-NP
     1.8/I-NP  billion/I-NP  in/B-PP  September/B-NP
```

ML Modeling

The following ML model configurations provide our best results.

BLS the baseline system assigns to each word the PC tag that was most frequently associated with the part-of-speech of that word in the training set. The only exception was the initial classification of prepositions in the SNR-CLIC Corpus. In this case, the initial classification is done on an individual basis: each preposition has its frequency individually measured and the NP tag is assigned accordingly, in a lexicalized method.

TBL in the TBL system we use a template set that contains the templates proposed by Ramshaw & Marcus [39]. The Ramshaw & Marcus's template set contains 100 handcrafted templates which make use of the features *word*, *POS* and *NP tags*, and use a context window of seven tokens. For the SNR-CLIC Corpus we extend the template set by adding the set of six special templates proposed by Santos & Oliveira [12]. The Santos & Oliveira's six templates are designed to reduce classification errors of preposition within the task of Portuguese noun phrase chunking. These templates use special handcrafted constraints that allow to efficiently check the feature *word* in up to 20 left side adjacent tokens.

ETL in the ETL learning, we use the features *word*, *POS* and *NP tags* in a context window of size 7. For the SNR-CLIC Corpus, we introduce the feature *left verb*. This feature assumes the word feature value of the nearest predecessor verb of the current token. In the DT learning step: only the 200 most frequent words are used; the PC tag of the word is the one applied by the initial classifier; and, the PC tag of neighbor words are the true ones. We report results for ETL trained with all the templates at the same time as well as using template evolution.

4.1 SNR-CLIC Corpus

Santos & Oliveira [12] present a TBL modeling that obtains state-of-the-art performance for Portuguese Noun Phrase Chunking. Our SNR-CLIC TBL system uses the same modeling presented by Santos & Oliveira. For the SNR-CLIC Corpus, we report the performance of ETL, TBL and BLS systems.

In Table 9, we summarize the performance results of the three systems. ETL increases the BLS $F_{\beta=1}$ by 31%. ETL system outperforms the TBL system's $F_{\beta=1}$ by 1.14. These results indicate that the templates obtained by entropy guided feature selection are very effective to learn transformation rules for this task.

Table 9 System performances for the SNR-CLIC Corpus

System	Accuracy (%)	Precision (%)	Recall (%)	$F_{\beta=1}$	# Templates
ETL	**97.97**	**88.77**	**88.93**	**88.85**	46
TBL	97.63	87.17	88.26	87.71	106
BLS	96.57	62.69	74.45	68.06	–

Using the template evolution strategy, training time is reduced by nearly 51%. On the other hand, there is a decrease of 0.4 in the ETL system's $F_{\beta=1}$.

4.2 Ramshaw and Marcus Corpus

Kudo & Matsumoto [27] present a SVM-based system with state-of-the-art performance for the Ramshaw & Marcus Corpus. Therefore, for this Corpus, we also list the SVM system performance reported by Kudo & Matsumoto.

In Table 10, we summarize the performance results of SVM, ETL, TBL and BLS systems. ETL increases the BLS $F_{\beta=1}$ by 16%. ETL system slightly outperforms the TBL system's. The ETL performance is competitive with the one of the SVM system.

Using the template evolution strategy, training time is reduced by nearly 62% and there is no performance loss.

Table 10 System performances for the Ramshaw & Marcus Corpus

System	Accuracy (%)	Precision (%)	Recall (%)	$F_{\beta=1}$	# Templates
SVM	–	**94.15**	**94.29**	**94.22**	–
ETL	97.52	92.49	92.70	92.59	106
TBL	97.42	91.68	92.26	91.97	100
BLS	94.48	78.20	81.87	79.99	–

Table 11 System performances for the CoNLL-2000 Corpus

System	Accuracy (%)	Precision (%)	Recall (%)	$F_{\beta=1}$	# Templates
SVM	–	**94.12**	**94.13**	**94.12**	
ETL	95.13	92.24	92.32	92.28	183
TBL	95.12	92.05	92.28	92.16	100
BLS	77.29	72.58	82.14	77.07	–

4.3 CoNLL-2000 Corpus

Wu et al. [44] present a SVM-based system with state-of-the-art performance for the CoNLL-2000 Corpus. Therefore, for this Corpus, we also list the SVM system performance reported by Wu et al.

In Table 11, we summarize the performance results of SVM, ETL, TBL and BLS systems. ETL increases the BLS $F_{\beta=1}$ by 20%. ETL system slightly outperforms the TBL system's. The ETL performance is competitive with the one of the SVM system.

Using the template evolution strategy, training time is reduced by nearly 81%. On the other hand, there is a slightly decrease of the ETL system's $F_{\beta=1}$.

4.4 SPSAL-2007 Corpus

PVS & Gali [36] present a state-of-the-art system for the SPSAL-2007 Corpus. Their system uses a combination of Hidden Markov Models (HMM) and Conditional Random Fields (CRF). Therefore, for this Corpus, we also list the HMM+CRF system performance reported by PVS & Gali.

In Table 12, we summarize the performance results of HMM+CRF, ETL, TBL and BLS systems. The results are reported in terms of chunking accuracy only, the same performance measure used in the SPSAL-2007 contest [2]. ETL increases the BLS $F_{\beta=1}$ by 12%. ETL and TBL systems achieved similar performance. The ETL performance is very competitive with the one of the HMM+CRF system.

Table 12 System performances for the SPSAL-2007 Corpus

System	Accuracy (%)	# Templates
HMM + CRF	**80.97**	–
ETL	78.53	30
TBL	78.53	100
BLS	70.05	–

We do not use template evolution for the SPSAL-2007 Corpus. Since the training corpus is very small, the training time reduction is not significant.

5 Named Entity Recognition

Named Entity Recognition (NER) is the problem of finding all proper nouns in a text and to classify them among several given categories of interest or to a default category called Others. Usually, there are three given categories: Person, Organization and Location. Time, Event, Abstraction, Thing, and Value are some additional, but less usual, categories of interest. In the example that follows, we use brackets to indicate the four Named Entities in the sentence.

[*PER* Wolff], currently a journalist in [*LOC* Argentina],
played with [*PER* Del Bosque] in the final years of the
seventies in [*ORG* Real Madrid]

This section presents the application of the ETL approach to language independent NER. We evaluate the performance of ETL over a Portuguese corpus, LearnNEC06 [32], a Spanish corpus, SPA CoNLL-2002 [43], and Dutch Corpus, DUT CoNLL-2002 [43]. Table 13 shows some characteristics of these corpora.

The LearnNEC06 Corpus [32] is annotated with only three categories: Person, Organization and Location. This corpus is already annotated with golden POS tags and golden noun phrase chunks. Both SPA CoNLL-2002 Corpus and DUT CoNLL-2002 Corpus were used in the CoNLL-2002[43] shared task. These two corpora are annotated with four named entity categories: Person, Organization, Location and Miscellaneous. The CONLL-2002 corpora are already divided into training and test sets. These two datasets also includes development corpora which have characteristics similar to the test corpora.

5.1 NER Modeling

We approach the NER task as a token classification problem, in the same way as in the CoNLL-2002 shared task [43]. For the LearnNEC Corpus, we use the $IOB1$ tagging style, where: O, means that the word is not a NE;

Table 13 Named Entity Recognition Corpora

Corpus	Language	Training Data		Test Data	
		Sentenc.	Tokens	Sentenc.	Tokens
LearnNEC06	Portuguese	2100	44835	–	–
SPA CoNLL-2002	Spanish	8323	264715	1517	51533
DUT CoNLL-2002	Dutch	15806	202931	5195	68994

$I - XX$, means that the word is part of a NE type XX and $B - XX$ is used for the leftmost word of a NE beginning immediately after another NE of the same type. The $IOB1$ tagging style is shown in the following example.

```
Wolff/I-PER  ,/O  currently/O  a/O  journalist/O  in/O
Argentina/I-PLA  ,/O  played/O  with/O  Del/I-PER
  Bosque/I-PER  in/O  the/O  final/O  years/O  of/O
  the/O  seventies/O  in/O  Real/I-ORG  Madrid/I-ORG
```

For both SPA CoNLL-2002 Corpus and DUT CoNLL-2002 Corpus, we use the $IOB2$ tagging style, which is the default tagging style of the CoNLL-2002 shared task. In the $IOB2$, a $B - XX$ tag is always used for the first word of a NE. The $IOB2$ tagging style is shown in the following example.

```
Wolff/B-PER  ,/O  currently/O  a/O  journalist/O  in/O
Argentina/B-PLA  ,/O  played/O  with/O  Del/B-PER
  Bosque/I-PER  in/O  the/O  final/O  years/O  of/O
  the/O  seventies/O  in/O  Real/B-ORG  Madrid/I-ORG
```

5.2 ML Modeling

The following ML model configurations provide our best results.

BLS   for the LearnNEC06 Corpus, we use the same baseline system proposed in [32], which makes use of location, person and organization gazetteers, as well as some simple heuristics.

for both SPA and DUT CoNLL 2002 corpora, the baseline system assigns to each word the named entity (ne) tag that was most frequently associated with that word in the training set. If capitalized, an unknown word is tagged as a person, otherwise it is tagged as non entity.

ETL   in the ETL learning, we use the basic features *word*, *pos*, and *ne tags*. Additionally, we introduce two new features: *capitalization information* and *dictionary membership*. The capitalization information feature provides a token classification, assuming one the following categorical values: First Letter is Uppercase, All Letters are Uppercase, All Letters are Lowercase, Number, Punctuation, Number with "/" or "-" inside or Other. Similarly, the dictionary membership feature assumes one the following categorical values: Upper, Lower, Both or None. In the DT learning step, only the 100 most frequent words are used, the named entity tag of the word is the one applied by the initial classifier, and the named entity tags of neighbor words are the true ones.

TBL   for the LearnNEC06 Corpus, the reported results for the TBL approach refer to TBL trained with the 32 handcrafted template set proposed in [32].

for both SPA and DUT CoNLL 2002 corpora, we use the Brill's template set [6].

Since the LearnNEC06 Corpus is very small, in our experiments we use 10-fold cross validation to measure the performance of the systems. We only report the ETL model that gives the best cross-validation result. For both SPA CoNLL-2002 Corpus and DUT CoNLL-2002 Corpus, we use the development sets to tune the context window size used by ETL.

5.3 *LearnNEC06 Corpus*

Milidiú et al. [32] present a SVM system with state-of-the-art performance for the LearnNEC06 Corpus. Therefore, for the LearnNEC06 Corpus, we also list the SVM system performance reported by Milidiú et al.

In Table 14, we summarize the performance results of the four systems. The best ETL system uses a context window of size 9. Both ETL and TBL systems increase the BLS $F_{\beta=1}$ by at least 15%. The ETL system slightly outperforms the TBL system. ETL results are very competitive with the ones of SVM.

Using the template evolution strategy, training time is reduced by nearly 20% and the learned transformation rules maintain the same performance. In this case, the training time reduction is not very significant, since the training set is very small.

Although the ETL's $F_{\beta=1}$ is only slightly better than the one of TBL with handcrafted template, it is an impressive achievement, since the handcrafted template set used in [32] contains many pre-instantiated rule tests that carry a lot of domain specific knowledge.

In Table 15, we show the ETL system results, broken down by named entity type, for the LearnNEC06 Corpus.

Table 14 System performances for the LearnNEC06 Corpus

System	Accuracy (%)	Precision (%)	Recall (%)	$F_{\beta=1}$	# Templates
SVM	**98.83**	**86.98**	**89.27**	**88.11**	–
ETL	98.80	86.89	88.54	87.71	102
TBL	98.79	86.65	88.60	87.61	32
BLS	97.77	73.11	80.21	76.50	–

Table 15 ETL results by entity type for the LearnNEC06 Corpus

Entity	Precision (%)	Recall (%)	$F_{\beta=1}$
Location	93.96	81.78	87.45
Organization	84.00	89.77	86.79
Person	85.75	91.93	88.73
Overall	86.89	88.54	87.71

5.4 SPA CoNLL-2002 Corpus

Carreras et al. [8] present a AdaBoost system with state-of-the-art performance for the SPA CoNLL-2002 Corpus. Their AdaBoost system uses decision trees as a base learner. Therefore, for the SPA CoNLL-2002 Corpus, we also list the AdaBoost system performance reported by Carreras et al.

In Table 16, we summarize the performance results of the four systems. The best ETL system uses a context window of size 13. The ETL system increase the BLS $F_{\beta=1}$ by 44%. In this corpus, we can see that ETL significantly outperforms the TBL system. The ETL system does not have a very competitive result when compared to AdaBoost system. On the other hand, for the SPA CoNLL-2002 Corpus, ETL is in top 5 when compared with the 12 CoNLL-2002 contestant systems.

Table 16 System performances for the SPA CoNLL-2002 Corpus

System	Accuracy (%)	Precision (%)	Recall (%)	$F_{\beta=1}$	# Templates
AdaBoost	–	**79.27**	**79.29**	**79.28**	–
ETL	96.76	75.23	77.24	76.22	154
TBL	96.32	71.62	74.32	72.95	26
BLS	93.92	46.84	61.14	53.04	–

In Table 17, we show the ETL system results, broken down by named entity type, for the SPA CoNLL-2002 Corpus.

Table 17 ETL results by entity type for the SPA CoNLL-2002 Corpus

Entity	Precision (%)	Recall (%)	$F_{\beta=1}$
Location	77.38	76.38	76.88
Organization	75.67	78.43	77.03
Person	78.61	88.03	83.06
Miscellaneous	56.77	51.76	54.15
Overall	75.23	77.24	76.22

5.5 DUT CoNLL-2002

The Carreras et al. [8] AdaBoost based system is also a state-of-the-art system for the DUT CoNLL-2002 Corpus. Therefore, for the DUT CoNLL-2002 Corpus, we also list the AdaBoost system performance reported by Carreras et al.

In Table 18, we summarize the performance results of the four systems. The best ETL system uses a context window of size 13. The ETL system increase the BLS $F_{\beta=1}$ by 51%. In this corpus, we can see that ETL significantly outperforms the TBL system. The ETL system does not have a very competitive result when compared to AdaBoost system. On the other hand, for the DUT CoNLL-2002 Corpus, ETL is in top 6 when compared with the 12 CoNLL-2002 contestant systems.

Table 18 System performances for the DUT CoNLL-2002 Corpus

System	Accuracy (%)	Precision (%)	Recall (%)	$F_{\beta=1}$	# Templates
AdaBoost	–	**77.83**	**76.29**	**77.05**	–
ETL	97.37	72.33	71.62	71.97	58
TBL	96.99	68.37	67.94	68.15	26
BLS	95.11	43.35	52.78	47.60	–

In Table 19, we show the ETL system results, broken down by named entity type, for the DUT CoNLL-2002 Corpus.

Table 19 ETL results by entity type for the DUT CoNLL-2002 Corpus

Entity	Precision (%)	Recall (%)	$F_{\beta=1}$
Location	76.81	79.90	78.32
Organization	75.57	57.13	65.06
Person	68.68	81.25	74.44
Miscellaneous	71.48	67.99	69.69
Overall	72.33	71.62	71.97

6 Conclusions and Discussions

Entropy Guided Transformation Learning is a new machine learning algorithm for classification tasks. ETL generalizes Transformation Based Learning by automatically solving the TBL bottleneck: the construction of good template sets. ETL uses the information gain in order to select the feature combinations that provide good template sets.

We describe the application of ETL to two language independent Text Mining preprocessing tasks: part-of-speech tagging and phrase chunking. We also report our findings on one language independent Information Extraction task: named entity recognition. Overall, we successfully apply it to six different languages: Dutch, English, German, Hindi, Portuguese and Spanish.

ETL modeling is simple. It only requires the training set and no hand-crafted templates. ETL also simplifies the incorporation of new input features such as capitalization information, which are successfully used in the ETL based NER models.

ETL training is reasonably fast. The observed ETL training time for the Mac-Morpho Corpus, using template evolution, is bellow one hour running on an Intel Centrino Duo 1.66GHz laptop. We also show that by using the *template evolution* strategy, one accelerates transformation learning by a factor of five for the English Text Chunking.

Using the ETL approach, we obtain state-of-the-art competitive performance results in nine out of the eleven corpus-driven tasks. For Spanish and Dutch named entity recognition, ETL achieves promising results. In all experiments, ETL shows better results than TBL with hand-crafted templates.

ETL templates are restricted to the combination of basic input features. On the other hand, TBL can process templates with *derived* features that are generated at training time. For instance, the rule

$$\text{pos[0]=DT} \quad \text{pos[1,3]=JJR} \quad \rightarrow \quad \text{post=RB}$$

uses the derived feature pos[1,3]. This feature checks the pos value of the three tokens following the current one. TBL derived features are very rich and complex [12], providing a powerful representation mechanism. Extending ETL to extract templates that use derived features is an open problem.

ETL inputs are just the training set and the Initial Classifier. Hence, Ensemble Modeling strategies such as Bagging [5] and Boosting [22] can use ETL as a base learner.

Our extensive experimental results demonstrate that ETL is an effective way to learn accurate transformation rules. We believe that by avoiding the use of handcrafted templates, ETL enables the use of transformation rules to a greater range of Text Mining applications, such as Shallow Semantics and Semantic Role Labeling [29].

References

1. Aluísio, S.M., Pelizzoni, J.M., Marchi, A.R., de Oliveira, L., Manenti, R., Marquiafável, V.: An account of the challenge of tagging a reference corpus for brazilian portuguese. In: PROPOR, pp. 110–117 (2003)
2. Bharati, A., Mannem, P.R.: Introduction to shallow parsing contest on south asian languages. In: Proceedings of the IJCAI and the Workshop On Shallow Parsing for South Asian Languages (SPSAL), pp. 1–8 (2007)

3. Brants, S., Dipper, S., Hansen, S., Lezius, W., Smith, G.: The TIGER treebank. In: Proceedings of the Workshop on Treebanks and Linguistic Theories, Sozopol (2002)
4. Brants, T.: Tnt – a statistical part-of-speech tagger. In: ANLP, pp. 224–231 (2000)
5. Breiman, L.: Bagging predictors. Machine Learning 24(2), 123–140 (1996)
6. Brill, E.: Transformation-based error-driven learning and natural language processing: A case study in part-of-speech tagging. Comput. Linguistics 21(4), 543–565 (1995)
7. Carberry, S., Vijay-Shanker, K., Wilson, A., Samuel, K.: Randomized rule selection in transformation-based learning: a comparative study. Natural Language Engineering 7(2), 99–116 (2001)
8. Carreras, X., Màrques, L., Padró, L.: Named entity extraction using adaboost. In: Proceedings of CoNLL 2002, Taipei, Taiwan, pp. 167–170 (2002)
9. Corston-Oliver, S., Gamon, M.: Combining decision trees and transformation-based learning to correct transferred linguistic representations. In: Proceedings of the Ninth Machine Tranlsation Summit, New Orleans, USA, pp. 55–62. Association for Machine Translation in the Americas (2003)
10. Curran, J.R., Wong, R.K.: Formalisation of transformation-based learning. In: Proceedings of the ACSC, Canberra, Australia, pp. 51–57 (2000)
11. Dash, M., Liu, H.: Feature selection for classification. Intelligent Data Analysis 1, 131–156 (1997)
12. dos Santos, C.N., Oliveira, C.: Constrained atomic term: Widening the reach of rule templates in transformation based learning. In: Bento, C., Cardoso, A., Dias, G. (eds.) EPIA 2005. LNCS, vol. 3808, pp. 622–633. Springer, Heidelberg (2005)
13. dos Santos, C.N., Milidiú, R.L., Rentería, R.P.: Portuguese part-of-speech tagging using entropy guided transformation learning. In: Proceedings of 8th Workshop on Computational Processing of Written and Spoken Portuguese, pp. 143–152 (2008)
14. Elming, J.: Transformation-based corrections of rule-based mt. In: Proceedings of the EAMT 11th Annual Conference, Oslo, Norway (2006)
15. Finger, M.: Técnicas de otimização da precisão empregadas no etiquetador tycho brahe. In: Proceedings of PROPOR, São Paulo, pp. 141–154 (November 2000)
16. Florian, R.: Named entity recognition as a house of cards: Classifier stacking. In: Proceedings of CoNLL 2002, Taipei, Taiwan, pp. 175–178 (2002)
17. Florian, R., Henderson, J.C., Ngai, G.: Coaxing confidences from an old friend: Probabilistic classifications from transformation rule lists. In: Proceedings of Joint Sigdat Conference on Empirical Methods in NLP and Very Large Corpora, Hong Kong University of Science and Technology (October 2000)
18. Forman, G., Guyon, I., Elisseeff, A.: An extensive empirical study of feature selection metrics for text classification. Journal of Machine Learning Research 3, 1289–1305 (2003)
19. Francis, W.N., Kucera, H.: Frequency analysis of english usage. Lexicon and grammar (1982)
20. Freitas, M.C., Duarte, J.C., dos Santos, C.N., Milidiú, R.L., Renteria, R.P., Quental, V.: A machine learning approach to the identification of appositives. In: Proceedings of Ibero-American AI Conference, Ribeirão Preto, Brazil (October 2006)

21. Freitas, M.C., Garrao, M., Oliveira, C., dos Santos, C.N., Silveira, M.: A anotação de um corpus para o aprendizado supervisionado de um modelo de sn. In: Proceedings of the III TIL / XXV Congresso da SBC, São Leopoldo - RS - Brasil (2005)

22. Freund, Y., Schapire, R.E.: A decision-theoretic generalization of on-line learning and an application to boosting. Journal of Computer and System Sciences 55(1), 119–139 (1997)

23. Higgins, D.: A transformation-based approach to argument labeling. In: Ng, H.T., Riloff, E. (eds.) HLT-NAACL 2004 Workshop: Eighth Conference on Computational Natural Language Learning (CoNLL 2004), Boston, Massachusetts, USA, May 6 - 7, 2004, pp. 114–117. Association for Computational Linguistics (2004)

24. Hwang, Y.-S., Chung, H.-J., Rim, H.-C.: Weighted probabilistic sum model based on decision tree decomposition for text chunking. International Journal of Computer Processing of Oriental Languages (1), 1–20 (2003)

25. IEL-UNICAMP and IME-USP. Corpus anotado do português histórico tycho brahe, http://www.ime.usp.br/~tycho/corpus/ (accessed January 23, 2008)

26. Jurafsky, D., Martin, J.H.: Speech and Language Processing. Prentice-Hall, Englewood Cliffs (2000)

27. Kudo, T., Matsumoto, Y.: Chunking with support vector machines. In: Proceedings of the NAACL 2001 (2001)

28. Mangu, L., Brill, E.: Automatic rule acquisition for spelling correction. In: Proceedings of The Fourteenth ICML. Morgan Kaufmann, San Francisco (1997)

29. Màrquez, L., Carreras, X., Litkowski, K.C., Stevenson, S.: Semantic role labeling: an introduction to the special issue. Computational Linguistics 34(2), 145–159 (2008)

30. Megyesi, B.: Shallow parsing with pos taggers and linguistic features. Journal of Machine Learning Research 2, 639–668 (2002)

31. Milidiú, R.L., dos Santos, C.N., Duarte, J.C.: Phrase chunking using entropy guided transformation learning. In: Proceedings of ACL 2008, Columbus, Ohio (2008)

32. Milidiú, R.L., Duarte, J.C., Cavalcante, R.: Machine learning algorithms for portuguese named entity recognition. In: Proceedings of Fourth Workshop in Information and Human Language Technology, Ribeirão Preto, Brazil (2006)

33. Milidiú, R.L., Duarte, J.C., dos Santos, C.N.: Tbl template selection: An evolutionary approach. In: Proceedings of Conference of the Spanish Association for Artificial Intelligence - CAEPIA, Salamanca, Spain (2007)

34. Mitchell, T.M.: Machine Learning. McGraw-Hill, New York (1997)

35. Ngai, G., Florian, R.: Transformation-based learning in the fast lane. In: Proceedings of North Americal ACL, pp. 40–47 (June 2001)

36. Avinesh, P.V.S., Gali, K.: Part-of-speech tagging and chunking using conditional random fields and transformation based learning. In: Proceedings of the IJCAI and the Workshop On Shallow Parsing for South Asian Languages (SP-SAL), pp. 21–24 (2007)

37. Ross Quinlan, J.: Induction of decision trees. Machine Learning 1(1), 81–106 (1986)

38. Ross Quinlan, J.: C4.5: programs for machine learning. Morgan Kaufmann Publishers Inc., San Francisco (1993)

39. Ramshaw, L., Marcus, M.: Text chunking using transformation-based learning. In: Armstrong, S., Church, K.W., Isabelle, P., Manzi, S., Tzoukermann, E., Yarowsky, D. (eds.) Natural Language Processing Using Very Large Corpora. Kluwer, Dordrecht (1999)
40. Tjong Kim Sang, E.F., Buchholz, S.: Introduction to the conll-2000 shared task: chunking. In: Proceedings of the 2nd workshop on Learning language in logic and the 4th CONLL, Morristown, NJ, USA, pp. 127–132. Association for Computational Linguistics (2000)
41. Skut, W., Krenn, B., Brants, T., Uszkoreit, H.: An annotation scheme for free word order languages. In: Proceedings of ANLP 1997 (1997)
42. Su, J., Zhang, H.: A fast decision tree learning algorithm. In: AAAI (2006)
43. Tjong Kim Sang, E.F.: Introduction to the conll-2002 shared task: Language-independent named entity recognition. In: Proceedings of CoNLL 2002, Taipei, Taiwan, pp. 155–158 (2002)
44. Wu, Y.-C., Chang, C.-H., Lee, Y.-S.: A general and multi-lingual phrase chunking model based on masking method. In: Proceedings of 7th International Conference on Intelligent Text Processing and Computational Linguistics, pp. 144–155 (2006)

Artificial Development

Arturo Chavoya

Abstract. Artificial Development is a field of Evolutionary Computation inspired by the developmental processes and cellular growth seen in nature. Multiple models of artificial development have been proposed in the past, which can be broadly divided into those based on biochemical processes and those based on a high level grammar. Two of the most important aspects to consider when designing a cellular growth model are the type of representation used to specify the final features of the system, and the abstraction level necessary to capture the properties to be modeled. Although advances in this field have been significant, there is much knowledge to be gained before a model that approaches the level of complexity found in living organisms can be built.

1 Introduction

Artificial Development is the study of computer models of cellular growth, with the objective of understanding how complex structures and forms can emerge from a small group of undifferentiated initial cells. In biological systems, development is a fascinating and very complex process that involves following an extremely intricate program coded in the organism's genome. To present day, we still marvel at how from a single initial cell, the zygote, a whole functional organism of trillions of coordinated cells can develop.

Over the years, artificial models of cellular growth have been proposed with the objective of either understanding the intricacies of the development process or with the goal of finding new computational paradigms inspired by nature. Artificial evolutionary systems with a developmental component have

Arturo Chavoya

University of Guadalajara, Periférico Norte 799, Núcleo Belenes, Zapopan, Jal. Mexico CP 45000

e-mail: `achavoya@cucea.udg.mx`

A.-E. Hassanien et al. (Eds.): Foundations of Comput. Intel. Vol. 1, SCI 201, pp. 185–215.

been given different names, such as artificial embryology [13], morphogenesis [32], artificial ontogeny [5], computational embryology [43], computational development [44] and artificial embryogeny [66]. We chose here to use the more generic name *artificial development*, since not all multicellular organisms in nature develop as embryos and not every model is forced to be implemented in computers.

Recent research in biology has shown that gene regulatory networks play a central role in the development and metabolism of living organisms [10]. It has been discovered in the last years that the diverse cell patterns created during the developmental stages are mainly due to the selective activation and inhibition of very specific regulatory genes. Artificial regulatory networks are computer models that seek to emulate the gene regulatory networks found in nature. On the other hand, evolutionary computation techniques have been extensively used in the past in a wide range of applications, and in particular they have previously been used to evolve artificial regulatory networks to perform specific tasks.

Traditionally, artificial development systems have been roughly divided into those based on biochemical processes and those based on a grammatical approach. Models based on cell chemical processes are inspired by the natural mechanisms seen in development such as gene expression, cell signalling, gene/protein interaction, gene mutation and recombination, chemical gradients, cell differentiation and cell death. On the other hand, grammatical approaches are based on the evolution of a set of rewrite rules, where the grammar may or may not be context-free and the rules may be parameterized. More recently, Stanley and Miikkulainen have proposed a new taxonomy for artificial development models based on the dimensions of the natural development process, such as cell fate, cell targeting and heterochrony [66]. These authors suggest that this taxonomy can aid in defining the capabilities of a given artificial development system.

This chapter covers the main research areas pertaining to artificial development with an emphasis toward the systems based on artificial regulatory networks. To this end, a short introduction to biological gene regulatory networks and their relationship with development is presented first. The sections on the various artificial development models are followed by a section on the canonical problem in artificial and natural development known as *the French flag problem*. The chapter ends with a section of concluding remarks.

2   Development and Gene Regulatory Networks

The mechanism of biological development can be viewed conceptually as consisting of a series of concentric layers [10]. On the outer layer, development is achieved through the spatial and temporal regulation of expression of myriads of genes coding for all the different proteins of the organism, which catalyze the creation of other constituents. A deeper layer is characterized by a

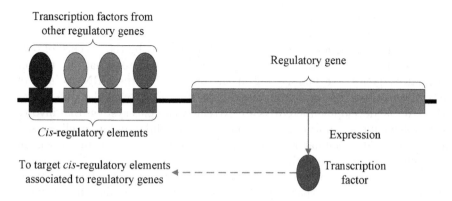

Fig. 1 Relationship between *cis*-regulatory elements, transcription factors and regulatory genes

dynamic progression of regulatory states, defined as the presence and activity state of the set of regulatory proteins that control gene expression. Finally, at the core layer is the genomic machinery consisting of all the modular DNA sequences that interact with the regulatory proteins in order to interpret the regulatory states. These DNA sequences are known as *cis*-regulatory elements (*cis* is Latin for "this side of"), as they refer to regulatory elements usually located on the same DNA molecule as the genes that they control. *Cis*-regulatory elements are the target of active diffusible proteins known as transcription factors or *trans* (Latin for "far side of") elements, which define the regulatory state.

Cis-regulatory elements read the information contained in the regulatory state of the cell, process that information, and interpret it into instructions that can be used by the cellular biochemical machinery to express genes contained in the genome. The term "regulatory genes" refer to those genes encoding the transcription factors that interact with *cis*-regulatory elements. Figure 1 shows a simplified illustration of the relationship between *cis*-regulatory elements, regulatory genes and transcription factors. Transcription factors bind to specific *cis*-regulatory elements, and this interaction can enhance or inhibit the expression of the associated regulatory gene into a transcription factor, which can in turn interact with its target *cis*-regulatory element on other genes.

The spatial and temporal expression of regulatory genes is central to development, as they determine to a great extent the fate and function of all cells in the developing organism. Developmental control systems take the form of gene regulatory networks (GRNs); when genes in a GRN are expressed, they produce transcription factors that can affect multiple target genes (through their associated *cis*-regulatory elements), which can in turn express transcription factors that affect their target genes. Each regulatory gene can have multiple inputs from other regulatory genes and multiple outputs to other regulatory genes. Thus a regulatory gene can be viewed as a node in a network of interactions.

The periphery of a developmental GRN is defined by the absence of outputs to other genes in the network, i.e. transcription factors that affect other regulatory genes. We mainly find at the outskirts of the GRN the sets of developmental genes that code for proteins that lead to cellular differentiation. Actual developmental GRNs are extremely complex systems that involve other elements not shown in Fig. 1, such as intracellular and intercellular signaling molecules, cellular receptors and lineage proteins that are present in one cell type and not in others.

Much of the knowledge on the development of organisms has been obtained from studying animals that are easy to maintain and reproduce in captivity. One of the most studied species in genetic and developmental research is the fruit fly *Drosophila melanogaster*. Over the years, researchers have been able to identify a number of proteins directly involved in the development of *D. melanogaster*. For instance, eight regulatory genes are responsible for the development of specific body segments on this insect. Mutation of these genes severely affects the development of the associated body segments.

Researchers continue to elucidate the mechanisms underlying development at the molecular level and much work on the subject remains to be done. However, it is increasingly evident that GRNs play an essential role in the development of multicellular organisms, from the simplest to the higher species of plants and animals.

3   Reaction-Diffusion Systems

It is usually attributed to Turing the founding of modern research on artificial development. He suggested in his seminal article on the chemical basis of morphogenesis [69] that an initially homogeneous medium might develop a structured pattern due to an instability of the homogeneous equilibrium, triggered by small random perturbations.

Using a set of differential equations, Turing proposed a reaction-diffusion model where substances called morphogens, or form generators, would react together and diffuse through a medium, which could be a tissue. In this model, one of the substances is autocatalytic and tends to synthesize more copies of itself. At the same time another substance also promotes synthesis of the first substance, but the latter inhibits the synthesis of the former. One key element of the model is that the two substances have very different diffusion coefficients, with one of them diffusing much more rapidly than the other. The system can be fine-tuned with the proper parameters such that at some point the slightest disruption in the equilibrium can be amplified and propagated through the medium generating unpredictable patterns.

For simplification purposes, Turing built his model using a few elemental cellular structures. He considered the cases of an isolated ring of cells, a hollow sphere of cells, and a 2D single layer of cells. He was particularly interested in investigating what caused the initial instability that led to the formation of patterns. He observed that there were six main patterns in the distribution of morphogens. One of the most interesting was the appearance of stationary waves on a ring of cells. He suggested that this could explain certain radial patterns that appeared during the morphogenesis of some organisms. Using the other cellular structures, he demonstrated how the gastrulation process could be generated in a sphere of cells by the reaction-diffusion mechanism. Other patterns, such as dappling, could be generated in a single layer of cells, which could account for the skin patterns seen in many animals.

Even though his model was based on an oversimplification of natural conditions, Turing succeeded in demonstrating how the emergence of a complex pattern could be explained in terms of a simple reaction and diffusion mechanism using well-known physical and chemical principles.

4 Self-activation and Lateral Inhibition Model

Experiments with biological specimens have demonstrated that development is a very robust process. Development can continue normally even after a substantial amount of tissue from certain parts has been removed from an embryo. However, there are small specialized regions that play a crucial role in the organization of the development process. Such organizing regions are usually very small and they control pattern generation in the surrounding tissues. When these organizing regions are transplanted to other parts of the embryo, they start to generate the structures that they would normally form in the original region [7, 8].

In order to explain the long range effect of these small organizing regions on the larger surrounding tissue and the robustness of their influence even after induced interferences, Wolpert introduced the concept of "positional information", whereby a local source region produces a signaling chemical [73, 74]. This theoretical substance was supposed to diffuse and decay creating a concentration gradient that provided cells with information regarding their position in the tissue.

Nevertheless, the problem remained as to how a local differentiated source region could be generated from a seemingly homogeneous initial cluster of developing cells. Even though many eggs have some predefined structure, all the patterns developed after a number of cell divisions cannot initially be present in the egg. A mechanism must exist that allows the emergence of heterogeneous structures starting with a more or less homogeneous egg.

Pattern generation from an almost homogeneous condition can often be observed in the inanimate world. Dunes, for example, can form from an initially homogeneous sand surface under the influence of wind. A small initial ripple in the sand can be amplified (positive feedback) until bigger and bigger sand deposits are formed.

Gierer and Meinhardt proposed that pattern formation was the result of local self-activation coupled with lateral inhibition [24, 23, 52]. In this model, which has some resemblance to Turing's model, a substance that diffuses slowly, called the activator, induces its own production (autocatalysis or self-activation) as well as that of a faster diffusing antagonist, the inhibitor. These authors suggest that pattern formation requires both a strong positive feedback (autocatalysis) and a long-ranging inhibitor to stop positive feedback from spreading indefinitely (lateral inhibition).

The concentration of the activator and the inhibitor can be in a stable state, since an increase in the activator concentration can be compensated by a corresponding increase in the inhibitor concentration, bringing the activator concentration back to the initial concentration. This equilibrium is nonetheless locally unstable. A local increase in the activator concentration will continue to increase by autocatalysis as the inhibitor diffuses more rapidly than the activator into the surrounding area. The smallest random fluctuations are sufficient for breaking the homogeneity so that pattern generation can be initiated.

One possible interaction between the activator a and the inhibitor h is suggested by the authors with the definition of the following equations:

$$\frac{\partial a}{\partial t} = \frac{\rho a^2}{h} - \mu_a a + D_a \frac{\partial^2 a}{\partial x^2} + \rho_a \tag{1}$$

$$\frac{\partial h}{\partial t} = \rho a^2 - \mu_h h + D_h \frac{\partial^2 h}{\partial x^2} + \rho_h \tag{2}$$

where t is time, x is the spatial coordinate, D_a and D_h are the diffusion coefficients and μ_a and μ_h are the decay rates of a and h, respectively. Parameter ρ is the source density that expresses the ability of cells to perform autocatalysis. A small activator production ρ_a can start the disruption of the homogeneous condition at low activator concentrations.

For the formation of stable patterns in this model, the diffusion of the activator has to be much slower than that of the inhibitor, i.e. $D_a \ll D_h$. Furthermore, the activator must have a longer time constant than the inhibitor, $\mu_a < \mu_h$, otherwise oscillations will be produced [55].

These equations were approximated by difference equations for discrete cells and used for computer simulations, which suggest how in an undifferentiated medium, a small perturbation could be amplified and then disrupt the homogeneous state. For instance, in an apparently homogenous egg, a small perturbation in the activator concentration at one end of the egg could lead to the accumulation of activator molecules at this end. The activator molecules could have other properties such as a signaling role that could induce cells near this zone to differentiated into the head (or tail) of the developing organism.

Gierer and Meinhardt have proposed many other examples of biological systems where their model could be applied [52, 53, 56, 55]. One well-known application of their model is in the simulated generation of shell patterns in seashells [54]. With the appropriate parameter values, these simulations generate patterns that are very close to those found in the shells of certain species of seashells.

These results suggest how a relatively simple mechanism of coupled biochemical interactions can account for the generation of very complex patterns. The components of the model are based on reasonable assumptions, since mutual activation and inhibition of biochemical substances and molecular diffusion actually exist in the real world. In recent years, molecular biology and genetics experiments have given support to many elements of the model. Possibly the weakest assumption the authors made for the model was the use of simple diffusion for cell signaling. Modern biological techniques have shown that intercellular communication is indeed very complex and it usually involves the expression of ligand-specific receptors at the cell's surface and intracellular transportation of certain signaling molecules to the cell's nucleus. Nevertheless, the logic behind the model seems to be sound.

5 Lindenmayer Systems

Lindenmayer systems, or L-systems, were originally introduced as a mathematical formalism for modeling development of simple multicellular organisms [48, 49]. The organism is abstracted as an assembly of repeating discrete structures or modules. The formalism is independent of the nature of the module, which can be an individual cell or a whole functional structure such as a plant branch. A module is represented by a symbol from an alphabet, and the symbol represents the module's type. Additionally, there can be parameters associated to a symbol in order to define its state, in which case the formalism is extended to what is known as Parametric L-systems [28].

An L-system is a formal grammar with a set of symbols and a set of rewriting rules. The rules are applied iteratively starting with the initial symbol. Unlike traditional formal grammars, rewriting rules are applied in parallel to simulate the simultaneous development of component parts of an organism.

L-Systems were initially conceived to derive theoretical results applicable to biological development. However, with the introduction of state-of-the-art computer graphics techniques and the widespread use of computer equipment, L-Systems were used to program and dynamically visualize development systems [64, 60].

One of the main applications of L-systems has been in the modeling of the development of higher plants [61]. The modeling does not take place at the cellular level. Instead, it is based on a modular construction of discrete structural units that are repeated during the development of plants, such as branches, leaves and petals [58, 59]. Initial models did not consider the influence of the environment on development. However, as organisms in nature are an integral part of an ecosystem, an extension to the modeling framework that considered interaction with the environment was introduced [51].

The use of L-Systems has been extremely fruitful in modeling the development of organisms at a high structural level. Implemented models of plant development that use L-systems are visually striking because of their resemblance to growth seen in real-life plants and trees.

6 Biomorphs

Richard Dawkins' well-known Biomorphs were first introduced in his famous book "The Blind Watchmaker" to illustrate how evolution might induce the creation of complex designs by means of micro-mutations and cumulative selection [11]. Dawkins intended to find a model to counteract the old argument in biology that a finished complex structure such as the human eye could not be accounted for by Darwin's evolution theory.

Biomorphs are the visible result of the instructions coded in a genome that can undergo evolution. The original genome consists of nine genes coded as integers, where the first eight genes determine the length and branching direction of growing lines in a 2D plane, whereas the last gene controls the depth of branching. Dawkins introduced a constraint of symmetry around an axis so that the resulting forms would show bilateral symmetry, as in many biological organisms.

The construction algorithm that grows biomorphs from their genome is recursive in nature, as Dawkins considered that actual embryological processes could to a large extent be considered recursive. The idea being that the shape of an adult individual emerged after a number of local cellular interactions in the whole developing body and that these effects consisted of simple divergencies such as binary cellular divisions.

Initially Dawkins thought that the forms produced would be limited to tree-like structures. However, to his surprise, the forms generated were extremely varied in shape and detail. There were biomorphs that roughly resembled insects, crustaceans or even mammals.

This author proposed next an "interactive" evolutionary algorithm, where the user played the part of the selection force. The algorithm implementation presents the user with an individual biomorph and the eighteen biomorphs that result from the addition of subtraction of one unity (micro-mutation) on each of the nine genes in the genome. Initially the user has to decide which form he/she wants to evolve, such as a spider or a pine tree, and in each step of the algorithm he/she chooses the biomorph that best resembles the target form (cumulative selection).

The algorithm was further improved by adding genes that could activate or disable the symmetrical growth of the generated forms in the 2D plane, both horizontally and vertically. The algorithm was also extended to take into account the segmentation process, which is considered one of the greatest innovations in biological evolution. Segmented bodies are present in at least three of the major phyla: vertebrates, arthropods and annelids.

Dawkins showed with his models that the evolution of complex structures was indeed feasible in a step by step manner by means of the cumulative selection of the individual that best approached the final structure.

7 Artificial Embryology

Hugo de Garis worked on the creation of a self-assembly process that he called "artificial embryogenesis". His motivation was that he believed that in the future, machines would have so many components that a sequential mechanical assembly would not be feasible. He theorized that highly complex machines should be self-assembled in a similar way as biological organisms are developed.

He worked on artificial "embryos" as 2D shapes formed by a colony of cells using the cellular automata paradigm. The basic idea is that a genetic algorithm can evolve the cellular automaton rule set to control reproduction of artificial cells [12]. The rule sets are encoded in the chromosomes used by the genetic algorithm and they can be switched on or off, depending on whether the state of a cell matches or not the gene's "condition field". If the action field is activated, then cells can reproduce.

De Garis used what he called "differentiable chromosomes", which consisted of several genes or "operons", where in general each operon contained a condition field C and an action field A. Each cell in a particular cellular automaton model contains the same chromosome and at every time step all cells calculate their current state by looking at the states of the neighboring cells. Next, each cell compares its present state with the condition fields C_i in the chromosome. If one of the condition C matches, then the corresponding action field A is activated and the reproduction instructions coded in this field are executed.

This author developed a model that evolved reproduction rules for cellular automata, with the goal that the final shape of a colony of cells was as close as possible to a predefined simple shape such as a square or a triangle [14, 13]. In this model, cells can only reproduce if there is at least one adjacent empty cell, i.e. only edge cells are allowed to reproduce. Assuming that only edge cells can reproduce and that no isolated cells are generated, then cells can be in one of fourteen possible states in a 2D lattice.

If one state is coded by one bit, then only fourteen bits are necessary to code all states, with a predefined position for each state in the bit string. If there is a "1" in the bit corresponding to a state, then all edge cells in that state are allowed to divide and generate a daughter cell, whereas a "0" means that cells in that state are not permitted to reproduce.

When edge cells are allowed to reproduce, it has to be determined the direction at which the daughter cell is to be placed. For states where there is only one empty adjacent cell, there is no option but to place the new cell at that position, so in this case there is no need to code the direction of reproduction. However, when there are two or three empty adjacent cells, a choice has to be made as to where to place the newly produced cell. For states with two empty adjacent cells, only one bit is necessary to determine the direction of reproduction, with a predefined direction for each bit value and each state. For states with three empty cells, two bits are sufficient to code the direction of reproduction, again with a predefined direction of reproduction for each state and each combination of the two bits. As a result, since there are 6 states with 2 empty adjacent cells, and 4 states with 3 vacant adjacent cells, it is necessary to use $6 \times 1 + 4 \times 2 = 14$ bits to specify the direction of reproduction for all states.

The combination of the two 14-bit strings determines for one reproduction cycle which cells are allowed to divide and at which relative position to place their daughter cells. In order to evolve the reproduction rules for more than one cycle with a possibly different reproduction rule for each cycle, it is necessary to introduce in the differentiable chromosome as many instances of the two 14-bit strings as iteration steps are desired. Even the number of iterations necessary to create a predefined shape can be evolved.

These chromosomes were evolved by a genetic algorithm using the following fitness function:

$$Fitness = \frac{ins - \frac{1}{2}outs}{des}, \tag{3}$$

where ins is the number of filled cells inside the desired shape, $outs$ is the number of filled cells outside the desired shape, and des is the total number of cells inside the desired shape. Thus, a fitness value of 1 represents a perfect match.

Using this setup, the generation of several target shapes, such as a triangle and a rectangle, was tried. Since isolated cells are not allowed in the model, reproduction started with a 2×2 cluster of cells. In both cases, the number of iterations evolved to be 4.

Several other target shapes, both convex and non-convex were tested. Results showed that convex shapes could be obtained with a fitness value around 95%, but non-convex shapes evolved poorly, with low fitness values.

After these initial results, de Garis concluded that evolving an artificial embryo implies a type of sequential, synchronized unfolding of shapes. For example, after the main body is grown, then the head and limbs can be grown, followed by the emergence of more detailed shapes, such as those corresponding to fingers and toes [13].

To put this idea to the test, de Garis attempted the generation of an L-shaped form, which is a non-convex shape. The basic idea is to use two genes to generate the L shape, where the first gene would generate the vertical component of the L, and it then would be shut off to let cells in the lower portion express the second gene in order to produce the horizontal region.

For this approach to work, it is necessary that cells can somehow determine their position in the grid. This author decided to use a concentration gradient of a theoretical chemical to provide cells with positional information. Each cell has a certain quantity of this chemical, which is replicated and a fraction of it is transmitted to the daughter cell. Thus a concentration gradient is formed with its peak at the position of the initial cells. Each cell also has an 8-bit storage for determining the direction of the highest concentration of the chemical. At each time step, concentration is measured at each cell position averaging the concentrations from the cell and the three cells towards the direction of the eight main "cardinal points" (N, NE, E, SE, S, SW, W and NW), thus obtaining eight average concentration values for each cell. The bit corresponding to the direction with the highest concentration is then given a value of 1, and the other bits are set to 0. In this manner, by measuring the concentration values from neighboring cells, a cell can determine the concentration gradient and have an estimate of its relative position. For example, cells in the south-east of the vertical region would have their highest gradients towards the north-west direction and would therefore have their north-west bit set to 1.

On the other hand, it is necessary that cells know when to switch the gene for the vertical region off so that the gene for the horizontal region can be expressed. The solution proposed by de Garis was to let cells carry an internal generation count, so that a parent cell with a count value of g would produce a daughter cell with a count of $g + 1$.

With this approach, de Garis obtained limited results when trying to produce an L shape. The fitness value of the resulting shape was about 80%, which was not as good as when generating convex shapes.

In an attempt to obtain better results, the author used a technique that he termed "shaping", which consisted in dividing the evolutionary process in phases with intermediate targets. In this case, the second operon was disabled and evolution was initially conducted with the objective of producing chromosomes that could generate the vertical region of the L shape. The resulting chromosomes were then fed as a starting chromosome population for the genetic algorithm, this time with both operons fully functional.

Despite the bias towards chromosomes that were already conditioned to start with the correct region, the shaping technique did not produce better result, as the final fitness value of the most successful experiments was again about 80%. This author went on and tried to generate other non-convex shapes, such as a snowman shape and a turtle shape with a combination of circle shapes, but he again ran into limitations in the maximum fitness values that he was able to obtain.

Even though the approach used by de Garis proved the potential of the application of evolutionary techniques to the growth of artificial cells in order to generate desired shapes, his results were of limited success. However, he was one of the first researchers to use the concept of sequential gene activation for the production of artificial cellular structures using the cellular automata paradigm.

8   Evolutionary Neurogenesis and Cell Differentiation

Kitano was another of the first researchers that conducted experiments towards evolving an artificial development system. This author was successful at evolving large neural networks using genetic algorithms [38]. He encoded into the genetic algorithm chromosome the neural network connectivity matrix using a graph generating grammar. Instead of using a direct encoding of the connectivity matrix, a set of rules was created by a grammar overcoming the scalability problem on the cases tested. Previous attempts saw how convergence performance was greatly degraded as the size of the neural network grew larger. The grammar used was an augmented version of Lindenmayer's L-System and worked with matrices as symbols.

Kitano later developed a model of neurogenesis and cell differentiation based on a simulation of metabolism [39]. The idea was to determine if artificial multicellular organisms could be created using genetic algorithms evolving the metabolic rules in the genome of the cell. Although all cells carry the same set of rules, individual cells can express different rules because of differences in their local environment, thus producing a sort of cell differentiation. Metabolic rules define which kind of metabolite can be transformed into another kind and under what conditions of metabolite concentration and enzyme presence. These rules are coded in the genome of the cell, and are of the form "if the level of metabolite A is less (or greater) than n, then start a reaction that converts metabolite B into C using the enzymatic properties of metabolite D." These rules are applied to all the metabolites in every cell of a developing organism at each iteration of the simulation. One of the metabolites represents DNA molecules, and when its concentration is above a specified threshold, cells can divide and produce daughter cells with a certain amount of random fluctuation. This randomness provides the base of a chaotic process that can lead to symmetry breaking, which is a widespread feature of developing organism. In addition, cell death can occur in the model when the overall metabolism is too low or too high.

Additionally, Kitano et al. developed a project to simulate the development of the soil nematode *C. elegans* [40]. They chose to model the embryogenesis of *C. elegans* because of its relatively simplicity in structure and because it is one of the best studied multicellular organism in biology. They used data on cell lineage and cell location published in [68] in order to generate a 3D computer graphics image from the division of the first cell to approximately 600 minutes after the first cellular cleavage. They tried to match as much as possible the data from the actual organisms and the results from the simulation. When there was missing simulation data regarding the actual position of cells, their system could calculate forces between cells such as the force that pushes adjacent cells. Their long-term goal was to produce a complete synthetic model of *C. elegans* cellular structure and function.

9   Evolutionary 2D/3D Morphogenesis

Fleischer and Barr presented a simulation framework and computational testbed for the study of 2D multicellular pattern formation [21]. Their initial motivation was the generation of neural networks using a developmental approach, but their interest soon shifted towards the study of the multiple mechanisms involved in morphogenesis.

Their approach combined several developmental mechanisms that they considered important for biological pattern formation. Previous work from other researchers had individually considered chemical factors, mechanical forces, and cell-lineage control of cell division to account for some aspects of morphogenesis. These authors decided to combine these factors into one modeling system in order to determine how the interactions between these components could affect cell pattern development.

The modeling framework consists of discrete cells capable of independent movement and controlled by an artificial genome. The latter is a set of differential equations that depend on the cell's current state and its local environment. The changes in the environment are in turn determined by differential equations which implement mechanical forces and the diffusion of extracellular substances. The computer implementation of the model was able to simulate a number of simple multicellular behaviors, such as cells following gradients, cell clustering, cell differentiation, pattern formation, and network generation.

Fleischer and Barr emphasized that it was the interactions between the developmental mechanisms that were at the core of the determination of multicellular and developmental patterns, and not the individual elements of the model.

On the other hand, Eggenberger used an evolutionary approach for studying the creation of neural network and the simulated morphogenesis of 3D organisms based on differential gene expression [16, 17]. His model for simulating morphogenesis includes a genome with two types of elements: regulatory units and structural genes. The regulatory units act as switches to turn genes on and off, while structural genes code for specific substances that are used to modulate developmental processes. Every gene is defined as having the same number of integers, with the last integer, called the marker, indicating the type of gene. The integers composing the genome are taken from the set $\{1, 2, 3, 4, 5, 6\}$. The fist gene of the genome is assumed to always be a regulatory unit. All genes from the first to the one ending in the marker 5 are defined as regulatory units. Next, all genes after the last regulatory unit and until the gene with the marker 6, are defined as structural genes, and the activity of the latter depends on the regulatory units that precede them. After the last marker 6, the next marker 5 is searched and all the genes between these markers are again considered regulatory units, which control the structural genes found next, until the gene with marker 6, and so on until the end of the genome (see Fig. 2).

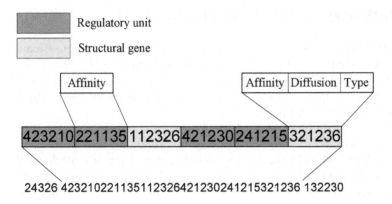

Fig. 2 Example of a segment of the genome used in Eggenberger's model

From the above definition, it follows that one or more adjacent regulatory units can control expression of one or more adjacent structural genes. Gene expression control is based on its biological counterpart, where sequences in the genome, called *cis*-elements, can show affinity with soluble factors, typically proteins, and the degree of matching determines the strength of the effect, either as enhancement or as inhibition of gene expression.

Structural gene expression is regulated by the concentration and affinity of transcription factors. Each cell contains a list of transcription factors, which consist of string of integers that can be compared for matching with the regulatory units in the genome. Affinity is calculated subtracting in the appropriate base the first n integers (from 6 to 8 in the implementation) of both strings. This difference represents the degree of affinity and its sign designates its role: positive for enhancement and negative for inhibition. On the other hand, every transcription factor in the cell has a concentration value. The product of the affinity and the concentration of each transcription factor at a regulatory unit is calculated and the resulting values are added. The same procedure is performed for every regulatory unit of a gene. The resulting sum is then fed to a sigmoidal function and if predefined thresholds are crossed, the corresponding gene is activated or inhibited. The associated equations are shown next:

$$r_j = \sum_{i=1}^{n} aff_i \times conc_i \tag{4}$$

$$a_k = \frac{1}{1 + e^{-\sum_{j=1} r_j}} \tag{5}$$

$$g_k = \begin{cases} -1.0 \ if \ a_k < 0.2 \\ 1.0 \ \ if \ a_k > 0.8 \\ 0.0 \ \ otherwise \end{cases} \tag{6}$$

where aff_i is the affinity of transcription factor i with regulatory unit j, $conc_i$ is the concentration of transcription factor i, r_j is the activity of regulatory unit j of a structural gene, a_k is the total sum of the activities of all regulatory units of gene k, and g_k is the activity of gene k [17].

An active structural gene can perform a number of functions, depending on its type, which is determined by some of its integers. A structural gene can be translated into a transcription factor, a cell adhesion molecule that can connect cells with the corresponding adhesion molecule, or a receptor used to regulate communication between cells. A structural gene can also elicit a function such as cell division or cell death.

Eggenberger implemented cell signaling in several ways: with intracellular substances which regulate the activity of its own gene, with specific receptors on the cell surface which can be stimulated by other substances, and with substances that can penetrate cell walls and diffuse to neighboring cells. These diffusing substances can provide cells with positional information.

In order to direct morphogenesis towards a 3D shape with desired properties, the artificial genome was evolved by means of a genetic algorithm, with mutation and single-point crossover as genetic operators. In the implementation of the model, a series of 8 genetic elements with 2 regulatory units and 2 structural genes each were used. Results from these simulations shows that a spherical shape tends to be formed due to the modulating effect of the morphogen concentration gradient on cell replication.

In other series of experiments, a fitness function that rewarded bilateralism was used. Before the initial cell is allowed to reproduce, the artificial environment is conditioned with morphogen sources on each of the three axes at varying distances to the origin. All three morphogens are different, so that they can regulate growth independently. Cells are able to read and respond to the varying morphogen concentrations. The fitness function was dependant on the total number of cells and their position with respect to one of the axes.

In order to gain more flexibility, this author extended his model later on. In the extended model, structural genes have seven parameters that encode their properties. Among the new additions there is a field that stores the probability of interaction with ligand molecules, a field for storing threshold levels of activation, and a field that contains the decay rate of the expressed molecule. Regulatory units are also endowed with a field that stores the threshold level at which the associated structural genes should be activated [18, 19].

The new model used floating point numbers, and for this reason Eggenberger decided to use the evolutionary strategy developed by Rechenberg as evolutionary algorithm, instead of the genetic algorithm used in the previous model. Results from these simulations show forms representing different stages of a simulated invagination, induced by placing a source of morphogen at a random position in the cell cluster. This invagination process is similar to the gastrulation process seen at the initial stages of development in higher animals.

Eggenberger's models showed that a number of mechanisms central to development such as cellular growth, cell differentiation, axis definition, and dynamical changes in shape could be simulated using a framework not based on a direct mapping between a genome and the resulting cellular structure. The shapes that emerge in the models are the result of the interaction among cells and their environment.

10 METAMorph

METAMorph, which stands for *Model for Experimentation and Teaching in Artificial Morphogenesis*, is an open source software platform for the simulation of cellular development processes using genomes encoded as gene regulatory networks. The design is made by hand and it allows visualization of the resulting morphological cellular growth process [67].

As in higher organisms, cellular growth starts in METAMorph with a single cell (the zygote) and is regulated by gene regulatory networks in interaction with proteins. All cells have the same genome consisting of a series of genes. Each gene can produce exactly one protein, although the same protein can be produced by different genes.

A protein is defined by a unique name, a type (internal or external) and two constants (decay and diffusion). Internal proteins can only diffuse inside a cell, while external proteins can travel through cell membranes and can have a signaling function for communication among cells.

The concentration of each protein is stored in 12 sites inside every cell. As a result, proteins may not be uniformly distributed within the cytoplasm. On the other hand, genes can be expressed differently at each of these sub-cellular sites based on local protein levels. The diffusion constant determines the amount of protein that migrates from one site to another at each time step.

Protein production by a gene is dependent on the promoter sequences located next to the gene. The influence of the promoter is calculated using the sum of the products of the weight and concentration of all proteins. The resulting value is then fed to a sigmoid function that determines the next protein concentration level.

In the model, cells are represented by spheres of a fixed radius and each cell occupies a position on an isospatial 3D grid, so that every cell can have up to 12 equidistant neighboring cells. Cell actions can be triggered when a specific protein concentration level rises above a threshold value. These actions are:

Cell division.   If there is an empty space available, a dividing cell produces a daughter cell placed in the space located in the direction of the mitotic spindle.

Mitotic spindle movement.   Cell orientation is achieved through the definition of a "mitotic spindle" pointing towards one of the 12 possible adjacent cell positions. The spindle orientation can be varied when specific proteins reach a threshold level.

Programmed cell death (apoptosis).   The cell is removed from its position leaving an empty space.

Differentiation.   Cell type is visualized in the model as a distinct external color. The type of a cell is not related to its function.

The main disadvantage of this simulation platform is that the cellular development model has to be designed through a trial and error process that is limited by the designer's ability to introduce the appropriate parameter values. By the authors' account, this trial and error process typically involves a considerable amount of time, since simulation times are usually high due to the parallel nature of the morphogenetic process. To compound the problem, small changes in design can have substantial consequences on the final shape caused by "the butterfly effect."

METAMorph would greatly benefit from introducing in the platform a search process (possibly an evolutionary algorithm) so that the system could find by itself a suitable design for a desired cellular growth pattern.

11 Random Boolean Networks

Random Boolean Networks (RBNs) are a type of discrete dynamical networks that consist of a set of Boolean variables whose state depends on other variables in the network. In RBNs, time and state values take only integer values. The first Boolean networks were proposed by Kauffman in 1969 as a randomized model of a gene regulatory network [33, 35, 36].

RBNs are also known as $N - K$ models or Kauffman networks. They consist of a set of N binary-state nodes, where each node represents a gene that can be on or off, and a set of K edges between nodes that represent relationships between nodes. The connections between nodes are randomly selected and remain fixed thereafter. The dynamics of the RBN is determined by the particular network configuration and by a randomly generated binary function, defined as a lookup table for each node.

RBNs are a generalization of cellular automata, but unlike the latter, the state for each node is determined by nodes that are not necessarily in the immediate vicinity. However, in cellular automata a node can take up any number of states, while a RBN is constrained to only two states. As in canonical cellular automata, updating is synchronous in RBNs, so that the state of all nodes at time $t + 1$ depends on the state of nodes at time t and are all updated at the same time.

Figure 3(a) presents an example of a RBN where all three nodes are connected to all the other nodes ($N = K = 3$). A possible update function is shown in 3(b), while the corresponding state space diagram is presented in 3(c) [22].

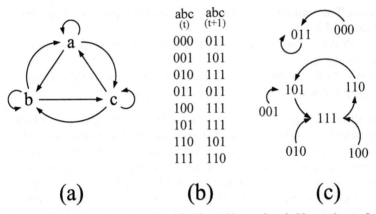

Fig. 3 Example of a Random Boolean Network. a) Network configuration; b) Lookup table for state transitions; c) State space diagram

Since the state space is finite, a state can eventually be visited more than once, and when that happens it is said that an *attractor* as been reached. If the attractor consists of a single state, it is called a *point attractor*, while if it contains two or more states, it is called a *cycle attractor*. The states that lead to an attractor are called the *attractor basin*. The RBN shown in Fig. 3 contains both a point and a cycle attractor, as can be seen in Fig. 3(c).

Depending on the behavior of the network dynamics, three different phases or regimes can be distinguished: ordered, chaotic and critical [37]. In order to identify these phases, a plot can be generated with a small black square for a node in state '1' and with all the nodes lined up at the top of the plot with time flowing downwards. Early studies of RBNs quickly revealed that parameter K had a great influence on the type of RBN generated. Broadly speaking, networks with $K \leq 2$ correspond to the ordered type, while the chaotic type is usually seen in networks with $K \geq 3$.

The critical type of behavior is usually considered by researchers as the most interesting of the three types. The ordered type is too static to derive useful observations applicable to dynamic systems, whereas the chaotic type is too random to study any kind of reproducible property. Of particular interest is what has been termed "the edge of chaos" [31], which in RBNs means a condition in networks of the critical type where dynamically changing states are in a phase transition between ordered and chaotic.

Kauffman also discovered that these transitional phases were related to diverse aspects of the stability of a network. He conducted experiments where he manually perturbed or damaged nodes in a RBN in order to study the effect of these alterations. He found that in the ordered regime, damage to a node does not usually spread to other nodes. In the chaotic phase a small change could have large consequences in the network dynamics. Finally, at the edge of chaos, changes can disseminate, but not necessarily affecting the whole network [22].

It has been suggested that living systems evolve more naturally at "the edge of chaos", since they need certain stability in order to survive, while at the same time they need flexibility to explore their space of possibilities [47, 37]. Furthermore, Kauffman suggested that biological entities could have originally been generated from random elements, with no absolute need of precisely programmed elements [33]. This conjecture was derived from his observations of the complex behavior of some of these randomly generated networks and the inherent robustness he found in them.

12 Artificial Regulatory Networks

Over the years, many models of artificial regulatory networks have emerged in an attempt to emulate the gene networks found in nature. Torsten Reil was one of the first researchers to propose an artificial genome with biological plausible properties based on template matching on a nucleotide-like sequence [63]. The genome is defined as a string of digits and is randomly created. Genes in the genome are not predefined, but are identified by a "promoter" sequence that precedes them. The string arbitrarily chosen for the promoter sequence is 0101, similar to TATA boxes found in biological genomes. The N digits immediately following the promoter sequence constitute the gene (Fig. 4).

After a gene is identified, it is translated using a simple transformation of the sequence. In the implementation used by Reil, the gene product is simply generated by the addition of one digit to the sequence, with the corresponding modulo operation. For each identified gene product, all direct matching sequences in the genome are searched and stored (see Fig. 4). These matching sequences are termed "*cis*-elements" in the model, as their biological counterparts, and

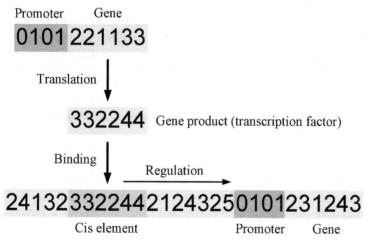

Fig. 4 Gene expression and regulation in the artificial genome. Gene length N is equal to 6

they constitute the regulatory units of the genome. A regulatory sequence can behave either as an enhancer to activate a gene, or as an inhibitor to block its activity. The role of the gene product is defined by the value of its last digit. For instance, all gene products ending in '1' are inhibitors. Regulation is not concentration dependent; it suffices to have one enhancer to activate a gene. However, inhibition is defined as having precedence over enhancement.

In order to test the model, after a genome was generated and all the above rules were applied, gene expression over time was studied. Initially all genes but one were set as turned off. It was then determined which genes were regulated by the initial active gene, and they were labeled as *on* or *off*, depending on their role as enhancers or inhibitors. In the next step, the same procedure was applied using the newly activated genes, and so on for all subsequent time steps until a specified number of cycles was reached. The resulting pattern of expression is visualized in what the author called an *expression graph*, which is a 2D graph with time in the X-axis and a black dot for every active gene in the Y-axis.

The author varied the following model parameters: genome size, gene length, base (range of digits), and degree of inhibition measured as a fraction of the range of digits that determined the role of inhibitor in *cis*-elements. It was found that the behavior of the model was highly dependent on the parameter values.

As with RBNs and other dynamical systems, three basic types of behavior were identified: ordered, chaotic, and complex. Gene expression was called ordered if genes were continuously active or inactive throughout the run. If gene expression seemed to be random with no apparent emerging pattern, it was called chaotic. If the expression of genes was considered to be between ordered and chaotic with the formation of identifiable patterns, then it was called complex. The author broadly identified which ranges of parameter values gave rise to each type of behavior.

For genomes with a behavior of the complex type, it was found that gene expression converged to the same pattern from a number of different start genes. This was viewed as the cycle attractors found in the RBNs described in Section 11. Every genome of this kind typically contained several such attractors. This finding supported the notion proposed by Kauffman that cell types could be viewed as attractors of gene expression in natural gene networks [34]. Thus, cell differentiation could be viewed as a dynamic system that moved from one attractor to another.

On the other hand, Reil observed that even after manual perturbations in the model, gene expression usually returned to the attractors that emerged previously. It must be emphasized that the artificial genomes endured no evolution. The behaviors observed were the result of the properties of genomes entirely generated at random. Reil hypothesized that robustness in natural genomes might be an inherent property of the template matching system, rather than the result of the natural selection of the most robust nucleotide sequences [63].

An important advancement in the design of an artificial genome model was made by Banzhaf, who designed a genetic representation based on artificial regulatory networks [2]. His genome consists of a randomly generated binary string. Similarly to other models, the "promoter" is a particular sequence that signals the beginning of a gene in the genome. The promoter was arbitrarily chosen as the string 'XYZ01010101', with 'XYZ' being any 3-bit combination. In a randomly generated bit string, the expected frequency of the pattern '01010101' is $2^{-8} \approx 0.0039 = 0.39\%$. The gene following the promoter was defined as a series of five 32-bit strings, for a total of 160 bits per gene. Immediately before the promoter sequence, two special 32-bit sites were defined: an enhancer and an inhibitor.

The five 32-bit regions after the promoter are translated into a protein using a majority rule, i.e. the first bit in the translated protein corresponds to the bit that is in majority in the first position of the five protein-coding regions, and so on until the end of the 32-bit sequence. In this model the transcription process seen in nature is completely disregarded. There is no intermediary element –such as the messenger RNA sequences found in biological systems– between the gene and the translated protein.

After a protein has been produced, it is then compared on a bit by bit basis with the enhancer and inhibitor sequences on all genes in the genome. The comparison is achieved through the use of an XOR operation, which renders a '1' if the bits compared are complementary. It is expected that a Gaussian distribution is found when measuring the match between a particular protein and all the 32-bit sequences of a randomly generated genome. Thus there will be few sequences with a high degree of matching and likewise there will be few poor-matching sequences. The majority of 32-bit strings will be average-matching sequences. Banzhaf confirmed through simulations that randomly generated genomes of various sizes contained the expected number of genes. For example, a genome consisting of 100,000 bits contained 409 genes, which is consistent with the 0.39% rule.

Each translated protein is compared with the inhibition and enhancer sites of all the regulatory genes in order to determine the degree of interaction in the regulatory network. The influence of a protein on an enhancer or inhibitor site is exponential with the number of matching bits. The strength of enhancement en or inhibition in for gene i with $i = 1, ..., n$ is defined as

$$en_i = \frac{1}{n} \sum_{j=1}^{n} c_j e^{\beta(u_{ij}^+ - u_{max}^+)} \qquad (7)$$

$$in_i = \frac{1}{n} \sum_{j=1}^{n} c_j e^{\beta(u_{ij}^- - u_{max}^-)}, \qquad (8)$$

where n is the total number of regulatory genes, c_j is the concentration of protein j, β is a constant that fine-tunes the strength of matching, u_{ij}^+ and u_{ij}^- are the number of matches between protein j and the enhancer and inhibitor sites of gene i, respectively, and $u_{\max}^+$ and $u_{\max}^-$ are the maximum matches achievable between a protein and an enhancer or inhibition site, respectively [2].

Once the en and in values are obtained for all regulatory genes, the corresponding concentration change for protein i in one time step is found using

$$\frac{dc_i}{dt} = \delta \left(en_i - in_i\right) c_i, \tag{9}$$

where δ is a constant that regulates the degree of protein concentration change. Protein concentrations are updated and normalized so that total protein concentration is always the unity.

After observing the dynamics of proteins from genomes that had experienced no evolution, Banzhaf used genetic programming in an attempt to drive the dynamics of gene expression towards desired behaviors. He started by evolving the genome to obtain a target concentration of a particular protein. He found out that in general the evolutionary process quickly converged towards the target state.

In the search of other applications for this model, Banzhaf and his colleagues evolved genomes where the protein concentrations were used to modulate output functions such as sinusoids, exponentials and sigmoids [46]. A randomly selected 62-bit string in the genome was chosen to function as an enhancer and an inhibitor site, and they were allowed to freely interact with expressed proteins. However, instead of controlling gene expression, these sites were used to calculate the output value of a function mapped to the $[-1, 1]$ interval. These authors found that it was feasible to evolve genomes for generating time-series for function optimization. They also used evolution of the artificial genome model to produce networks with small-world and scale-free topologies [45].

Another author that evolved an artificial regulatory network in order to perform a specific task was Bongard [4]. He designed virtual modular robots that were evaluated for how fast they could travel over an infinite horizontal plane during a time interval previously specified. The robots are composed of one or more morphological units and zero or more sensors, motors, neurons and synapses. Each morphological unit contains a genome, and at the beginning of the evolution a genome and a motor neuron are inserted into the initial unit. As in similar genome models, a gene is preceded by a promoter sequence. Transcription factor sources are placed at the poles of the initial unit to allow the artificial regulatory network to establish its anterior/posterior axes. The unit then starts producing transcription factors that activate expression of genes in the genome. Transcription factors can have a direct influence in the external features of the developing unit. They can activate one of 23 predefined phenotypic transformations, such as increasing the length of the unit, causing a unit to divide in two, or adding, deleting or modifying neurons or synapses. The unit's behavior is dependent on the real-time propagation of sensory information through its neural network to motor neurons, which can actuate the unit's joints to generate movement.

Using this model, Bongard demonstrated that mobile units could be evolved in a virtual environment. His results suggest that a similar model might be applied in the design of physical robots.

Other authors have performed research on artificial regulatory networks using a number of approaches. Willadsen and Wiles designed a genome based on the model proposed by Reil [63]. As in other models, the genome consists of a string of randomly generated integers where a promoter precedes a fixed-length gene. Gene products are generated, which can regulate expression of other genes [71]. While their genome model offered no major improvement over previous models, these authors succeeding in showing that there was a strong relationship between gene network connectivity and the degree of inhibition with respect to generating a chaotic behavior. Low connectivity gene networks were found to be very stable, while in higher connectivity networks there was a significantly elevated frequency of chaotic behavior. The same research group suggested that the synchronous updating of the network dynamics regularly used in genome models was not the most adequate since it was not biologically plausible [27]. They suggested that asynchronous updating of the dynamics was more realistic since biological cells do not work synchronously. They found that using asynchronous updating, dynamics converged to an attractor under almost all conditions [26].

Flann *et al.* used artificial regulatory networks to construct 2D cellular patterns such as borders, patches and mosaics [20]. They implemented the artificial regulatory network as a graph, where each node represents a distinct expression level from a protein, and each edge corresponds to interactions between proteins. A protein is influenced when its production or inhibition is altered as the function of other protein concentration levels. A set of differential equations was used to define the rate of production or inhibition. In their model, cell to cell contact signaling was sufficient to form a number of global patch patterns. On the other hand, they found it difficult to produce certain patterns with a single artificial regulatory network, but they solved the problem by using disjoint networks run in parallel and combining their protein concentration levels. These authors conjectured that complex regulatory networks in nature might have evolved by combining simpler regulatory networks.

Nehaniv's research group has worked on artificial regulatory networks aiming at evolving a biological clock model [42, 41]. They studied the evolvability of regulatory networks as active control systems that responded with appropriate periodic behaviors to periodic environmental stimuli of several types. Their genome model is based on the one proposed in [62], from the same research group. Unlike the model on which it is based, the genome in the biological clock model contains an evolvable number of complex *cis*-regulatory control sites. Each regulatory site in turn contains a number of activating or inhibitory binding factors. Although their model only considered the evolution of the genome of one single cell, their results with the biological clock model could be used to synchronize reproduction of cells in an artificial development model.

13   Evolutionary Development Model

Kumar and Bentley designed a developmental testbed that they called the Evolutionary Development System (EDS). It was intended for the investigation of multicellular processes and mechanisms, and their potential application to computer science. The EDS contains the equivalent of many key elements involved in biological development. It implements concepts such as embryos, cells, cell cytoplasm, cell wall, proteins, receptors, transcription factors, genes and *cis*-regulatory regions [43, 44].

In the EDS, proteins are implemented as objects. Each protein has a field to identify it as one of the eight types defined. Protein objects contain a current and a new state object that are used to simulate parallelism in protein behavior. These state objects include information such as the protein diffusion coefficient. Protein diffusion in the medium is implemented by means of a Gaussian function centered at the protein source. It is assumed that proteins diffuse uniformly in all directions.

Unlike other models, there are two types of genomes in the EDS. The first genome stores protein-related parameter values such as the rates of synthesis, decay and diffusion. The second genome encodes the architecture to be used for development by describing which proteins have a role in the regulation of the different genes. The second genome is contained inside every cell during the simulation of the developmental process. The first genome is only used to initialize proteins with their respective values.

The genome inside cells is represented by an array of genes, where each gene consists of a *cis*-regulatory region and a protein-coding region. The *cis*-regulatory region contains in turn an array of target sites for binding matching transcription factors. As in other models, the binding of transcription factors in the *cis*-regulatory region modulates the activity of the gene. By summing the product of concentration and interaction weight of each transcription factor, a value is obtained that is fed to a sigmoid function to determine whether or not the associated protein-coding region should be translated.

Cells in the EDS are autonomous agents that have sensors in the form of surface receptors capable of binding to substances in the environment. Depending on their current state, cells can exhibit a number of activities such as division, differentiation shown as an external color, and apoptosis or programmed cell death.

A genetic algorithm with tournament selection was used to evolve the genomes. One of the morphogenesis experiments consisted in evolving spherical embryos using the equation of a sphere as a fitness function. Results showed that evolution did not make use of many proteins and the evolved regulatory networks were not very complex. The authors considered that it was likely that their system had a natural tendency to produce almost spherical cell clusters and that it did not take much to achieve the goal desired.

The design of the EDS was probably too ambitious by involving many elements that introduced more variables and interactions in the system than desired. Results obtained with the EDS are meager considering the number of concepts involved. The system might prove its true potential with a more complex target cellular structure.

14   Compositional Pattern Producing Networks

A developmental model called Compositional Pattern Producing Networks (CPPNs) has recently been proposed by Stanley [65]. This model is independent of local interactions and temporal unfolding, which are traditionally considered as important to artificial development systems. This author suggests that local interactions and temporal unfolding are essential to natural development due to physical constraints in the developing organism, but that they can be abstracted away in an artificial development system without loss of expressivity.

In this model, the pattern producing process is achieved through the composition of mathematical functions, each of which is based on simpler gradient patterns similar to those found in nature. CPPNs are structurally similar to artificial neural networks (ANNs), as they consist of nodes interconnected by weighted edges. However, the activation functions in CPPNs are not necessarily sigmoid functions as in ANNs; instead they can be any function that represents a basic pattern, such as a linear gradient or a Gaussian distribution.

Starting with simpler networks, CPPNs are evolved by adding or deleting nodes and edges in the network and by modifying the weights of connecting edges. In a graphic implementation made to test the model, evolution is guided interactively by a human operator, who manually selects the parent shapes to mate in order to obtain offspring that best resembles a predefined 2D target shape. This selection process is similar to the selection of biomorphs in Dawkins' evolutionary system presented in Section 6.

In his model, Stanley was able to demonstrate the existence of intrinsic properties found in natural development, such as bilateral symmetry and repeating patterns with and without variation. This author emphasizes that the role of local interactions and temporal unfolding may not be as important in an abstract developmental model, where physical constraints might not have to be followed rigorously.

15   The French Flag Problem

The problem of generating a French flag pattern was first introduced by Wolpert in the late 1960s when trying to formulate the problem of cell pattern development and regulation in living organisms [72]. This formulation has been used since then by some authors to study the problem of artificial pattern development.

Lindenmayer and Rozenberg used the French flag problem to illustrate how a grammar-based L-System could be used to solve the generation of this particular pattern when enunciated as the production of a string of the type $a^n b^n c^n$ over the alphabet $\{a, b, c\}$ and with $n > 0$ [50]. On the other hand, Herman and Liu developed an extension of a simulator called CELIA [1] and applied it to generate a French flag pattern in order to study synchronization and symmetry breaking in cellular development [30].

Miller and Banzhaf used what they called Cartesian genetic programming to evolve a cell program that would construct a French flag pattern [57]. They tested the robustness of their programs by manually removing parts of the developing pattern. They found that some of their evolved programs could repair to some extent the damaged patterns. Bowers also used this problem to study the phenotypic robustness of his embryogeny model, which was based on cellular growth with diffusing chemicals as signaling molecules [6].

Gordon and Bentley proposed a development model based on a set of rules that described how development should proceed [25]. A set of rules evolved by a genetic algorithm was used to develop a French flag pattern. The morphogenic model based on a multiagent system developed by Beurier et al. also used an evolved set of agent rules to grow French and Japanese flag patterns [3]. On the other hand, Devert et al. proposed a neural network model for multicellular development that grew French flag patterns [15]. Even models for developing evolvable hardware have benefited from the French flag problem as a test case [70, 29].

More recently, Chavoya and Duthen proposed an artificial development model for cell pattern generation based on an artificial regulatory network [9]. In this model, cellular growth is controlled by a genome consisting of an artificial regulatory network and a series of structural genes. The genome was evolved by a genetic algorithm in order to produce 2D cell patterns through the selective activation and inhibition of genes. Morphogenetic gradients were used to provide cells with positional information that constrained cellular replication. After a genome was evolved, a single cell in the middle of the CA lattice was allowed to reproduce until a cell pattern was formed. The model was applied to the problem of growing a French flag pattern with and without a flagpole.

16   Concluding Remarks

The artificial development models presented here have to a varying degree been based on the biological development process. Despite their intrinsic limitations, these models have proven in general to be useful either in proposing a computational paradigm or in solving specific problems. Additionally, these models could shed light on the mechanisms involved in the natural development process.

One question that is frequently raised in designing models inspired by nature is the level of abstraction needed. For certain applications, it may not be necessary to rigorously mimic the biological processes down to the molecular level. A more abstract model may be more useful in finding solutions to particular problems by removing constraints not found in artificial systems. However, it is not always easy to determine the right level of abstraction to achieve a goal, and in general this determination remains an open question.

As for artificial regulatory networks, it is expected that research will continue on the development of models based on this paradigm. If the goal is to propose more realistic development models, the role of the environment on the developing cells has also to be considered. It would be desirable that artificial cells have an individual program –possibly a separate regulatory network– for reacting to local unexpected changes in their environment. Morphogenetic fields provide a means to extract information from the environment, but an independent program would lend more flexibility and robustness to a developing organism. After all, living organisms do contain a series of gene regulatory networks for development and metabolism control. One could even envision either a hierarchy of regulatory networks, where some networks could be used to regulate others networks, or a network of regulatory networks, where all networks could influence and regulate each other.

Additional work is needed in order to explore structure formation of more complex forms, both in 2D and 3D. Furthermore, in order to increase the usefulness of some of the models, interaction with other artificial entities and extraction of information from a more physically realistic environment may be necessary.

One of the long-term goals of this exciting field is to study the emergent properties of the artificial development process. It can be envisioned that one day it will be feasible to build highly complex structures arising mainly from the interaction of myriads of simpler entities.

References

1. Baker, R.W., Herman, G.T.: Celia - a cellular linear iterative array simulator. In: Proceedings of the fourth annual conference on Applications of simulation, pp. 64–73. Winter Simulation Conference (1970)
2. Banzhaf, W.: Artificial regulatory networks and genetic programming. In: Riolo, R.L., Worzel, B. (eds.) Genetic Programming Theory and Practice, ch. 4, pp. 43–62. Kluwer, Dordrecht (2003)
3. Beurier, G., Michel, F., Ferber, J.: A morphogenesis model for multiagent embryogeny. In: Rocha, L.M., Yaeger, L.S., Bedau, M.A., Floreano, D., Goldstone, R.L., Vespignani, A. (eds.) Proceedings of the Tenth International Conference on the Simulation and Synthesis of Living Systems (ALife X), pp. 84–90 (2006)
4. Bongard, J.: Evolving modular genetic regulatory networks. In: Proceedings of the 2002 Congress on Evolutionary Computation (CEC 2002), pp. 1872–1877. IEEE Press, Piscataway (2002)

5. Bongard, J.C., Pfeifer, R.: Repeated structure and dissociation of genotypic and phenotypic complexity in artificial ontogeny. In: Proceedings of The Genetic and Evolutionary Computation Conference, pp. 829–836. Morgan Kaufmann, San Francisco (2001)
6. Bowers, C.: Simulating evolution with a computational model of embryogeny: Obtaining robustness from evolved individuals. In: Capcarrère, M.S., Freitas, A.A., Bentley, P.J., Johnson, C.G., Timmis, J. (eds.) ECAL 2005. LNCS, vol. 3630, pp. 149–158. Springer, Heidelberg (2005)
7. Carroll, S.B.: Endless Forms Most Beautiful: The New Science of Evo Devo. W. W. Norton (2006)
8. Carroll, S.B., Grenier, J.K., Weatherbee, S.D.: From DNA to Diversity: Molecular Genetics and the Evolution of Animal Design, 2nd edn. Blackwell Science, Malden (2004)
9. Chavoya, A., Duthen, Y.: An artificial development model for cell pattern generation. In: Randall, M., Abbass, H.A., Wiles, J. (eds.) ACAL 2007. LNCS, vol. 4828, pp. 61–71. Springer, Heidelberg (2007)
10. Davidson, E.H.: The Regulatory Genome: Gene Regulatory Networks in Development And Evolution, 1st edn. Academic Press, London (2006)
11. Dawkins, R.: The Blind Watchmaker: Why the Evidence of Evolution Reveals a Universe Without Design. W. W. Norton (1996)
12. de Garis, H.: Genetic programming: artificial nervous systems artificial embryos and embryological electronics. In: Schwefel, H.-P., Männer, R. (eds.) PPSN 1990. LNCS, vol. 496, pp. 117–123. Springer, Heidelberg (1991)
13. de Garis, H.: Artificial embryology - the genetic programming of cellular differentiation. In: Artificial Life III Workshop, Santa Fe, New Mexico, USA (1992)
14. de Garis, H.: Artificial embryology: The genetic programming of an artificial embryo. In: Souček, B. (ed.) Dynamic, Genetic, and Chaotic Programming, pp. 373–393. John Wiley, New York (1992)
15. Devert, A., Bredeche, N., Schoenauer, M.: Robust multi-cellular developmental design. In: GECCO 2007: Proceedings of the 9th annual conference on Genetic and evolutionary computation, pp. 982–989. ACM Press, New York (2007)
16. Eggenberger, P.: Creation of neural networks based on developmental and evolutionary principles. In: Gerstner, W., Hasler, M., Germond, A., Nicoud, J.D. (eds.) ICANN 1997. LNCS, vol. 1327, pp. 337–342. Springer, Heidelberg (1997)
17. Eggenberger, P.: Evolving morphologies of simulated 3D organisms based on differential gene expression. In: Harvey, I., Husbands, P. (eds.) Proceedings of the 4th European Conference on Artificial Life, pp. 205–213. Springer, Heidelberg (1997)
18. Eggenberger Hotz, P.: Combining developmental processes and their physics in an artificial evolutionary system to evolve shapes. In: Kumar, S., Bentley, P.J. (eds.) On Growth, Form and Computers, ch. 16, pp. 302–318. Academic Press, New York (2003)
19. Eggenberger Hotz, P.: Asymmetric cell division and its integration with other developmental processes for artificial evolutionary systems. In: Proceedings of the 9th International Conference on Artificial Life IX (2004)
20. Flann, N., Hu, J., Bansal, M., Patel, V., Podgorski, G.: Biological development of cell patterns: Characterizing the space of cell chemistry genetic regulatory networks. In: Capcarrère, M.S., Freitas, A.A., Bentley, P.J., Johnson, C.G., Timmis, J. (eds.) ECAL 2005. LNCS, vol. 3630, pp. 57–66. Springer, Heidelberg (2005)

21. Fleischer, K., Barr, A.H.: A simulation testbed for the study of multicellular development: The multiple mechanisms of morphogenesis. In: Langdon, C. (ed.) Proceedings of the Workshop on Artificial Life ALIFE 1992, pp. 389–416. Addison-Wesley, Reading (1992)

22. Gershenson, C.: Introduction to random boolean networks. In: Bedau, M., Husbands, P., Hutton, T., Kumar, S., Susuki, H. (eds.) Worshop and Tutorial Proceedings, Ninth International Conference on the Simulation and Synthesis of Living Systems (ALife IX), pp. 160–173 (2004)

23. Gierer, A.: Generation of biological patterns and form: Some physical, mathematical, and logical aspects. Prog. Biophys. Molec. Biol. 37, 1–47 (1981)

24. Gierer, A., Meinhardt, H.: A theory of biological pattern formation. Kybernetik 12, 30–39 (1972)

25. Gordon, T.G.W., Bentley, P.J.: Bias and scalability in evolutionary development. In: GECCO 2005: Proceedings of the 2005 conference on Genetic and evolutionary computation, pp. 83–90. ACM Press, New York (2005)

26. Hallinan, J., Wiles, J.: Asynchronous dynamics of an artificial genetic regulatory network. In: Artificial Life IX: Proceedings of the Ninth International Conference on the Simulation and Synthesis of Living Systems, pp. 399–403. MIT Press, Cambridge (2004)

27. Hallinan, J., Wiles, J.: Evolving genetic regulatory networks using an artificial genome. In: CRPIT 2004: Proceedings of the second conference on Asia-Pacific bioinformatics, pp. 291–296. Australian Computer Society, Inc., Darlinghurst (2004)

28. Hanan, J.S.: Parametric l-systems and their application to the modelling and visualization of plants. Ph.D. thesis, University of Regina (1992)

29. Harding, S.L., Miller, J.F., Banzhaf, W.: Self-modifying cartesian genetic programming. In: GECCO 2007: Proceedings of the 9th annual conference on Genetic and evolutionary computation, pp. 1021–1028. ACM Press, New York (2007)

30. Herman, G.T., Liu, W.H.: The daughter of celia, the french flag and the firing squad. In: WSC 1973: Proceedings of the 6th conference on Winter simulation, p. 870. ACM, New York (1973)

31. Heudin, J.C.: L'évolution au bord du chaos. Hermes, Paris, France (1998)

32. Jakobi, N.: Harnessing morphogenesis. In: International Conference on Information Processing in Cells and Tissues, pp. 29–41 (1995)

33. Kauffman, S.A.: Metabolic stability and epigenesis in randomly constructed genetic nets. Journal of Theoretical Biology 22, 437–467 (1969)

34. Kauffman, S.A.: Gene regulation networks: A theory for their global structure and behavior. Current Topics in Dev. Biol. 6, 145 (1971)

35. Kauffman, S.A.: The large-scale structure and dynamics of gene control circuits: An ensemble approach. Journal of Theoretical Biology 44, 167 (1974)

36. Kauffman, S.A.: The Origins of Order: Self-Organization and Selection in Evolution. Oxford University Press, Oxford (1993)

37. Kauffman, S.A.: Investigations. Oxford University Press, Oxford (2004)

38. Kitano, H.: Designing neural networks using genetic algorithms with graph generation system. Complex Systems 4, 461–476 (1990)

39. Kitano, H.: A simple model of neurogenesis and cell differentiation based on evolutionary large-scale chaos. Artificial Life 2(1), 79–99 (1994)

40. Kitano, H., Hamahashi, S., Kitazawa, J., Luke, S.: The perfect C. elegans project: an initial report. Artificial Life 4(2), 141–156 (1998)

41. Knabe, J.F., Nehaniv, C.L., Schilstra, M.J.: The essential motif that wasn't there: Topological and lesioning analysis of evolved generic regulatory networks. In: Proceedings of the 2007 IEEE Symposium on Artificial Life (CI-ALife 2007), pp. 47–53. IEEE Computational Intelligence Society (2007)

42. Knabe, J.F., Nehaniv, C.L., Schilstra, M.J., Quick, T.: Evolving biological clocks using genetic regulatory networks. In: Rocha, L.M., Yaeger, L.S., Bedau, M.A., Floreano, D., Goldstone, R.L., Vespignani, A. (eds.) Proceedings of the Artificial Life X Conference (Alife 10), pp. 15–21. MIT Press, Cambridge (2006)

43. Kumar, S., Bentley, P.J.: Computational embryology: past, present and future. In: Advances in evolutionary computing: theory and applications, pp. 461–477. Springer, New York (2003)

44. Kumar, S., Bentley, P.J.: An introduction to computational development. In: Kumar, S., Bentley, P.J. (eds.) On Growth, Form and Computers, ch. 1, pp. 1–44. Academic Press, New York (2003)

45. Kuo, P.D., Banzhaf, W.: Small world and scale-free network topologies in an artificial regulatory network model. In: Pollack, J., Bedau, M., Husbands, P., Ikegami, T., Watson, R. (eds.) Proceedings of Artificial Life IX (ALIFE-9), Boston, USA, pp. 404–409. MIT Press, Cambridge (2004)

46. Kuo, P.D., Leier, A., Banzhaf, W.: Evolving dynamics in an artificial regulatory network model. In: Yao, X., Burke, E.K., Lozano, J.A., Smith, J., Merelo-Guervós, J.J., Bullinaria, J.A., Rowe, J.E., Tiňo, P., Kabán, A., Schwefel, H.-P. (eds.) PPSN 2004. LNCS, vol. 3242, pp. 571–580. Springer, Heidelberg (2004)

47. Langton, C.: Computation at the edge of chaos: Phase transitions and emergent computation. Physica D 42, 12–37 (1990)

48. Lindenmayer, A.: Mathematical models for cellular interaction in development, Parts I and II. Journal of Theoretical Biology 18, 280–315 (1968)

49. Lindenmayer, A.: Developmental systems without cellular interaction, their languages and grammars. Journal of Theoretical Biology 30, 455–484 (1971)

50. Lindenmayer, A., Rozenberg, G.: Developmental systems and languages. In: STOC 1972: Proceedings of the fourth annual ACM symposium on Theory of computing, pp. 214–221. ACM Press, New York (1972)

51. Mech, R., Prusinkiewicz, P.: Visual models of plants interacting with their environment. In: Proceedings of SIGGRAPH 1996, pp. 397–410 (1996)

52. Meinhardt, H.: Models of Biological Pattern Formation. Academic Press, London (1982)

53. Meinhardt, H.: Models of biological pattern formation: common mechanism in plant and animal development. Int. J. Dev. Biol. 40, 123–134 (1996)

54. Meinhardt, H.: The algorithmic beauty of seashells. Springer, Heidelberg (1998)

55. Meinhardt, H.: Models for pattern formation and the position-specific activation of genes. In: Kumar, S., Bentley, P.J. (eds.) On Growth, Form and Computers, ch. 7, pp. 135–155. Academic Press, London (2003)

56. Meinhardt, H., Gierer, A.: Pattern formation by local self-activation and lateral inhibition. BioEssays 22(8), 753–760 (2000)

57. Miller, J.F., Banzhaf, W.: Evolving the program for a cell: from French flags to Boolean circuits. In: Kumar, S., Bentley, P.J. (eds.) On Growth, Form and Computers, pp. 278–301. Academic Press, London (2003)

58. Prusinkiewicz, P.: Modeling and Vizualization of Biological Structures. In: Proceeding of Graphics Interface 1993, pp. 128–137 (1993)

59. Prusinkiewicz, P.: Visual models of morphogenesis. In: Langton, C.G. (ed.) Artificial life: an overview, pp. 61–74. The MIT Press, Cambridge (1997)
60. Prusinkiewicz, P., Hanan, J., Mech, R.: An l-system-based plant modeling language. In: Münch, M., Nagl, M. (eds.) AGTIVE 1999. LNCS, vol. 1779, pp. 395–410. Springer, Heidelberg (2000)
61. Prusinkiewicz, P., Lindenmayer, A.: The Algorithmic Beauty of Plants. Springer, Heidelberg (1990)
62. Quick, T., Nehaniv, C.L., Dautenhahn, K., Roberts, G.: Evolving embodied genetic regulatory network-driven control systems. In: Banzhaf, W., Ziegler, J., Christaller, T., Dittrich, P., Kim, J.T. (eds.) ECAL 2003. LNCS (LNAI), vol. 2801, pp. 266–277. Springer, Heidelberg (2003)
63. Reil, T.: Dynamics of gene expression in an artificial genome - implications for biological and artificial ontogeny. In: Proceedings of the 5th European Conference on Artificial Life (ECAL), pp. 457–466. Springer, New York (1999)
64. Smith, A.R.: Plants, fractals, and formal languages. In: SIGGRAPH 1984: Proceedings of the 11th annual conference on Computer graphics and interactive techniques, pp. 1–10. ACM Press, New York (1984)
65. Stanley, K.O.: Compositional pattern producing networks: A novel abstraction of development. Genetic Programming and Evolvable Machines Special Issue on Developmental Systems 8(2), 131–162 (2007)
66. Stanley, K.O., Miikkulainen, R.: A taxonomy for artificial embryogeny. Artif. Life 9(2), 93–130 (2003)
67. Stewart, F., Taylor, T., Konidaris, G.: Metamorph: Experimenting with genetic regulatory networks for artificial development. In: Capcarrère, M.S., Freitas, A.A., Bentley, P.J., Johnson, C.G., Timmis, J. (eds.) ECAL 2005. LNCS, vol. 3630, pp. 108–117. Springer, Heidelberg (2005)
68. Sulston, J., Du, Z., Thomas, K., Wilson, R., Hillier, L., Staden, R., Halloran, N., Green, P., Thierry-Mieg, J., Qiu, L.: The C. elegans genome sequencing project: a beginning. Nature 356(6364), 37–41 (1992)
69. Turing, A.M.: The chemical basis of morphogenesis. Philosophical Transactions of the Royal Society of London. Series B, Biological Sciences 237(641), 37–72 (1952)
70. Tyrrell, A.M., Greensted, A.J.: Evolving dependability. J. Emerg. Technol. Comput. Syst. 3(2), 7 (2007)
71. Willadsen, K., Wiles, J.: Dynamics of gene expression in an artificial genome. In: Proceedings of the IEEE 2003 Congress on Evolutionary Computation, pp. 199–206. IEEE Press, Los Alamitos (2003)
72. Wolpert, L.: The French flag problem: a contribution to the discussion on pattern development and regulation. In: Waddington, C. (ed.) Towards a Theoretical Biology, pp. 125–133. Edinburgh University Press, New York (1968)
73. Wolpert, L.: Positional information and the spatial pattern of cellular differentiation. J. Theor. Biol. 25, 1–47 (1969)
74. Wolpert, L.: Positional information and pattern formation. Phil. Trans. R. Soc. Lond. B 295(1078), 441–450 (1981)

Robust Training of Artificial Feedforward Neural Networks

Moumen T. El-Melegy, Mohammed H. Essai, and Amer A. Ali

Abstract. Artificial feedforward neural networks have received researchers' great interest due to its ability to approximate functions without having a prior knowledge about the true underlying function. The most popular algorithm for training these networks is the backpropagation algorithm that is based on the minimization of the mean square error cost function. However this algorithm is not robust in the presence of outliers that may pollute the training data. In this chapter we present several methods to robustify neural network training algorithms. First, employing a family of robust statistics estimators, commonly known as M-estimators, in the backpropagation algorithm is reviewed and evaluated for the task of function approximation and dynamical model identification. As these M-estimators sometimes do not have sufficient insensitivity to data outliers, the chapter next resorts to the statistically more robust estimator of the least median of squares, and develops a stochastic algorithm to minimize a related cost function. The reported experimental results have indeed shown the improved robustness of the new algorithm, especially compared to the standard backpropagation algorithm, on datasets with varying degrees of outlying data.

1 Introduction

Artificial neural networks have been used with success in many diverse areas of scientific and technical disciplines including computer science, engineering, physics, medicine, and cognitive sciences. One of the practical areas in

Moumen T. El-Melegy
Electrical Engineering Department, Assiut University, Assiut 71516, Egypt
e-mail: moumen@aun.edu.eg

Mohammed H. Essai
Electrical Engineering Department, Al-Azhar University, Qena 83513, Egypt

Amer A. Ali
Electrical Engineering Department, Assiut University, Assiut 71516, Egypt

A.-E. Hassanien et al. (Eds.): Foundations of Comput. Intel. Vol. 1, SCI 201, pp. 217–242.
springerlink.com   © Springer-Verlag Berlin Heidelberg 2009

which feedforward neural networks have found extensive application is function approximation. This is due to its ability as a universal function approximator [20, 33]. Many examples of systems developed at an industrial and commercial level can be abstracted into the task of approximating an unknown function from a training dataset of input-output pairs. Most of these applications need real time processing, which can indeed benefit from the fact that neural networks are inherently parallel processors.

Multilayered feedforward neural networks (MFNNs) are often trained by using the popular backpropagation (BP) learning algorithm. Unfortunately, due to its use of the mean square error (MSE) as an error measure between the actual and desired output, this popular learning algorithm is not robust completely in the presence of outliers. Outliers are data severely deviating from the pattern set by the majority of the data. Examples of such deviations include the contamination of data by gross errors, such as blunders in measuring, wrong decimal points, errors in copying, inadvertent measurement of a member of a different population, rounding and grouping errors, and departure from an assumed sample distribution.

It has been noted [15] that the occurrence of outliers in routine data ranges from 1% to 10%, or even more [43]. Accordingly numerous studies have been conducted (e.g., [25, 34, 35]), which clearly show that the presence of outliers poses a serious threat to standard least squares analysis. When there is only a single outlier, some of these methods work quite well by looking at the effect of deleting one point at a time. Unfortunately, it is much more difficult to diagnose outliers when there are several of them, and diagnostics for such multiple outliers are quite involved and often give rise to extensive computations. In addition, it can be difficult or even impossible to spot outliers in multivariate or highly structured data.

To overcome the possibility of presence of outliers, many robust estimators which are not very sensitive to departure from the assumptions on which they depend are being used recently [34, 35], such as M-estimators (Maximum-likelihood estimators), R-estimators (estimates based on rank transformations), L-estimators (Linear combination of order statistics), and LMedS estimators (Least Median of Squares). In this chapter we address the issue of robust training of feedforward neural networks based on utilizing these robust estimators. There have been some previous efforts in that regard, mostly based on M-estimators. Liano [25] has used M-estimators to study the mechanism by which outliers affect the resulting network models. He investigated the use of the least mean log square (LMLS) estimator to reduce this effect. Chen and et. al [8] also have used M-estimators as a robust estimator in the presence of outliers. They used the Tukey, Huber, Talwar and Logistic estimators, and compared their performances with the non-robust MSE. Alpha and et.al [1] combined the benefits of the so-called TAO non-linear regression model and M-estimators to produce the TAO-robust learning algorithm. They compared the proposed algorithm results with the non-robust MSE and the robust LMLS estimators. Approaches based on the adaptive learning rate [44] and Least Trimmed Squares (LTS) [45] estimator were also proposed. The concept of initial data analysis by the MCD (Minimum Covariance Determinant) estimator was also investigated [48]. Some authors [46, 47] have tried to apply robust learning algorithms to radial basis function networks as well.

While these previous research efforts focused mainly on M-estimators to make neural network training more robust, these estimators sometimes do not provide enough resistance (robustness) to outliers. Therefore in this chapter we propose to train neural networks based on the more statistically robust estimator of the least median of squares (LMedS). It can theoretically tolerate up to 50% of outlying data [34]. Nevertheless, to the very best of our knowledge, this estimator has not been used before for training neural networks. One reason perhaps is that there is no explicit formula for the derivative of the LMedS estimator, the fact that makes deterministic algorithms cannot be used for minimization this estimator in a backpropagation-like fashion. To get around this problem, we resort to another family of optimization algorithms that is called stochastic algorithms. We develop a new stochastic simulated annealing algorithm for training neural networks based on the LMedS estimator. In neural network learning, simulated annealing algorithm has been used for Boltzmann learning in Boltzmann machines [17] and in the mean-field sense to optimize other connectionist architectures [10, 13]. However, it has not been used before to minimize more robust network measures.

We can outline the objectives of this chapter in the following:

- One main goal is to evaluate the very popular backpropagation learning algorithm in terms of its sensitivity to input data possibly-contaminated with outliers. This is in contrast to the majority of already published papers on the backpropagation algorithm, which have focused on its evaluation with respect to issues like accuracy, generalization ability and training time. We thus strive to emphasize the importance of robustness in neural network training algorithms in practical and real applications.
- As M-estimators have been the main strategy to achieve more robust neural network training by several researchers, the chapter surveys the theory required to apply them in backpropagation learning algorithms. In addition, it conducts a comprehensive comparative study of several popular M-estimators used for training neural networks for the tasks of function approximation and dynamical model identification. We hence seek some answers to the question on which M-estimator to use in these popular neural network applications.
- A new, stochastic simulated annealing algorithm for robust training of MFNNs based on the highly-robust LMedS estimator is proposed. Its performance on the tasks of function approximation and dynamical model identification is assessed and quantitatively compared to the usual method

of MFNN training by the MSE-based backpropagation algorithm and to M-estimator-based training.

The rest of this chapter is organized as follows. Section 2 reviews M-estimators as the most popular robust technique and shows how they can be used to robustify the traditional backpropagation learning algorithm. The performance of the various M-estimators is assessed and compared on several experiments for function approximation and dynamical model identification. Section 3 explains the least median of squares (LMedS) as a robust estimator, then develops a stochastic simulated annealing algorithm based on that estimator for neural network training. In section 4, we report all our experimental results to evaluate the various developed robust algorithms. Finally, we conclude in Section 5.

2 M-Estimators

M-estimators are one popular robust technique which corresponds to the maximum likelihood type estimate. They generalize straightforwardly to multi-parameter problems [34]. Let r_i be the residual of the ith datum, i.e. the difference between the ith observation and its fitted value. The standard least-squares method tries to minimize $\sum_i r_i^2$, which is unstable if there are outliers present in the data. Outlying data give an effect so strong in the minimization that the parameters thus estimated are distorted. The M-estimators try to reduce the effect of outliers by replacing the squared residuals r_i^2 by another function of the residuals, yielding

$$\min \sum_i \rho(r_i), \tag{1}$$

where $\rho(.)$ is a symmetric, positive-definite function with a unique minimum at zero, and is chosen to be less increasing than square. Instead of solving directly this problem, it can be implemented as an iterated reweighed least-squares one [43]. The derivative $\psi(r_i) = \partial \rho(r_i)/\partial r_i$ is called the influence function. It measures the influence of a datum on the value of the parameter estimate. For example, for the least-squares with $\rho(x) = x^2/2$ the influence function is $\psi(x) = x$. That is, the influence of a datum on the estimate increases linearly with the size of its error, which confirms the non-robustness of the least-squares estimate. When an estimator is robust, it may be inferred that the influence of any single observation (datum) is insufficient to yield any significant offset [40]. The ratio $\frac{\psi(r_i)}{r_i}$ is commonly known as the weight function, and reflects the rate at which an influence function is increasing in comparison to that of the least-squares. Fig. 1 illustrates some commonly used M-estimators, and their influence and weight functions.

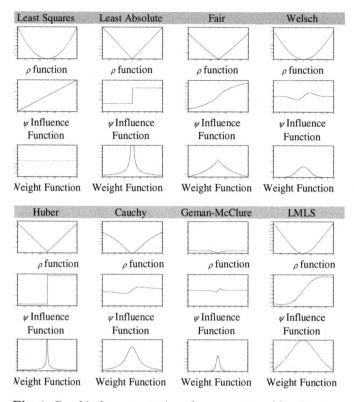

Fig. 1 Graphical representation of some common M-estimators, and their influence and weight functions (After [43])

2.1 Robust Backpropagation Algorithm

To robustify the traditional backpropagation learning algorithm based on the M-estimators concept in order to reduce the effect of outliers, all one should do is to replace the squared residuals r_i^2 in the network error by another function of the residuals, yielding

$$E = \frac{1}{N} \sum_i^N \rho(r_i), \qquad (2)$$

where N is the total number of samples available for network training. We are going to derive the updating of the network weights based on the gradient decent learning algorithm, but his can be extended to any other similar learning algorithm (e.g., conjugate gradient). In the following we assume, without loss of generality, a feedforward neural network with one hidden layer.

The weights from the hidden to output neurons W_{oh} are updated as

$$\triangle W_{oh} = -\alpha \frac{\partial E}{\partial W_{oh}} = -\frac{\alpha}{N} \sum_i^N \frac{\partial \rho(r_i)}{\partial W_{oh}}$$

$$= \frac{\alpha}{N} \sum_i^N \psi(r_i) \cdot \frac{\partial f_o}{\partial net_o} \cdot O_h, \tag{3}$$

where α is a user-supplied learning constant, O_h is the output of the hth hidden neuron, $O_o = f_o(net_o)$ is the output of the oth output neuron, $net_o = \sum_h W_{oh} O_h$ is induced local field produced at the input of the activation function associated with the output neuron (o), and f_o is the activation function of the neurons in the output layer. In this work, a linear activation function (purelin) is used in the output layer's neurons.

Analogously, the weights from the input to hidden neurons W_{hk} are updated as

$$\triangle W_{hk} = -\alpha \frac{\partial E}{\partial W_{hk}} = -\frac{\alpha}{N} \sum_i^N \frac{\partial \rho(r_i)}{\partial W_{hk}}$$

$$= \frac{\alpha}{N} \sum_i^N \sum_o \psi(r_i) \cdot \frac{\partial f_o}{\partial net_o} \cdot W_{oh} \cdot \frac{\partial f_h}{\partial net_h} \cdot I_k, \tag{4}$$

where I_k is the input to the kth input neuron, $net_h = \sum_k W_{hk} I_k$ is induced local field produced at the input of the activation function associated with the hidden neuron (h), and f_h is the activation function of the neurons in the hidden layer. We have assumed the log-sigmoid function as the activation function for the hidden layer's neurons in our experimental results.

2.2 Simulation Results

In this subsection, the performance of neural networks trained with the robust, M-estimators-based backpropagation algorithm is evaluated in several situations for the task of one-dimensional (1-D) and two-dimensional (2-D) function approximation and, for dynamical model identification.

1-D Function Approximation

A 1-D function is given by the following equation:

$$y = |\,x\,|^{2/3} . \tag{5}$$

This test function was first used in [25] and also used in [1, 8, 48]. We compare the performance of several M-estimators with that of a neural network trained by the usual MSE estimator. The so-called *Tao*-robust training

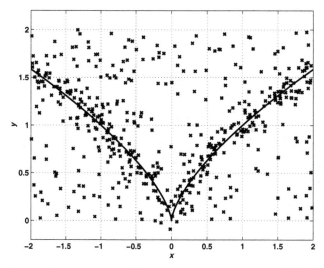

Fig. 2 Data points used in our experiments at $\varepsilon = 40\%$, along with true function shown in continuous line

algorithm of [1] (also based on using M-estimators) was also tested, but we found its results no better than those of the robust training algorithm based on the Cauchy M-estimator. Therefore its results are not included in our experimental results reported below.

The data used for this experiment consist of $N = 501$ points generated by sampling the independent variable in the range [-2 2], and using (5) to calculate the dependent variable. The data points are then corrupted in both the x- and y-coordinates with Gaussian noise of zero mean and 0.1 standard deviation. A variable percentage, ε, of the data points was selected at random and substituted with background noise, uniformly distributed on the range [-2 2]. Fig. 2 illustrates the data points used in our experiments at an outliers percentage of $\varepsilon = 40\%$, along with the true function shown in continuous line.

To assess the performance for each estimator, it is necessary to use a criterion. Therefore, we used the root mean square error (RMSE):

$$RMSE = \sqrt{\frac{\sum_{i=1}^{N}(t_i - y_i)^2}{N}}, \tag{6}$$

where the target t_i is the true value of the function at x_i, y_i is the output of the network given x_i as its input, and N is the number of dataset points.

The neural network architecture considered in all the methods is a MFNN with one hidden layer having ten hidden neurons. The network is trained with the backpropagation algorithm using the M-estimators mentioned above. In our implementation, we have used the Matlab neural network toolbox with the following settings: network training function is traincgf (conjugate gradient

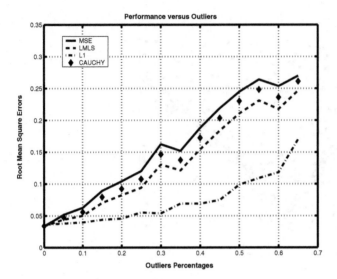

Fig. 3 Averaged Performance of MSE, LMLS, L1, and CAUCHY estimators versus varying degrees of outliers for 1-D function approximation

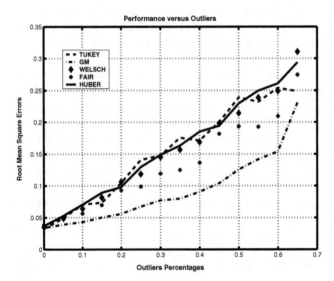

Fig. 4 Averaged Performance of GM, WELSCH, FAIR, TUKEY, and HUBER estimators versus varying degrees of outliers for 1-D function approximation

backpropagation with Fletcher-Reeves updates), maximum number of epochs to train is 500 epochs, and performance goal is 0.

The outliers percentage ε is changed from 0 to 65%. At each outliers percentage, the network was trained and the RMSE was calculated from the resultant network model. This procedure is repeated 10 times for each of M-estimator

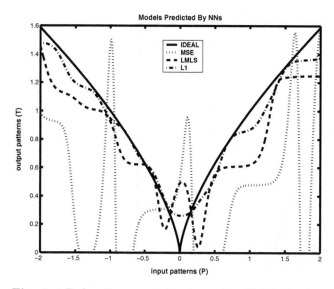

Fig. 5 1-D function approximation using LMLS, L1, and MSE estimators at $\varepsilon = 0.5$

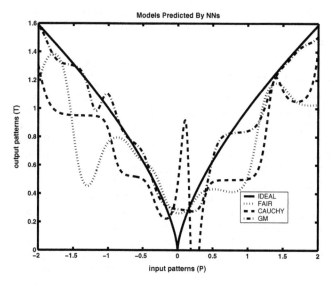

Fig. 6 1-D function approximation using FAIR, CAUCHY, and GM estimators at $\varepsilon = 0.5$

(to compensate for the random initialization of network weights and biases) and the *averaged* obtained RMSE scores are plotted versus ε in Fig. 3 and Fig. 4. The scores are also tabulated in Table 1. The resultant models (the approximated functions) at the case $\varepsilon = 50\%$ are graphed in Figs. 5-7.

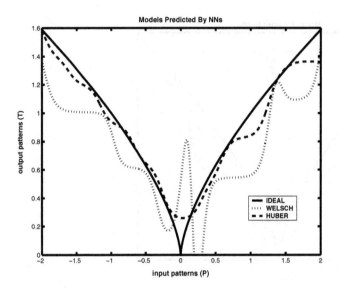

Fig. 7 1-D function approximation using WELSCH and HUBER estimators at $\varepsilon = 0.5$

Table 1 Averaged RMSE scores versus outliers percentages for networks trained with MSE, LMLS, L1, FAIR, CAUCHY, GM, WELSCH and HUBER estimators for 1-D function approximation

ε	MSE	LMLS	L1	CAUCHY	HUBER	TUKEY	GM	WELSCH	FAIR
0%	0.0332	0.0336	0.0362	0.0339	0.0369	0.0348	0.0342	0.0352	0.0387
5%	0.0517	0.0442	0.0377	0.0476	0.0531	0.0515	0.0393	0.0489	0.0511
10%	0.0627	0.0502	0.0396	0.0560	0.0705	0.0637	0.0432	0.0641	0.0566
15%	0.0892	0.0703	0.0439	0.0794	0.0892	0.0813	0.0503	0.0816	0.0697
20%	0.1041	0.0823	0.0458	0.0923	0.0979	0.0977	0.0560	0.1063	0.0928
25%	0.1199	0.0939	0.0552	0.1076	0.1296	0.1306	0.0674	0.1186	0.0988
30%	0.1625	0.1303	0.0539	0.1462	0.1488	0.1444	0.0772	0.1453	0.1194
35%	0.1520	0.1211	0.0691	0.1374	0.1632	0.1613	0.0798	0.1568	0.1250
40%	0.1881	0.1544	0.0690	0.1725	0.1850	0.1758	0.0905	0.1687	0.1365
45%	0.2188	0.1837	0.0748	0.2034	0.1942	0.1899	0.1041	0.1980	0.1817
50%	0.2446	0.2105	0.0987	0.2300	0.2294	0.2364	0.1257	0.2141	0.1931
55%	0.2642	0.2308	0.1091	0.2481	0.2495	0.2415	0.1418	0.2387	0.1926
60%	0.2537	0.2175	0.1181	0.2359	0.2608	0.2728	0.1540	0.2489	0.2094
65%	0.2701	0.2466	0.1704	0.2613	0.2939	0.2803	0.2309	0.3105	0.2744

It is clear that from the table and figures that MSE technique is not robust. Its RMSE starts to exceed 0.1 at $\varepsilon = 20\%$ whereas almost all other M-estimators have a RMSE well below 0.1 for the same outliers percentage. On the other hand, L1, GM, LMLS, FAIR, CAUCHY, and WELSCH estimators give better performance than that of the MSE. The best performance was obtained by the L1-estimator.

2-D Function Approximation

Another experiment is conducted to approximate a 2-D function defined as

$$y = x_1 e^{-\rho}, \rho = x_1^2 + x_2^2, x_1, x_2 \in [-2, 2].$$ (7)

The function is plotted in Fig. 8. The domain of the function is the square $x_1, x_2 \in [-2, 2]$. In order to form the training set, the function is sampled on a regular 16*16 grid. In order to quantify the robustness of estimators in this case, we have generated a data set for this function similar to the 1-D example. The neural network architecture considered in this case is a MFNN with 2 input neurons and one hidden layer having ten hidden neurons. The outlier percentage is varied within the range (0-65%). The results presented below are the average RMSE scores, defined analogously to (6), over ten runs for each estimator. This was done to take into account the different initial values of weights and biases at the beginning of each training. The RMSE scores versus varying percentages of outliers contaminated in the data are tabulated in Table 2, and plotted in Fig. 9 and Fig. 10.

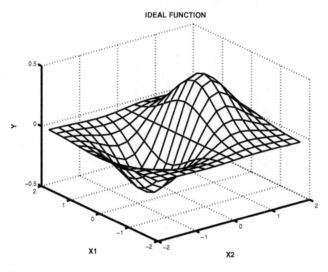

Fig. 8 2-D function to be approximated

From Fig. 9, we can note the better performance of CAUCHY-estimator over MSE, LMLS, and L1 estimators. From Fig. 10, we can observe clearly the superiority of the GM estimator over the HUBER, TUKEY, WELSCH, and FAIR estimators. On the whole, the GM estimator has exhibited the best overall on the 2-D function approximation case.

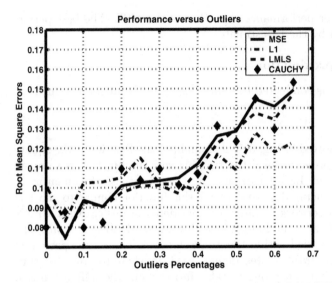

Fig. 9 Averaged RMSE of MSE, LMLS, L1 and CAUCHY estimators on 2-D function approximation versus varying outlier percentage

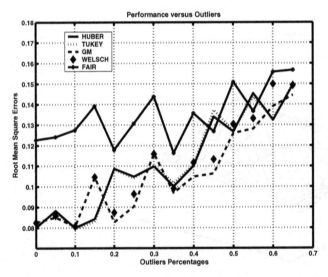

Fig. 10 Averaged RMSE of HUBER, TUKEY, GM, WELSCH, and FAIR estimators on 2-D function approximation versus varying outlier percentage

Model Identification Using M-Estimators

In this experiment, the simulation results of nonlinear plant identification are presented. A parallel model is assumed to identify the plant, and the backpropagation algorithm that uses the usual MSE cost function and the

Table 2 Averaged RMSE scores of networks trained with MSE, LMLS, L1, FAIR, CAUCHY, GM, WELSCH and HUBER estimators for 2-D function approximation versus outliers percentages

ε	MSE	LMLS	L1	CAUCHY	HUBER	TUKEY	GM	WELSCH	FAIR
0%	0.0921	0.0923	0.1011	0.0800	0.0801	0.0813	0.0826	0.0825	0.1227
5%	0.0748	0.0743	0.0831	0.0878	0.0883	0.0875	0.0851	0.0859	0.1242
10%	0.0937	0.0928	0.1024	0.0798	0.0800	0.0795	0.0806	0.0807	0.1275
15%	0.0905	0.0904	0.1029	0.0823	0.0843	0.0829	0.1045	0.1046	0.1391
20%	0.1010	0.0975	0.1051	0.1094	0.1087	0.1082	0.0827	0.0874	0.1179
25%	0.1025	0.1008	0.1150	0.1037	0.1048	0.1040	0.0903	0.0964	0.1306
30%	0.1035	0.1011	0.1017	0.1093	0.1098	0.1103	0.1162	0.1159	0.1438
35%	0.1049	0.0969	0.1018	0.1013	0.1000	0.1017	0.0969	0.0991	0.1163
40%	0.1120	0.1086	0.0981	0.1068	0.1101	0.1105	0.1048	0.1117	0.1356
45%	0.1261	0.1223	0.1167	0.1311	0.1340	0.1372	0.1063	0.1133	0.1268
50%	0.1284	0.1294	0.1086	0.1233	0.1267	0.1271	0.1260	0.1303	0.1509
55%	0.1445	0.1375	0.1277	0.1448	0.1454	0.1437	0.1281	0.1331	0.1366
60%	0.1409	0.1343	0.1178	0.1294	0.1323	0.1329	0.1390	0.1499	0.1558
65%	0.1491	0.1473	0.1227	0.1530	0.1495	0.1500	0.1446	0.1493	0.1567

Table 3 Averaged RMSE scores versus outliers percentages for networks trained with MSE, LMLS, L1, FAIR, CAUCHY, GM, WELSCH and HUBER estimators for model identification

ε	MSE	LMLS	L1	CAUCHY	GM	FAIR	WELSCH	HUBER	TUKEY
0%	0.0082	0.0111	0.0208	0.0111	0.0227	0.0742	0.0120	0.0116	0.0100
5%	0.0625	0.0596	0.0487	0.0643	0.0597	0.0616	0.0618	0.0629	0.0629
15%	0.0960	0.0883	0.0584	0.0938	0.0717	0.0891	0.0839	0.0863	0.0862
30%	0.1405	0.1288	0.0747	0.1408	0.1044	0.1335	0.1342	0.1374	0.1369
45%	0.2031	0.1700	0.1180	0.1822	0.1546	0.1697	0.1900	0.1862	0.1931

various M-estimators is used to train the neural network. The plant to be identified is governed by the difference equation [27]:

$$y_p(k + 1) = 0.3y_p(k) + 0.6y_p(k - 1) + f[u(k)], \qquad (8)$$

where the unknown function has a form $f(u) = u^3 + 0.3u^2 - 0.4u$.

The MFNN used to identify the model has a topology of (1-10-1). The network weights are adjusted at every $(T_i = 1)$ according to the backpropagation algorithm using the conjugate gradient updating rule. The input to the plant and the model is taken as the sinusoid $u(k) = \sin(2\pi k/250)$. The data that will be used by the neural network are disturbed by additive Gaussian noise

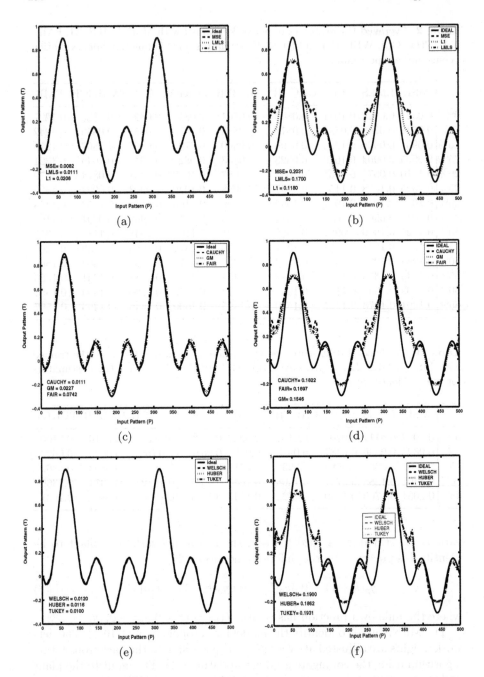

Fig. 11 Outputs of plant identification model experiment using various M-estimators (First column for $\varepsilon = 0\%$, and Second column for $\varepsilon = 45\%$): (a)-(b) MSE, LMLS, and L1. (c)-(d) CAUCHY, GM, and FAIR. (e)-(f) WELSCH, HUBER, and TUKEY

with 0 mean and 0.1 standard deviation. A variable percentage ε of the data are substituted by uniformly distributed background noise (that represents the outliers).

We have used various outliers percentages to carefully study its effects on the identification process. In order to assess the identification process performance, we use the root mean square error (RMSE) for each model. Table 3 shows the obtained errors, averaged over 10 trials, versus ε for the various M-estimators. For the sake of illustration, the outputs from the models identified by the neural network, in comparison with the ideal one, for the two cases ($\varepsilon = 0, 45\%$) are graphed in Fig. 11. It is clear that in the case of pure data, the usual MSE estimator outperformed all the other estimators with a RMSE score of 0.0082. Afterwards, the L1 estimator has achieved the best performance up to $\varepsilon = 30\%$. For higher outliers percentages, all estimators have demonstrated rather poor performance as one can note from Table 3 and the second column of Fig. 11.

3 Neural Network Training Based on Least Median of Squares Estimator

When a high level of contamination by outliers is expected, it is appropriate to use estimators with high breakdown point. Unfortunately M-estimators do not have sufficiently-higher breakdown point [34]. In the experiments of the previous section, M-estimators managed to tolerate outliers up to $20 - 30\%$. Robust statistics literature introduces a more robust estimator that is called the least median of squares (LMedS), which can tolerate up to 50% of outlying data [34, 35, 40]. Generally, the LMedS method estimates the parameters by solving the nonlinear minimization problem,

$$\min_{i} med(r_i^2). \tag{9}$$

That is, the estimator must yield the smallest value for the median of squared residuals (r_i^2) computed for the entire data set.

Unlike the M-estimators, however, the LMedS problem cannot be reduced to a weighted least problem; it is probably impossible to write down a straightforward formula for the derivative of LMedS estimator. Therefore deterministic algorithms cannot be used to minimize that estimator. Monte-Carlo technique [14] has been used for solving this problem in some non-neural applications. To get around this problem, we resort to another family of optimization algorithms that is called stochastic algorithms. These algorithms use random search to obtain a solution. Stochastic algorithms are thus relatively slow but more likely to find the global minimum. In this section, we use one such algorithm, the simulated annealing algorithm [26], to minimize an LMedS-based network error.

3.1 Overview of Simulated Annealing

Simulated annealing [26] is analogous to the physical behavior of annealing a molten metal. Above its melting temperature, a metal enters a phase
where atoms (particles) are positioned at random according to statistical
mechanics. As with all physical systems, the particles of the molten metal
seek minimum energy configurations (states) if allowed to cool. A minimum
energy configuration means a highly ordered state such as a defect-free crystal lattice. In order to achieve the defect-free crystal, the metal is annealed:
First, the metal is heated above its melting point and then cooled slowly
until the metal solidifies into a "perfect" crystalline structure. Slow cooling
(as opposed to quenching) is necessary to prevent dislocations of atoms and
other crystal lattice disruption. The defect-free crystal state corresponds to
the global minimum energy configuration [33].

The same principle can be applied to function optimization. Suppose we
want to find the global minimum of a function that is fraught with many of
local minima. In essence, simulated annealing (SA) draws an initial random
point to start its search. From this point, the algorithm takes a step within
a range predetermined by the user. This new point's objective function value
is then compared to the initial point's value in order to determine if the new
value is smaller. For the case of minimization, if the objective function value
decreases it is automatically accepted and it becomes the point from which
the search will continue. The algorithm will then proceed with another step.
Higher values for the objective function may also be accepted with a probability determined by the Metropolis criteria [18]. By occasionally accepting
points with higher values of objective function, the SA algorithm is able to
escape local minima. As the algorithm progresses, the length of the steps
declines, closing in on the final solution.

3.2 SA for Neural Network Training

The SA algorithm has been widely used for global optimization in many
fields, including training of artificial neural networks [14]. In these applications, a given network is trained minimizing a mean square error (MSE)
cost function. In our work here, we use the (SA) algorithm to minimize the
LMedS error computed over the whole training data. For the case of minimization, if the objective function value decreases, it is automatically accepted and it becomes the point from which the search will continue. The
algorithm will then proceed with another step. Higher values for the objective function that do not improve the score may also be accepted when
they serve to allow the algorithm to explore more of the possible space of
solutions with a probability determined by the Metropolis criterion [18, 32].

Algorithm 1. Simulated Annealing -Least Median of Squares Algorithm (SA-LMedS).

Objective: Train a MFNN with N training data $\mathbf{x}_i$, each of length d, which may contain outliers.

User-supplied parameters: MFNN topology, the start temperature *starttemp*, stop temperature *stoptemp*, the number of temperature reduction steps *ntemps*, the number of iterations per temperature reduction loop *niters*, and scaling factor *ratio*.

1. Initialize:
   - $Curnet=$ network with randomly-initialized weights.
   - $Bestnet = Curnet$;
   - Compute network outputs for $Curnet$ and error residual r_i for each input $\mathbf{x}_i$.
   - $Curcost = median_i^N(r_i^2)$;
   - $Bestcost = Curcost$; $Temp = starttemp$;
   - $tempmul = exp(log(stoptemp/starttemp)/(ntemp - 1))$;
2. **for** itemp= 0 To ntemps **do**
   for iter= 0 To niters **do**
     Let Newnet= Curnet + random disturbance proportional to Temp.
     Compute network outputs for $Newnet$ and error residual r_i for each input $\mathbf{x}_i$.
     $Newcost = median_i^N(r_i^2)$;
     $\Delta cost = Newcost - Curcost$;
     if $\Delta cost < 0$ **then**
       $Curnet = Newnet$;
       $Curcost = Newcost$;
     else
       P=random number$\in [0, 1]$;
       $prop = exp(-\Delta cost/(T * ratio))$;
       if $P < prop$ **then**
         $Curnet = Newnet$;
         $Curcost = Newcost$;
       end if
     end if
     if $Newcost < Bestcost$ **then**
       $Bestnet = Newnet$;
       $Bestcost = Newcost$;
     end if
   end for
   Temp=Temp*tempmul;
   end for
3. Calculate the robust standard deviation [35]:
   $\sigma = 1.4826[1 + 5/(N - d)]\sqrt{BestCost}$.
4. Identify inlier data points that have $r_i^2 < (2.5 * \sigma^2)$ and throw away the remaining outlying points.
5. Re-train the network using only the inliers data using the usual MSE cost function.

This Metropolis acceptance probability criterion can be described through the following equation:

$$P_{accept} = e^{\frac{\Delta f}{T}} > R(0,1), \tag{10}$$

where Δf denotes the change of the score implied by objective function f, T is a temperature, $R(0,1)$ is a random number in the interval $[0,1]$, and $f = median_{i=1}^{N}(r_i^2)$ is our interest objective function. Algorithm 1 summarizes our robust SA algorithm for feedforward neural network training (SA-LMedS).

It is important to note here that unlike the MSE estimator, the LMedS estimator has rather a poor performance in case of Gaussian noise [34, 35]. To compensate for this deficiency, Algorithm 1 starting from step 3 tries to identify outlying data by comparing the residuals to a robust estimate of residuals σ. The constant 1.4826 is a coefficient to achieve the same efficiency as MSE in the presence of only Gaussian noise. The factor $5/(N-d)$ is to compensate the effect of a small set of data. The reader is referred to [35] for the details of these magic numbers. Having identified and removed outliers, Algorithm 1 next re-trains the network using the (MSE) this time, which will not be influenced by outliers as they have been discarded and in the same time it has good performance with Gaussian noise.

4   Experimental Results

In this section we will perform the same experiments carried out in Section 2.2 for both 1-D and 2-D function approximation, and for model identification. The same datasets and network topologies are assumed here. The goal here is to assess the performance of the simulated annealing algorithm minimizing the Least Median of Squares estimator (Algorithm:SA-LMedS), in comparison to the traditional backpropagation algorithm minimizing the usual MSE estimator and the best members of the robust M-estimators as obtained in Section 2.2.

It is important to note that the performance of the (SA) algorithm depends on a number of user defined parameters, such as start temperature, stop temperature, number of temperatures, and temperature reduction factor. Therefore we have tried to carefully tune these parameters in order to give the SA algorithm a full opportunity to obtain a global solution.

4.1   1-D Function Approximation

A experiment similar to that of Section 2.2 is conducted to evaluate the performance of the SA-LMedS algorithm for 1-D function approximation versus varying degrees of outliers contaminated in the data. Its behavior is compared with the backpropagation algorithm minimizing the MSE estimator and the L1 estimator (as it outperformed the other M-estimator

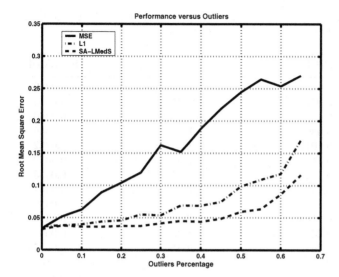

Fig. 12 Averaged RMSE scores versus outliers percentage for MSE, L1 and SA-LMedS for 1-D function approximation

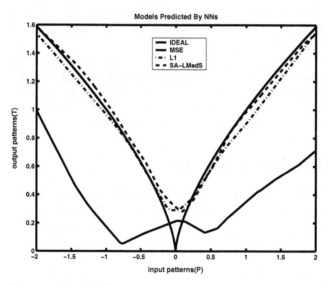

Fig. 13 The ideal function used in the training and the predictions of MSE, L1 and SA-LMedS networks at $\varepsilon = 50\%$

on 1-D function approximation). After some tuning experimentation, we used $starttemp = 5, stoptemp = 0.01, ntemps = 200, niters = 25$, and $ratio = 0.001$ for the SA-LMedS algorithm.

The outliers percentage is varied from 0% to 65%. At each outliers percentage, the network is trained with the three methods and the RMSE from

Table 4 Averaged RMSE scores versus outliers percentage for MSE, L1, and SA-LMedS methods for 1-D function approximation

ε	MSE	L1	SA-LMedS
0%	0.0332	0.0362	0.0320
5%	0.0517	0.0377	0.0375
10%	0.0627	0.0396	0.0360
15%	0.0892	0.0439	0.0359
20%	0.1041	0.0458	0.0372
25%	0.1199	0.0552	0.0372
30%	0.1625	0.0539	0.0413
35%	0.1520	0.0691	0.0451
40%	0.1881	0.0690	0.0439
45%	0.2188	0.0748	0.0485
50%	0.2446	0.0987	0.0595
55%	0.2642	0.1091	0.0637
60%	0.2537	0.1181	0.0864
65%	0.2701	0.1704	0.1162

(6) was calculated from the resultant network model. This was repeated 10 times and the average obtained RMSE scores are tabulated in Table 4 and graphed in Fig. 12. Fig. 13 shows the resultant model (the approximated functions) at the case $\varepsilon = 50\%$.

We see from Fig. 13 that in spite of the presence of such corrupted points, L1 and SA-LMedS were able to estimate a good model. The MSE estimator shows significant deviations from the ideal model. It is clear from Fig. 12 and Table 4 that the SA-LMedS algorithm is consistently better than the L1 estimator. Its RMSE is below 0.06 even at a high outlier contamination of 50%.

4.2 2-D Function Approximation

Here we will evaluate the performance of the SA-LMedS algorithm versus varying degrees of outliers contaminated in the data for a 2-D function. It is compared to the GM estimator-based algorithm as it has witnessed the best performance among all the M-estimators for such a task in Section 2.2. After several tuning trials, we used $starttemp = 5, stoptemp = 0.01, ntemps = 50, niters = 100$, and $ratio = 0.001$ for the SA-LMedS algorithm. The obtained results are represented graphically in Fig. 14. It is so clear that SA-LMedS has better performance compared with GM estimator.

4.3 Model Identification

Here we report the results that have been obtained by using the proposed SA-LMedS algorithm, compared to the MSE and L1 estimators. The results are

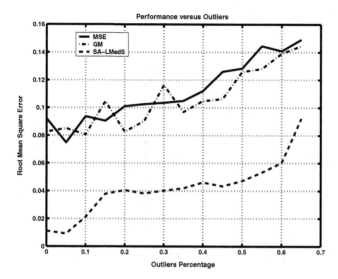

Fig. 14 Averaged RMSE scores versus outliers percentage for MSE, GM and SA-LMedS for 2-D function approximation

Table 5 Averaged RMSE values of networks trained with MSE, L1, SA-LMedS for model identification versus outliers percentages

ε	MSE	L1	SA-LMedS
0%	0.0082	0.0208	0.0363
5%	0.0625	0.0487	0.0479
15%	0.0960	0.0584	0.0502
30%	0.1405	0.0747	0.0143
45%	0.2031	0.1180	0.0391

tabulated in Table 5 and graphed in Fig. 15. It is clear that SA-LMedS outperformed considerably the other estimators for all outliers percentages. Even at 45% of outlying data, the RMSE of SA-LMedS is below 0.04, while it is over 0.1 for the L1 estimator, and even above 0.2 for the MSE estimator.

4.4   Time Performance

Another measure we have used to evaluate the performance of the various MFNN training algorithms for 2-D function approximation is the processing time. Table 6 summarizes the average run times (in seconds) for all the M-estimators-based backpropagation algorithms as well as the SA-LMedS algorithm. All the times are recorded for the algorithms implemented in Matlab and run on a pc with 2.8GHz Intel Pentium 4 processor with 1MB cache and

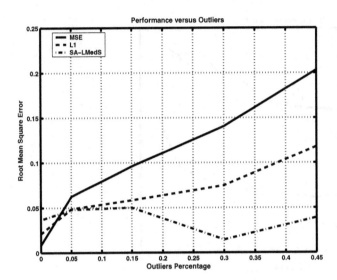

Fig. 15 Averaged RMSE scores versus outliers percentage for plant identification model, when MFNN trained by MSE, L1 and SA-LMedS

Table 6 Averaged run-time performance in seconds for all the M-estimators-based backpropagation algorithms and the SA-LMedS algorithm for 2-D function approximation

MSE	LMLS	L1	FAIR	CAUCHY	GM	WELSCH	HUBER	TUKEY	SA-LMedS
8.88	14.59	5.03	1.1090	14.59	13.23	13.70	4.29	5.23	485.64

256MB RAM. The reported times are merely indicative of the algorithms' time performance as no efforts were done to optimize them for speed.

From Table 6 we can note that the FAIR estimator has the smallest processing time. The SA-LMedS algorithm, similar to other stochastic algorithms, generally consumes more processing time than the M-estimators, and this is due to its random search nature and the large number of iterations required to find a good solution.

5 Conclusions

The chapter has addressed the issue of training feedforward neural networks in the presence of outlying data. While outliers are sample values that cause surprise in relation to the majority of samples, the presence of outliers is a common feature in many real data sets. Training feedforward neural networks by the backpropagation algorithm, when data are corrupted with these outliers, is not accurate and produces wrong models. In other words, the popular MSE cost function is not robust.

To tackle the problem of outlying data and to increase the robustness of the backpropagation algorithm, the chapter has demonstrated the use of M-estimators, borrowed from the field of robust statistics. At first the non-robust MSE cost function is replaced by one of several available M-estimators in order to make the backpropagation learning algorithm more robust. The new equations updating the network weights as needed by the backpropagation algorithm have been derived. Several M-estimators have been evaluated and compared quantitatively for the tasks of function approximation and dynamical model identification using neural networks. Our comprehensive comparative study has indeed shown that M-estimators have higher insensitivity to outlying data when compared with the MSE estimator. Among the nine M-estimators we have examined, the L1 and GM estimators have shown the best performance on the tasks considered in this work.

Unfortunately M-estimators sometimes do not have sufficient insensitivity against the outliers presence. In our experiments, M-estimators managed to tolerate outliers up to $20 - 30\%$. So the chapter has proposed to employ the more robust, Least-median-of-squares estimator that has theoretically the maximum insensitivity to outliers as it can tolerate up to 50% of outlying data. Nevertheless, unlike the M-estimators, there is no straightforward method (explicit formula) to minimize the least median of squares estimator. Therefore the chapter resorts to the simulated annealing algorithm from the family of stochastic techniques in order to minimize this LMedS estimator.

The new proposed robust stochastic algorithm based on the LMedS estimator has demonstrated even improved robustness over that of M-estimators. It managed to tolerate up to about 45% outliers while maintaining an accurate model within a RMSE of 0.048 on both 1-D and 2-D function approximation.

Whereas the stochastic algorithm has rendered better robustness than M-estimators, it is several orders of magnitudes slower. As a remedy, the algorithm can be optimized for speed making use of current parallel programming concepts. Another direction worthy of future research is to try to improve the speed of the stochastic algorithm by combining both fast deterministic algorithms with robust stochastic algorithms to obtain a hybrid algorithm. The most basic idea of hybridizing stochastic and deterministic algorithms is to alternate them. Anneal for a while, then efficiently descent for a while, then anneal some more, and so forth.

The proposed robust algorithms for training neural networks can also be used in applications other than function approximation and system identification, such as pattern classification, detection of spurious patterns, financial risk management, and computer network congestion control.

References

1. Pernia-Espinoza, A.V., Ordieres-Mere, J.B., Martinez-de-Pison, F.J., Gonzalez-Marcos, A.: TAO-robust backpropagation learning algorithm. Neural Networks 18, 1–14 (2005)
2. Annema, A.-J.: Feed-forward neural networks: Vector decomposition analysis, Modelling and Analog Implementation. Kluwer Academic Publishers, Boston (1995)
3. Baxt, W.G.: Use of an artificial neural network for data analysis in clinical decision making: The diagnosis of acute coronary occlusion. Neural Computation 2, 480–489 (1990)
4. Baxt, W.G.: Use of an artificial neural network for the diagnosis of myocardial infarction. Annals of Internal Medicine 115, 843–848 (1991)
5. Bishop, C.M.: Neural networks for pattern recognition. Clarendon Press, Oxford (1995)
6. Goodall, C.: M-estimators of location: An outline of the theory. In: Hoaglin, Mosteller, Turkey (eds.) Understanding Robust and Exploratory Data Analysis, pp. 339–403 (1983)
7. Peterson, C., Andrson, J.: A mean field theory learning algorithm for neural networks. Complex Systems 1(1), 995–1019 (1987)
8. Chuang, C.C., Su, S.F., Hsiao, C.C.: The annealing robust backpropagation (ARBP) learning algorithm. IEEE Trans. on Neural Networks 11(5), 1067–1077 (2000)
9. Churchland, P.S., Sejnowski, T.J.: The computational brain in deutscher sprache. In: Comutational intelligence. Vieweg Verlag (1997)
10. Corana, A., Marchesi, M., Martini, C., Ridella, S.: Minimizing multimodal functions of continuous variables with the simulated annealing algorithm. ACM Trans. on Mathematical Software 13(3), 262–280 (1987)
11. Cowan, J.D.: Neural networks: The early days. In: Touretzky, D. (ed.) Advances in neural information processing systems 2 (NIPS), pp. 828–842. Morgan Kaufmann, San Francisco (1990)
12. Rumelhart, D.E., McClelland, J.L.: Parallel distributed processing. MIT Press, Cambridge (1986)
13. Van den Bout, D.E., Miller, T.K.: Graph partitioning using annealed networks. IEEE Trans. on Neural Networks 1, 192–203 (1990)
14. Aarts, E.H., Korst, J.: Simulated annealing and Boltzmann machines: stochastic approach to combinatorial optimization and neural computing. John Wiley and Sons, Inc., New York (1989)
15. Hampel, F.R., Ronchetti, E.M., Rousseeuw, P.J., Stahel, W.A.: Robust Statistics, The approach based on influence functions. Wiley, NewYork (1986)
16. Dreyfus, G.: Neural networks methodology and applications. Springer, Heidelberg (2005)
17. Bibro, G.L., Synder, W.E., Garnier, S.J., Gault, J.W.: Mean field annealing: A formalism for constructing GNC-like algorithms. IEEE Trans. on Neural Networks 3, 131–138 (1992)
18. Goffe, W.L., Ferrier, G.D., Rogers, J.: Global optimization of statistical functions with simulated annealing. Journal of Econometrics 60, 65–99 (1994)
19. Gupta, P., Sinha, N.: An improved approach for nonlinear system identification using neural networks. Journal of the Franklin Institute 336(4), 721–734 (1999)

20. Haykin: Neural Networks: A comprehensive foundation, 2nd edn. Macmillan College Publishing, New York (1994)
21. Hertz, J., Krogh, A., Palmer, R.G.: Introduction to the theory of Neural Computation. Addison Vesley, New York (1991)
22. Hornik, K.: Multi-layer Feed-Forward Networks are Universal Approximators. In: White, H., et al. (eds.) Artificial Neural Networks: approximation and Learning Theory. Blackwell publishers, Cambridge (1992)
23. Hutchinson, J.M.: A radial basis function approach to financial time series analysis. Ph.D. dissertation, Massachusetts Institute of Technology (1994)
24. Moody, J., Darken, C.: Fast learning in networks of locally-tuned processing units. Neural Computa 1, 281–284 (1989)
25. Liano, K.: Robust error measure for supervised neural network learning with outliers. IEEE Trans. Neural Networks 7, 246–250 (1996)
26. Kenneth, D., Kahng, A.B.: Simulated annealing of neural networks: The cooling strategy reconsidered. Technical report CA90024, UCLA Computer science Dept., Los Angeles (1965)
27. Kumpati, S., Narendra: Identification and control of dynamical systems using neural networks. IEEE Trans. on Neural Networks 1(1), 4–27 (1990)
28. Hornik, K.: Approximation capabilities of multilayer feedforward networks. Neural Networks 4(2), 251–257 (1991)
29. Huang, L., Zhang, B.L., Huang, Q.: Robust interval regression analysis using neural network. Fuzzy Sets Syst., 337–347 (1998)
30. Le Cun, Y., Boser, B., Denker, J.S., Henderson, D., Howard, R.E., Hubbard, W., Jackel, L.D.: Handwritten digit recognition with a back-propagation network. In: Advances in Neural Information Processing Systems, vol. 2, pp. 248–257 (1990)
31. Leung, M.T., Engeler, W.E., Frank, P.: Fingerprint processing using backpropagation neural networks. In: Proceedings of the International Joint Conference on Neural Networks I, pp. 15–20 (1990)
32. Masters, T.: Advanced algorithms for neural networks: A C++ source book. John Wiley and Sons, Inc., New York (1995)
33. Hassoun, M.H.: Fundamentals of artificial neural networks. MIT Press, Cambridge (1995)
34. Huber, P.J.: Robust Statistics. John Wiley and Sons, New York (1981)
35. Rousseeuw, P.J., Leroy, A.M.: Robust regression and outlier detection. Wiley, New York (1987)
36. Pomerleau, D.A.: Neural network perception for mobile robot guidance. Kluwer, Boston (1993)
37. Rosenblatt, F.: The perceptron: A probabilistic model for information storage and organization in the brain. Psychological Review 65, 386–408 (1959)
38. Rumelhart, D.E., Hinton, G.E., Williams, R.J.: Learning representations by propagating errors. Nature 323, 533–546 (1986)
39. Welch, R.M., Sengupta, S.K., Goroch, A.K., Rabindra, P., Rangaraj, N., Navar, M.S.: Polar cloud and surface classification using AVHRR imagery: An intercomparison of methods. Journal of Applied meteorology 31, 405–420 (1992)
40. William, J.J.: Introduction to robust and quasi-robust statistical methods. Springer, Heidelberg (1983)
41. Zamarreno, J.M., Vega, P.: State space neural network: Properties and application. Neural Networks 11(6), 1099–1112 (1998)

42. Zaprains, A.D., Refenes, A.P.: Principles of neural model identification, selection, and adequacy with applications to financial econometrics. In: Perspective in Neuro Computing. Springer, London (1999)
43. Zhang, Z.: Parameter estimation techniques: A tutorial with application to conic fitting. Image and Vision Computing Journal 15(1), 59–76 (1997)
44. Rusiecki, A.L.: Robust learning algorithm with the variable learning rate. In: Rutkowski, L., Tadeusiewicz, R., Zadeh, L.A., Żurada, J.M. (eds.) ICAISC 2006. LNCS, vol. 4029, pp. 83–90. Springer, Heidelberg (2006)
45. Rusiecki, A.L.: Robust LTS backpropagation learning algorithm. In: Sandoval, F., Prieto, A.G., Cabestany, J., Graña, M. (eds.) IWANN 2007. LNCS, vol. 4507, pp. 102–109. Springer, Heidelberg (2007)
46. Chuang, C., Jeng, J., Lin, P.: Annealing robust radial basis function networks for function approximation with outliers. Neurocomputing 56, 123–139 (2004)
47. Lee, C., Chung, P., Tsai, J., Chang, C.: Robust radial basis function neural networks. IEEE Trans. Systems, Man, and Cybernetics – Part B: Cybernetics 29(6), 674–685 (1999)
48. Rusiecki, A.L.: Robust MCD-based backpropagation learning algorithm. In: Rutkowski, et al. (eds.) ICAISC 2008. LNCS (LNAI), vol. 5097, pp. 154–163. Springer, Heidelberg (2008)

Workload Assignment in Production Networks by Multi Agent Architecture

Paolo Renna and Pierluigi Argoneto

Abstract. Distributed production networks are considered organizational structures able to match agility and efficiency necessary to compete in the global market. Performances of such organization structures heavily depend on the ability of the network actors of coordinating their activities. This chapter concerns the low level of the production planning that is aim is to allocation the orders to the distributed plant. The orders assigned by the medium level have to be allocated to the plant able to manufacture the product. A this level the coordination among the plant is crucial activity to purse an high level of performance. In this context, three approaches for coordinating production planning activities within production networks are proposed. Moreover, the proposed approaches are modeled and designed by a Multi Agent Technology and a negotiation model is proposed. In order to test the functionality of the proposed agent based distributed architecture for distributed production planning, a proper simulation environment has been developed. Moreover, a benchmark model is proposed in order to evaluate the performances of the proposed approaches. The Multi Agent System and simulation environment can be utilized to learn the optimal coordination policy to adopt in distributed production planning problem.

Keywords: Order allocation, operative production planning, Multi Agent Systems, Coordination Policies, Discrete event simulation.

1 Introduction

Market globalization requires companies to operate in a wide and complex international market by matching agility and efficiency. This can be achieved either by splitting geographically the production capacity or by working together in supply chain organization involving several independent entities. In both cases, companies need to be able to design, organize and manage distributed production networks where the actions of any entity affect the behavior and the available alternatives of any other entity in the network [Wiendahl and Lutz, 2002].

Basically, two approaches are available for managing complex distributed production networks: a centralized approach, where a unique entity (the planner

Paolo Renna and Pierluigi Argoneto
Dipartimento di Ingegneria e Fisica dell'Ambiente, Università degli Studi dellaBasilicata,
Via dell'Ateneo Lucano,10, 85100, Potenza - Italy

A.-E. Hassanien et al. (Eds.): Foundations of Comput. Intel. Vol. 1, SCI 201, pp. 243–277.
springerlink.com   © Springer-Verlag Berlin Heidelberg 2009

for instance) has got all the necessary information to make planning decisions for the entire network; On the other hand, a decentralized approach can be used; in this case, each entity in the network has the necessary information and knowledge to make autonomous planning decisions, while, the common goal is reached through a cooperation among all the network actors.

It has been quite acknowledged that, while centralized approaches are theoretically better in pursuing global system performance, they have several drawbacks concerning operational costs, reliability, reactiveness, maintenance costs and so forth [Ertogral and Wu, 2000].

This is the reason why several authors propose to use decentralized approaches for distributed production planning. However, when it comes to manage complex distributed systems, questions may arise about what kind of technology and approaches should be used to assure the necessary effectiveness and efficiency of distributed systems. As far as the technology is concerned, multi-agent system (MAS) technology seems to have demonstrated its suitability. MAS, indeed, as branch of distributed artificial intelligence (DAI), is a technology based on autonomous and intelligent agents cooperation; the term agent represent an hardware or (more usually) software based computer system that has the properties of autonomy, reactivity, pro-activeness, and proper knowledge belief. Then, MAS is the logical framework for developing distributed applications, and this is particularly true in Distributed Production Planning problems.

A multi-facility production planning problem can be formulated as it follows: given external demand for items over a time horizon and a set of facilities able to produce those items, the problem is to find a production plan over multiple facilities that maximize customer and company satisfaction. Of course this problem is subject to some constraints: customer constraints such as required production volumes, due dates, item quality and so forth and internal constraints due to the limited capacity of facility resources.

The chapter is organized as it follows: in section 2 a literature review is explained; section 3 presents a general description of the Multi Agent; in section 4 the distributed production planning problem is described; the competitive and cooperative approaches are discussed in sections 5 and 6; in section 7 the fixed price environment is presented; The negotiation approach is described in section 8 while in section 9 the centralized approach as benchmark is presented; the performance measures, the simulation case study and simulation results are discussed in sections 10,11 and 12; the simulation of negotiation approach is discussed in section 13 and ,finally, conclusions and future development are reported in section 14.

2 Literature Review

Bhatnagar and Chandra (1993) have presented a survey on multi-plant coordination. In their paper, they divide coordination into two categories: (i) coordination among different functional areas such as production planning, distribution, and marketing; and (ii) coordination of the same function across multiple layers of the organization. Kutanoglu and Wu (1999) have proposed an auction-based mechanism to complex resource allocation problems, and Ertogral and Wu (2000) have proposed to use auction-based mechanism to address multi facility production planning coordination. Brandolese et al. (2000) also addressed the problem of allocating common resources requested by a multitude of system actors. In particular, they proposed a new approach for the problem of allocating production capacity to multiple requirements, based on the MAS paradigm. They implemented the proposed approach to support a distributed decision making process in companies where several product lines compete for the same resources in a scenario with swiftly changing and scarcely foreseeable demand. The authors presented a medium- term capacity allocation mechanism based on negotiations among the managers of different divisions. Timpe and Kallrath (2000), have faced with a multi sites, multi products production environment and their model considers production, distribution and marketing aspects dealing substantially with a lot sizing problem in a distributed scenario. However, their model maximizes the contribution margin or the satisfied demand in a priori known demand scenario; such a case is quite difficult to happen in a real e-marketplace scenario.

In Tharumarajah (2001) a survey of resource allocation methods in distributed manufacturing environment is provided: the survey considers many issues assuming different points of view: in particular problem decomposition (resource, task and hybrid view), type of problem solving organization (hierarchical and heterarchical), methods used for solving local problem, strategy used for solving conflict (coordination and control), inter-entity communications and measurement of performance.

Dhaenens-Flipo and Finke (2001) also consider production and distribution issues of a multi facilities, multi plant and multi period industrial problem supposing as known the demand over the planning period.

Kanyalkar and Adil (2005) considered production planning of multi-site production facilities with substitutable capacities serving multiple selling locations where the supplying plant is dynamically determined. A linear programming model is developed to produce the time and capacity aggregated plan and the detailed plan simultaneously to overcome the drawback of the hierarchical planning approaches. The model developed concerns the coordination in production and distribution planning of a company with multi-location parallel manufacturing facilities in one or several countries, in several selling locations and in using the make-to-stock product staging strategies.

Alvarez (2007) discusses some specific characteristics of the planning and scheduling problem in the extended enterprise including an analysis of a case study, and reviews the available state-of-the-art research studies in this field. Most studies suggest that integrated approaches can have a significant impact on the system performance, in terms of lower production costs, and less inventory levels. Agent-based and auction based systems can be considered as some of the most promising techniques for implementing collaborative planning and scheduling systems in the supply chain, although they have not been able to provide a good enough global optimal solution that is well accepted in industry. Other important challenges are the definition of the interactions and development of negotiation protocols between the different echelons of the manufacturing network at the planning and scheduling level. In an extended manufacturing network it is more difficult to manage all the interactions that are necessary to ensure that disruptions and related changes in one plant are taken into account by other plants avoiding unnecessary waits and inventory. This issue has not been sufficiently addressed.

Lin and Chen (2007) presented a monolithic model of the multi-stage and multi-site production planning problem is proposed in this paper. The contribution of this planning model is to combine two different time scales, i.e., monthly and daily time buckets. In addition, we also consider other practical planning characteristics and constraints. An example from a thin film transistor- liquid crystal display (TFT-LCD) manufacturer case in Taiwan is illustrated for explaining the monolithic planning model.

Tsiakis and Papageorgiou (2008) investigate the optimal configuration of a production and distribution network subject to operational and financial constraints. Operational constraints include quality, production and supply restrictions, and are related to the allocation of the production and the work-load balance. Financial constraints include production costs, transportation costs and duties for the material flowing within the network subject to exchange rates. As a business decision the out-sourcing of production is considered whenever the organization cannot satisfy the demand. A mixed integer linear programming (MILP) model is proposed to describe the optimization problem.

As the reader can notice, many authors have addressed the multi-facility production-planning problem as a distributed problem by using agent techniques. However, none of them addresses the problem in terms of comparison of different coordination strategies from a performance point of view. Most of the papers concerning performance comparison are addressed by comparing centralized approaches with decentralized one. In the authors belief, such problem, even if very relevant as a benchmark for decentralized system, it is not critical from a strategic point of view. Indeed, the choice of a distributed approach is based on the advantages it can lead into a distributed enterprise (lower investing and operating cost, reactiveness, reliability and so forth). Most important, from a strategic point of view, is to understand what kind of coordination policy can lead a better global result for the enterprise within a distributed framework. This is the main motivation of the proposed research.

3 Multi Agent Systems for Distibuted Production Planning

The scientific literature has acknowledged the use of Multiple Agent Systems (MASs) as a very promising technology to support the workload allocation in distributed environment. The agent systems respond to the changes in their environment in a timely fashion. The task of the agent is to learn from indirect, delayed reward, to choose sequences of actions that produce the greatest cumulative rewards. The simulation tool based on MAS is a valid decision support system in order to learning the optimal policy to adopt in distributed production planning problem. Therefore, the coordination policies will be tested by the simulation MAS for each operative conditions and the optimal policy is selected to coordinate the plants. The proposed approaches are based on MAS; in particular for the specified problem the following agents have been developed:

- *plant agent*; the agent manages a local database about the information of the plant (production capacity, costs, etc.). Moreover, the agent holds a set of algorithms to support the decision at local level.
- *customer agent*; the agent manages the input order process and it holds a set of algorithms to support the decision process to accept or refuse a proposal submitted by the supplier.
- *master plant agent*; the agent organizes the coordination process among the plant that constitute the network enterprises of the supplier. The agent performs the coordination strategy supported by the proposed approaches.

The workload assignment process starts with the order submission by the customer. The order is processed through the customer agent and it is delivered to the master plant agent. The order o consists of the array (V^o, dd^o, p^o) being V^o the required quantity, dd^o, the requested delivery date, and p^o, the asked price. The activity diagram of Figure 1 carries out the following actions:

Transmits order; the Customer agent transmits the order array $(V, dd, p)_o$ to the master plant agent.

Provides customer order proposal; the master plant agents transmits the customer order proposal to the plants of the network.

Proposal production planning; each plant of the network runs the local production planning algorithm and computes the volume, due date and price.

Transmits proposal; the plant agent transmits the proposal to the master plant agent;

Performs the coordination strategy; the master plant agent gathers the proposals from each plant agent, implements the selected coordination strategy and transmits the volume that needs to be produced to the plant agents of the network.

Eventual new production planning; the plant agent compares its volume proposal with the volume transmitted by the master plant agent. If the volume is the same, the plant agent keeps the production planning computed, otherwise it run again the production planning matching the volume transmitted by the master plant agent and computes the due date and the price.

Transmits final order proposal; the master plant agent computes the terms of the order to supply and it transmits the data to the customer.

Computes utility; the customer agent computes its utility resulting by final supply solution offered by the network of plants.

4 Problem Formalization

The problem is formalized by using a mathematical model whose symbolism is reported in this section (Lo Nigro et al., 2003). Each generic customer agent periodically puts an order proposal within the network. The order is marked with a progressive number o ($o = 1,...,O$) and it is characterised by an arrival date t^o_a, a due date t^o_c and a requested volume V^o. The order o with the related characteristic vector (t^o_a, t^o_c, V^o) enters the network; the master Plant P, the agent in charge of negotiating with the customer, forwards order data to each plant and remains waiting for information needed to assign orders to the plants.

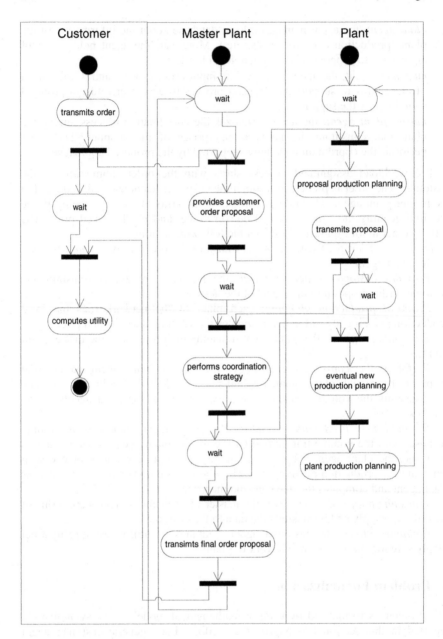

Fig. 1 Activity Diagram MAS

In order to reply to P, each local agent runs a local optimisation algorithm, that has been assumed to be a genetic algorithm, whose decision variables for each $t \in \left[t_a^o, t_c^o \right]$, where the interval $\left[t_a^o, t_c^o \right]$ is planning period, and the optimization algorithm perform in this interval, because the master Plant transmits the order data to the plants in real time, are:

$W_j^o(t)$: number of hours of direct and regular manpower allocated to order o in the period t in the plant j;

$S_j^o(t)$:   number of hours of direct and overtime manpower allocated to order o in the period t in the plant j.

Both variables determine the planned production volume for that order.
Specific characteristics of each plant j are expressed through the following parameters:

h_j — manpower hours needed to produce one item in the plant j;
w_j — unit cost per hour of regular manpower in the plant j;
s_j — unit cost per hour of overtime manpower in the plant j;
m_j — unit direct production cost (energy + raw part) in the plant j;
i_j — unit inventory cost in the plant j;
$MAXW_j$ — hours of direct and regular manpower available in the unit time period in the plant j;
$MAXS^j$ — hours of direct and overtime manpower available in the unit time period in the plant j.

Each plant agent j is called to schedule its production activities over each planning period t, whose collection determines the planning horizon T, by setting $W_j^o(t)$ and $S_j^o(t)$ which are indeed the optimisation problem variables.
The model constraints for each o, j and t are:

$$X_j^o(t) = \frac{W_j^o(t) + S_j^o(t)}{h_j} \tag{1}$$

$$VR_j^o = \sum_{t=t_a^o}^{t_c^o} X_j^o(t) \le V^o \tag{2}$$

$$0 \le \sum_{o=1}^{O} W_j^o(t) \le AW_j(t) \tag{3}$$

$$AW_j^o(t) = MAXW_j(t) - \sum_{i=1}^{o-1} W_j^i(t) \tag{3bis}$$

$$0 \le \sum_{o=1}^{O} S_j^o(t) \le AS_j(t) \tag{4}$$

$$AS_j^o(t) = MAXS_j(t) - \sum_{i=1}^{o-1} S_j^i(t) \tag{4bis}$$

$$\forall t \in \left[t_a^o, t_c^o - 1 \right] \quad E_j^o(t) = X_j^o(t) + E_j^o(t-1) \tag{5}$$

assuming $E_j^o(t_a-1)=0$

$$\forall t \in \left[t_a^o, t_c^o\right] E_j^o(t) \geq 0 \tag{6}$$

Constraint (1) links the decision variables to the number of products to be produced in the period t for the order o in the plant j; constraint (2) expresses that the volume, VR_j^o, produced by plant j for order o cannot exceed the customer demand; (3) and (4) express respectively the regular and overtime manpower hours capacity constraints over each period; (5) provides inventory at the end of the period t for the order o in the plant j,($E_j^o(t)$); finally, (6) expresses that stock out is not allowed, this expression is consequential from the expression (2), (3) and (5).

Constraints (3bis) and (4bis) just express that a plant can allocate to an order only the capacity that has not be already allocated to orders already planned.

It has been assumed that the volume produced is shipped to the customer in unique solution, tardiness and earliness are not allowed.

In setting the above decision variables, agent j computes the order production cost by the following expression:

$$C_j^o = \sum_{t=t_a^o}^{t_c^o}\left[i_j \cdot E_j^o(t) + w_j \cdot W_j^o(t) + s_j \cdot S_j^o(t)\right] + m_j \cdot VR_j^o(t) \tag{7}$$

while the corresponding cost unit is:

$$cunit_j^o = \frac{C_j^o}{VR_j^o} \tag{8}$$

At this point, the crucial issue concerns the definition of the objective function. This problem is particularly complex, because the goal of the supplier firm consists in contemporarily maximizing customer satisfaction in term of demand, price, due date reliability and its own satisfaction in term of profit, cost, losses in ordered production calculated over the planning period for all the orders. This goal must be reached globally by operating locally; furthermore, the problem is even more complex because at each planning step t, the supplier company does not know what kind of order it will collect in the next period. Therefore, the optimization problem cannot be approached as a global optimization problem in the whole planning horizon T. Within this framework, the following problems must be solved:

i.   what kind of objective function each local optimization algorithm has to face with;
ii. what kind of coordination policy must be followed in order to get a global efficiency by acting locally.

In this chapter three different coordination strategies are proposed.

The first two policies aim at giving autonomy to the plants, that is each plant decides autonomously about its production planning. However, the second approach has a cooperative nature, because it pursues also an equal distribution of the plant workload in order to balance plant utilization. The third approach implements a negotiation protocol. Furthermore, in order to test the robustness of the coordination policies, the objective function definition has been adapted to two different competition conditions of the supplier firm. The first situation regards a price based market in which the supplier can fix the price to propose to the customer, through a mark-up strategy. The second situation concerns a market in which the supplier is a price taking firm; this case is typical of very competitive markets. Next paragraphs clarify how problems i) and ii) are solved in the previously mentioned cases.

In a mark-up pricing strategy, order prices are computed by marking-up the order cost with a mark-up constant. Since it has been assumed that the company pursues customer and its own satisfaction, following this pricing strategy the company aims at minimizing production cost and maximizing production volume. In this case, the profit of the plant is assured by the mark-up, the minimizing of the production cost is needed to achieve more competitiveness to acquire a major volume of the orders.

In a fixed pricing strategy the order price is fixed then the firm will maximize the profit through the cost minimization. In this case the profit is variable and the plant should reach a trade off between the profit and the competitiveness to acquire a major share of the orders.

5 Competitive Approach

The first approach assumes that each plant can decide autonomously about how to make its production, that is, it decides when and how much to produce. The objective function, i.e. the fitness function of the local genetic algorithm, aims at minimizing the unit production cost and maximizing the produced volume. It is possible to normalize the terms of the objective function rating them to their optimal values. In particular, assuming for a plant j the minimum unit cost as:

$$c\min_j = m_j + h_j * w_j \qquad (9)$$

the FF (Fitness Function) can be expressed as it follows:

$$\frac{VR_j^o}{V^o} + \frac{c\min_j}{cunit_j^o} \qquad (10)$$

It should be noticed that, FF achieves is maximum value, equal to 2, when the plant j produces the entire requested volume with the minimum unit cost. When the optimal plan for the order o is found out, each plant agent sends the related output vector $\left(cunit_j^o, VR_j^o\right)$ to P; the master plant agent, P, computes the following performance index:

$$IP_j^o = \frac{cunit_j^o - cunit\max^o}{cunit\min^o - cunit\max^o} + \frac{VR_j^o - V\min^o}{V\max^o - V\min^o} \qquad (11)$$

where:

$$cunit\min^o = \min_j\left\{cunit_j^o\right\} \tag{12}$$

$$cunit\max^o = \max_j\left\{cunit_j^o\right\} \tag{13}$$

$$V^o\min = \min_j\left\{VR_j^o\right\} \tag{14}$$

$$V^o\max = \max_j\left\{VR_j^o\right\} \tag{15}$$

Afterwards, plants are ranked basing on IP_j^o and the order is progressively split among the L^o plants based on the above index until the desired volume is achieved; of course, if just one plant is able to realise the volume $L^o = 1$. It can happen that the plant with the lower IP_j^o among those belonging to the set L^o, could get a volume, NVR_j^o, that is different from the optimal one; this is because the plant obtains only the remaining part of the volume, i.e.:

$$NVR_j^o = V^o - \sum_{k=1}^{L^o-1} VR_k^o \tag{16}$$

In this case, the plant has to run again the local genetic algorithm to find out the new optimal production plan, substituting equation (2) with the following one:

$$NVR_j^o = \sum_{t=t_c^o}^{t_a^o} X_j^o(t) \tag{17}$$

As the reader can notice, IP_j^o takes into account both the order cost and the volume each plant can effectively offer; this minimizes the number of plants among which the order is split.

The activity diagram of figure 2 shows the interaction among the master Plant and the plants that implements the described process.

6 Cooperative Approach

The previous approach splits progressively the order to those plants that better satisfy the order with lower cost; in other words, each plant acquires a share of the order based on its efficiency. Therefore, the inefficient plants are assigned to product low volume and they will be more inefficient and lead to overload the workload of the efficient plants.

The cooperative approach wants to overcome this limit avoiding to penalize less efficient plants by a strategy that involves all the plants of the network in the production of the orders.

This is obtained through a proper coordination activity among the plants. In particular, this second approach differs from the first one because of the strategy used to split the order among the plants.

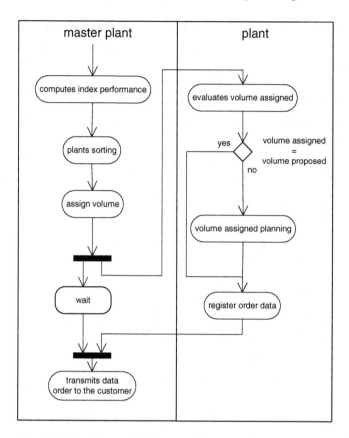

Fig. 2 Activity Diagram MAS competitive approach

After the master plant, P, receives the plan vector $\left(cunit_j^o, VR_j^o\right)$ from each plant j, it allocates the order in a proportional way as it follows:

$$NVR_j^o = \frac{VR_j^o}{\displaystyle\sum_{j=1}^{N} VR_j^o} \cdot V^o \tag{18}$$

By assigning a new production volume to the plants the following cases can happen:

a) $NVR_j^o = VR_j^o$, no further optimisation runs are required to the plant j agent;

b) the plant j is not able to produce NVR_j^o because it has not got the required production capacity; in this case the plant will produce the maximum possible volume and there will be a lost of production.

c) plant j is able to produce NVR_j^o, in this case it runs again the genetic algorithm, in order to find out the new optimal production plan, by substituting eq. (2) with (17)

The activity diagram of figure 3 shows the interaction among the master Plant and the plants that implements the cooperative approach.

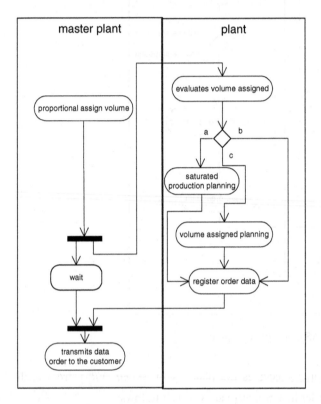

Fig. 3 Activity Diagram MAS cooperative approach

7 Fixed Price Policy

The proposed approaches are helpful in a cost driven pricing strategy, i.e. in those cases where the price can be fixed basing on the estimated cost.

However, in several other real situations the price is established in a different way: it can be the equilibrium point between demand and supply or it can be obtained by a negotiation process between customer and supplier.

In order to take into account such situations, the case of a price taking firm has also been considered. For sake of simplicity, the price has been fixed by incrementing of 50% the average unit cost obtained by the simulation carried out with the mark up strategy. Nevertheless, it is to be noticed that, the exact value of the price does not affect the result; moreover, it is also to be highlighted that the results obtained for the two techniques of pricing cannot be compared each other. Moreover, in case of a price taking firm whose aim is also the determination of the production volume to submit to the customer, the unit cost minimization does not necessarily coincide with profit maximisation. Because of that, the company profit maximisation has been considered as company goal.

In particular, in the fixed price approach, the fitness function of the local genetic algorithm includes the following terms: a) the ratio between the offered volume and the demanded one and b) the ratio between the profit related to the offered volume (pr_j^o) and an ideal maximum profit ($pr\max_j^o$), given by:

$$pr\max_j^o = \left(p - c\min_j \right) \cdot V^o \tag{19}$$

$$pr_j^o = \left(p - cunit_j^o \right) \cdot \sum_{t=t_a^o}^{t_c^o} X_j^o(t) \tag{20}$$

The maximum profit $pr\max_j^o$ represents a superior limit for pr_j^o because it is calculated supposing that all the demanded volume is realised under the best production conditions that is at the minimum cost.

Then, the local fitness function is given by:

$$\frac{\sum_{t=t_a^o}^{t_c^o} X_j^o(t)}{V^o} + \frac{\left(p - cunit_j^o \right) \cdot \sum_{t=t_a^o}^{t_c^o} X_j^o(t)}{pr\max_j^o} \tag{21}$$

In order to obtain the optimal production plan, each plant runs its local algorithm and, once finished, it sends to the master plant, P, the vector $\left(pr_j^o, VR_j^o \right)$. Based on the received information P computes the following expressions:

$$pr\min^o = \min_j \left\{ pr_j^o \right\} \tag{22}$$

$$pr\max^o = \max_j \left\{ pr_j^o \right\} \tag{23}$$

$$V^o \min = \min_j \left\{ VR_j^o \right\} \tag{24}$$

$$V^o \max = \max_j \left\{ VR_j^o \right\} \tag{25}$$

and it uses them to assign to each plant the following performance index for the current order o:

$$IP_j^o = \frac{pr_j^o - pr\min^o}{pr\max^o - pr\min^o} + \frac{VR_j^o - V\min^o}{V\max^o - V\min^o} \tag{26}$$

Finally, plants are ranked using this new index and the order is split among them as illustrated in paragraph 3. In the cooperative approach the information sent to P are the same of the previous competitive approach and the order assignment follows the criterion illustrated in paragraph 4

8 Negotiation Approach

Negotiation is a decision making theory activity in which several independent subjects with owns objectives and knowledge tray to undertake a particular course of action, modify a planned course of action or come to an agreement on a common course of action. According with Kraus, negotiation is a crucial activity in multi agent environment. Several methodologies have been proposed to address negotiation in MAS. However, no studies have addressed the impact of the negotiation on the efficiency of distributed production planning performed through MAS networks. This paper focuses such an issue. In particular, a new negotiation approach for a distributed production planning process is proposed in the paper and it is tested within a proper simulated multi-agent production planning environment.

The problem is formalized by using a mathematical model whose symbolism is here reported.

Let us suppose O orders coming from various customers; a generic customer is allowed to input its order o $(o=1,..,O)$ along with its related attributes, that are the desired quantity (V), the expected due date (DD) and price (P). The customer is also called to provide an evidence of the importance of each order's attribute.

Such information is available for the entire supplier company and each plant agent needs to find out a production plan able to assure supplier and customer satisfaction; customer satisfaction is evaluated in terms of offered quantity, estimated due date and price having considered the importance evidences. It has been assumed that the supplier can fix the price by using a mark-up pricing strategy, that is the following relation gives the price of the product:

$$P = mu \cdot cunit \tag{27}$$

where $cunit$ is the production unit cost and mu the mark up.

In particular, each generic NAi periodically puts an order proposal within the network. The order is marked with a progressive number o $(o = 1,...,O)$ and the negotiation protocol establishes that the order is structured according to the following array $\left(V^o, R_V^o, DD^o, R_{DD}^o, P^o, R_P^o \right)$ where:

- $\left(V^o, R_V^o \right)$, is the asked volume and the relative importance;

- $\left(DD^o, R_{DD}^o \right)$, is the expected due date and the relative importance;

- $\left(P^o, R_P^o \right)$, represents the desired price and the relative importance.

The customer satisfaction in meeting the order requirements is measured through a set of utility functions UA, $A = \{V, DD, P\}$, mapping in the interval [0, 1] the customer satisfaction in front of a given value of the attribute A. That is the satisfaction function form is: $U_A : D_A \rightarrow [0,1]$ where D_A is the domain of the attribute A and $U_A \in [0,1]$. The utility function form depends on the attribute type, the expected value of the attribute and on the importance given to the attribute.

Figure 4 shows the utility function for the volume and the due date attributes; in particular, bA, is equal to V_o e DD_o respectively when A is V and DD, and $bA - aA = cA - aA$.

Figure 5 shows the utility function for the price attribute; in this case, $bA = Po$.

The attribute importance is expressed through linguistic expressions whose values are: Very Important (VI), Important (I) and Not so Important (NI). Depending on the linguistic importance assigned to the attribute A, the spread, s, of UA changes. The value of s is defined in the following way, depending on the attribute type:

$$s = cA - aA, \text{ for } A = \{V, DD\} \tag{28}$$

$$s = 2 \cdot (cP - bP), \text{ for } A = \{P\} \tag{29}$$

Table 1 provides the spread variation corresponding to the linguistic values *VI*, *I* and *NI* for each utility function.

Fig. 4 *V* and *DD* utility function

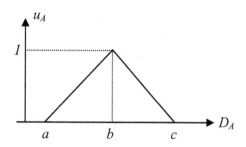

Fig. 5 *P* utility function

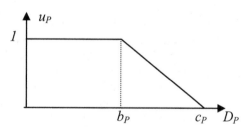

The reader should notice that, while the *s* value is expressed as function of *b* in the case of the attributes V and P, it is expressed directly in terms of unit of time for the *DD* attribute.

Table 1 Utility function spreads

	A		
S	V	DD	P
VI	$0,1 \cdot b$	4	$0,1 \cdot b$
I	$0,2 \cdot b$	6	$0,2 \cdot b$
NI	$0,3 \cdot b$	8	$0,3 \cdot b$

8.1 Negotiation Process

By indicating with r the negotiation step index, for $r = 1$, the process works through the following steps:

1. *NAi* enters the order o through the network with the related attribute array, $\left(V^o, R_V^o, DD^o, R_{DD}^o, P^o, R_P^o \right)$;

2. *NAP* forwards the order data to each *PPAj* and it remains waiting for those information needed to co-ordinate plants activities and to elaborate the counterproposal to announce to the customer;

3. In order to reply to *NAP*, each *PPAj* runs a local evolutionary algorithm whose aim is to find out how much to produce for each period with the available production capacity; the algorithm used in this step optimises a multi objective function whose terms are the customer utility function and the plant profit. This optimisation process is constrained by the plant production capacity and by the linguistic values concerning the order attributes. At the end of the optimisation process, each PPAj provides *NAP* with the following array $\left(\overline{V}_j^{o,r}, \overline{DD}_j^{o,r}, \overline{cunit}_j^{o,r} \right)$ being $\overline{V}_j^{o,r}$ the quantity that *PPAj* would like to produce, $\overline{DD}_j^{o,r}$ the due date and $\overline{cunit}_j^{o,r}$ the production unit cost.

4. In order to compensate the customer advantage of the first move, *NAP*, generates its first counter proposal by using a cooperative-coordination approach described in paragraph 5.4 has been proved to be able to maximize the profit of a multi plant distributed company. Such a coordination strategy consists in constraining the *PPAj* to plan again their production activities by producing a fixed quantity that is equal to

$$\max \left\{ \overline{V}_j^{o,r} \div \sum_{j=1}^{N} \overline{V}_j^{o,r} \cdot V^o, V_j^o \max \right\} \tag{30}$$

where V_j^o max is the maximum volume that plant j can produce when the order o arrives.

5. Each *PPAj* runs again its own evolutionary algorithm as in step 3 with the new volume constraint. At the end of this process each *PPAj* provides *NAP* with an attribute array, that is effectively used to formulate the first proposal, i.e. $\left(V_j^{o,r}, DD_j^{o,r}, cunit_j^{o,r} \right)$

6. At this point, *NAP* elaborates the proposal array to submit to *NAi*, $\left(V^{o,r}, DD^{o,r}, P^{o,r} \right)$, in the following way:

$$V^{o,r} = \sum_{j=1}^{N} V_j^{o,r} \tag{31}$$

$$DD^{o,r} = \sum_{j=1}^{N} \left(DD_j^{o} \cdot V_j^{o} \right) \div V^{o,r} \tag{32}$$

$$P^{o,r} = mu\left(r, pu \right) \cdot \sum_{j=1}^{N} \left[\left(cunit_j^{o,r} \cdot V_j^{o,r} \right) \div V^{o,r} \right] \tag{33}$$

where *mu* is the mark up, whose definition is the following:

$$mu\left(r, pu \right) = mu\_min \cdot g^{(1-pu)} \cdot \left(r\,max - r \right) \div \left(r\,max - 1 \right) +$$
$$+ mu\_max \cdot l^{pu} \cdot \left(r - 1 \right) \div \left(r\,max - 1 \right) \tag{34}$$

mu_min, mu_max define the mark up range limits and *pu* is the plants average utilization. Finally, *g* and *l* define the compromise attitude of *NAP* with *g, l*∈ℜ+ and $\left(mu\_max \div mu\_min \right) \le g < 1; l \le 1$. Once the first counter proposal is transmitted to the *NAi,* the negotiation process starts.

8.2 Negotiation Process Characterization

The negotiation is carried out by *NAP* and *NAi* and can be dealt with as a bilateral, multi issue and integrative negotiation [Raiffa, 1982]. The negotiation process is fully characterized once the following issues are defined:

     I. maximum length of the negotiation process;
    II. the *NAi* proposal evaluation strategy;
    III. the strategies *NAP* adopts to elaborate counter proposal during the negotiation;
    IV. the negotiation scheme.

Maximum length of the negotiation process
The negotiation is assumed as iterative, that is *r** rounds, with *r** ≤ *rmax*, of offers and counteroffers will occur before an agreement is reached or the negotiation fails.

The *NAi* evaluation strategy

Being $\left(V^{o,r}, DD^{o,r}, P^{o,r}\right)$ the generic *NAP* proposal at *r-th* negotiation step, *NAi* computes its utility U_r^o and the corresponding utility threshold $ThU(r)$.

$$U_r^o = u_V\left(V^{o,r}\right)^{h\left(R_V^o\right)} + u_{DD}\left(DD^{o,r}\right)^{h\left(R_{DD}^o\right)} + u_P\left(P^{o,r}\right)^{h\left(R_P^o\right)} \tag{35}$$

$$ThU(r) = \left(ThU\_min - ThU\_max\right)\cdot(r-1)\div(r\max-1)^e + ThU\_max \tag{36}$$

where
$h(VI) = 3;\ h(I) = 2;\ h(NI) = 1,$ ThU_min, ThU_max are respectively the thresholds for the last and for the first round and, finally, e is a parameter indicating the compromise attitude of *NAi* varying *r*.

NAi accepts the proposal if the following condition is verified:

$$U_r^o \geq ThU(r) \tag{37}$$

If the (37) is not verified and *r < rmax*, *NAi* evaluates each proposal attribute by computing the following unsatisfactory indexes:

$$II\_V^{o,r} = abs\left(V^{o,r} - V^o\right) \div spread\_V^o \tag{38}$$

$$II\_DD^{o,r} = abs\left(DD^{o,r} - DD^o\right) \div spread\_DD^o \tag{39}$$

$$\begin{cases} If\ P^{o,r} > P^o\ Then\ II\_P^{o,r} = \left(P^{o,r} - P^o\right)\div spread\_V^o \\ If\ P^{o,r} \leq P^o\ Then\ II\_P^{o,r} = 0 \end{cases} \tag{40}$$

and it compares them with the following unsatisfactory threshold ($ThI(r)$):

$$ThI(r) = \left(ThI\_max - ThI\_min\right)\cdot(r-1)\div(r\max-1)^f + ThI\_min \tag{41}$$

where ThI_min, ThI_max are respectively the unsatisfactory threshold for the first and the last round and *f* is a parameter indicating the compromise attitude of *NAi*. For each attribute whose the corresponding unsatisfactory index is greater than the unsatisfactory threshold, the *NAi* will communicate to the *NAP* to improve the attribute in the next proposal.

If *r = rmax*, *NAi* ends unsuccessfully the negotiation.

The strategy NAP adopts to elaborate counter proposal during the negotiation
For *r > 1*, *NAP* elaborates its counter proposal by the following steps:
1. it pushes each *PPAj* to improve the attribute requested by the customer; this is done by adding the attribute constraints (42), (43) or (44) to the model;

$$V_j^{o,r} > V_j^{o,r-1} \tag{42}$$

$$DD_j^{o,r} > DD_j^{o,r-1} \quad \text{or} \quad DD_j^{o,r} < DD_j^{o,r-1} \tag{43}$$

$$cunit_j^{o,r} < cunit_j^{o,r-1} \tag{44}$$

2. each *PPAj* runs its evolutionary algorithm and provides *NAp* with a new attribute array $\left(V_j^{r,o}, DD_j^{r,o}, cunit_j^{r,o}\right)$;

3. it coordinates the plants using a competitive approach described in paragraph 5.3 by ranking them according to the performance index reported in (45) – (48), and assigning the workload on the base of the obtained rank;

$$IP_j^{o,r} = \frac{pr\_unit_j^{o,r}}{pr\_unit\_max^{o,r}} + \frac{U_j^{o,r}}{U\_max^{o,r}} \tag{45}$$

where:

$$pr\_unit_j^{o,r} = pr_j^{o,r} \div VR_j^{o,r} \tag{46}$$

$$pr\_unit\_max^{o,r} = \max_j \left\{ pr\_unit_j^{o,r} \right\} \tag{47}$$

$$U^{o,r} \max = \max_j \left\{ U_j^{o,r} \right\} \tag{48}$$

and $U_j^{o,r}$ is the *NAi* utility when $\left(V^{o,r}, DD^{o,r}, P^{o,r}\right) \equiv \left(V_j^{o,r}, DD_j^{o,r}, P_j^{o,r}\right)$.

4. it calculates the new proposal array, $\left(V_j^{r,o}, DD_j^{r,o}, cunit_j^{r,o}\right)$ as in (31) (32) and (33), to submit to *NAi*;

5. it calculates the current value of U_r^o;

6. finally, if $U_r^o \geq U_{r-1}^o$ it submits the new proposal to *NAi*, otherwise it submits again the previous one.

The negotiation scheme

For $r \leq rmax$ the negotiation process evolves according with the following scheme:

Step 1: NAi evaluates $\left(V^{o,r}, DD^{o,r}, P^{o,r}\right)$; if $U_r^o \geq ThU(r)$ is true *NAi* accepts the offer and the negotiation ends successfully; if $U_r^o \geq ThU(r)$ is false and $r = rmax$, the negotiation ends unsuccessfully; if both the above conditions are not verified, *NAi* indicates to *NAP* the attributes to improve, $r = r+1$ and it goes to the step 2;

Step 2: *NAP* elaborates the new proposal $\left(V^{o,r}, DD^{o,r}, P^{o,r}\right)$ and submits it to *NAi*;

The above described process is shown in figure 6 by an activity diagram that explains the interaction among the customer, master Plant and plant agent.

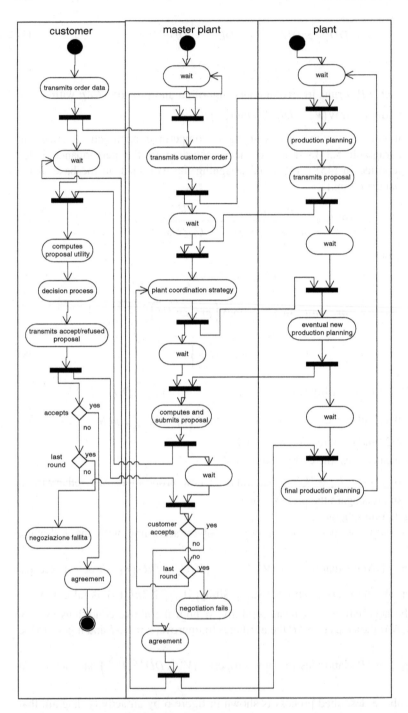

Fig. 6 Activity Diagram MAS negotiation approach

9 Centralized Approach (Benchmark)

In order to test such affirmations in the previously described environment, the competitive approach, has been modified to allow a centralized optimisation. In this case, it has been supposed that, the master plant P knows all the information regarding the single plant; P has to make decision about the production plan of all the plants, trying to pursue a common global goal. In this case, the optimisation problem has a number of decision variables equal to N-times the number of the decision variables in the distributed hypothesis; analogously, the constraints are given by expressions (1)-(6) (paragraph 2) for j=1,2,…,N substituting equation 2 with the following:

$$\sum_{j=1}^{N} VR_j^o = \sum_{j=1}^{N} \sum_{t=t_a^o}^{t_c^o} X_j^o(t) \le V^o \tag{2bis}$$

The fitness function needs to be modified in order to pursue the minimum global cost as shown in the following expression:

$$\frac{\sum_{j=1}^{N} VR_j^o}{V^0} + \frac{\overline{c \min}}{\overline{c}^{-o}} \tag{49}$$

being

$$\overline{c \min} = \sum_{j=1}^{N} \left(\frac{MAXW_j}{h_j} \cdot c \min_j \right) \div \sum_{j=1}^{N} \frac{MAXW_j}{h_j} \tag{50}$$

the minimum average (weighted average) unit cost depending on the single $c \min_j$ and on the maximum volumes each plant can produce with the minimum cost, and

$$\overline{c}^{-o} = \sum_{j=1}^{N} \left(VR_j^o \cdot c_j^o \right) \div \sum_{j=1}^{N} VR_j^o \tag{51}$$

the average unit cost for the produced volume.
In the case of fixed price policy the fitness function of the master Plant is the following:

$$\frac{\sum_{j=1}^{N} VR_j^o}{V^0} + \frac{\sum_{j=1}^{N} \left(p - cunit_j^o \right) \cdot VR_j^o}{\left(p - \overline{c \min} \right) \cdot V^o} \tag{52}$$

The centralized algorithm is linked with local algorithm of each plant; the centralized algorithm transmits the volume assigned to the plant that runs the local algorithm with objective to minimize the production planning cost (equation 7).

After that, the master Plant computes the fitness function eq. (1) and (4) of the centralized algorithm.

The proposed centralized approach is performed in the same conditions of the decentralized approach, that is:

- demand not known at priori;
- an hybrid approach centralized/decentralized to distribute the computational load to the agent in order to obtained the results in time comparable with distributed approach.

The activity diagram of figure 7 shows the interaction among the Customer, master Plant and the plants that implements the centralized approach.

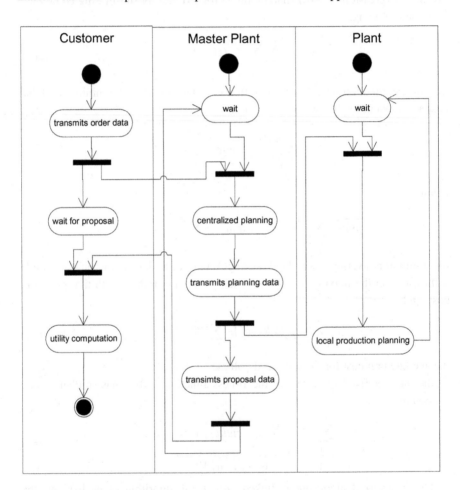

Fig. 7 Activity Diagram MAS centralized approach

10 Performance Measurement

In order to compare the proposed approaches and to evaluate them, a set of performance indexes are utilised. The purpose of such analysis is to understand the ability of the three approaches to achieve the goals. For such a reason, the performance indexes include the terms of the fitness function. Concerning the mark-up pricing technique, the indexes are:

    i)   a cost performance index (p_cost);
    ii)   a volume performance index (p_vol);
    iii) a system utilization performance index (p_ut)
    iv)   a cooperation index (p_div);

In order to allow an easy interpretation of the obtained results the performance indexes p_cost, p_vol, p_ut are normalised by rating them to the best (sometimes ideal) corresponding values; in particular, the unit cost is rated to the minimum unit cost, the produced volume to the asked one, while the system utilisation index considers a balanced indicator of the plant utilisation. The following expressions illustrate the mentioned indexes:

$$p\_cost = \frac{\sum_{o=1}^{O}\sum_{j=1}^{N} c\min_j \cdot V_j^o}{\sum_{o=1}^{O}\sum_{j=1}^{N} cunit_j \cdot V_j^o} \tag{53}$$

$$p\_vol = \frac{\sum_{o=1}^{O}\sum_{j=1}^{N} V_j^o}{\sum_{o=1}^{O} V^o} \tag{54}$$

$$p\_ut = \frac{\sum_{j=1}^{N} U_j}{N \cdot U^*}, \ \ \text{being} \ \ U^* = \max_j \{U_j\} \tag{55}$$

$$p\_div = \frac{O}{\sum_{o=1}^{O} L^o} \tag{56}$$

Where

$$U_j = \frac{\sum_{t=1}^{T}\sum_{0=1}^{O} W_j^o(S_j^o)}{T \cdot MAXW_j(MAXS_j)} \tag{57}$$

is the utilization rate of the plant j in regular (W) or overtime (S) activity.

The latest index, *p_div*, takes into account the degree of cooperation among the plants; in particular, it rates the number of the orders to the number of the plants involved in all the orders. If only one plant obtains the order, as in the pure distributive approach proposed in Amico et al. (2001) *p_div = 1*, meanwhile if the order is always split among all the plants *p_div* will assume the lowest value, depending on the number of plants.

The index related to the system utilization can be separated in regular (*p_ut reg*) and overtime utilisation (*p_ut over*).

The index *p_vol* expresses both a supplier and customer satisfaction, while *p_div* and *p_ut* refer exclusively to firm goals. Those indexes can be indifferently used for both pricing strategies.

In case of the mark-up pricing strategy, being the price based on the cost, also the index p_cost reflects both supplier and customer satisfaction.

In case of fixed price policy the performance measure of costs equation 53 must be substitute by a measurement of profit:

$$p\_pr = \frac{\sum\limits_{o=1}^{O}\sum\limits_{j=1}^{N}\left(p - cunit_j\right)\cdot V_j^o}{\sum\limits_{o=1}^{\overline{=}}\left(p - \overline{c\min}\right)\cdot V^o} \qquad (58)$$

11 The Simulated Case Study

The described simulation environment has been used to implement a distributed network consisting of 4 and 8 plants. The simulation test cases consist of 60 classes of experiments: three approaches (cooperative, competitive, centralized), two network configurations (4 and 8 plants), two pricing techniques and five mixes (representing different market scenario). Such classes of experiments have been divided into four clusters of 15 experiments each, as depicted in   table 2; 5 experiments (different mix) for each of the three approaches. The experimental plan reported in table 2 has been replicated for the 4 and 8 plants network in case of mark-up and fixed price strategy .

Table 2 Test case data

Simulation n.	Model	mix	Simulation n.	Model	mix
1	Cooperative	1	9		4
2		2	10		5
3		3	11	Centralized	1
4		4	12		2
5		5	13		3
6	Competitive	1	14		4
7		2	15		5
8		3			

Table 3 Plant characteristics

	Measure unit	Plant 1	Plant 2	Plant 3	Plant 4
h_j	hr/product	2	1.6	2.2	1.9
w_j	Euro/hr	15,5	18,1	13,9	15,5
s_j	Euro/hr	18,6	26,2	18,1	18,6
m_j	Euro/product	21,7	14,5	23	21,7
i_j	Euro/product per week	2,1	1,3	2,6	1,5
$MAXW(t)$	hr/week	40	96	40	96
$MAXS(t)$	hr/week	80	48	128	72

Let us consider the network with four plants first. Table 3 reports the production data characterizing the plants. The distributed production system is called to reply to 12 orders inputted by customers. In order to test the three approaches under different system workload situations, five mixes of orders have been simulated. Table 4 shows the orders mix characteristics; there, the parameter r indicates the percentage of orders that need a production capacity per period higher than 75% of the available one (average plant capacity). Therefore the higher the number associated with the mix is, the higher is the workload required to the whole distributed system.

The planning unit is the week, the planning period is 60 weeks, and the resulting production capacity is equal to 18.887 products. The number of iterations of the local evolutionary algorithm is 5000, while it is 100 for the centralized one (some attempts have been conducted with 200 iterations without significant improvement).

The network with 8 plants is obtained from the previous one duplicating the number of plants: 2 plants for each type of plant. The orders are 24 (12 + 12) and each order of each mix has double volumes.

12 Numerical Results

Figure 8 - 14 show the simulation results. As previously reported, five performance indexes have been considered to compare the proposed approaches (Fig. 8,9,10 and 11): p_cost or p_pr depending on the price technique used, p_vol, p_ut reg (p_ut in regular time), p_ut over (p_ut in over time) and p_div. Figure 1 reports the average value of the mentioned indexes obtained for the five mixes. Figure 12 and 13 report the plants utilization rate for each price technique and each network. Finally, Figure 14 shows the rate between the number of products inventoried in all the plants and the produced volume during the production planning horizon. In all the figures "1° price" stands for mark-up pricing strategy, while "2° price" for fixed pricing strategy.

Table 4 Order characteristics

| Order | Mix 1 0 | | | Mix 2 0.25 | | | Mix 3 0.5 | | |
	t_a^o	t_c^o	V^o	t_a^o	t_c^o	V^o	t_a^o	t_c^o	V^o
1	1	21	1200	1	21	1200	1	21	1200
2	5	15	700	5	15	700	5	11	700
3	7	30	1500	7	30	1500	7	30	1500
4	12	25	1000	12	21	1000	12	21	1000
5	17	28	800	17	28	800	17	28	800
6	19	35	1200	19	30	1200	19	30	1200
7	22	45	1500	22	45	1500	22	45	1500
8	24	39	1000	24	39	1000	24	33	1000
9	27	45	1200	27	38	1200	27	38	1200
10	30	55	1600	30	55	1600	30	55	1600
11	34	60	1000	34	60	1000	34	60	1000
12	35	54	1500	35	54	1500	35	44	1500
		Total	14200		Total	14200		Total	14200

| Order | Mix4 0.75 | | | Mix 5 1 | | |
	t_a^o	t_c^o	V^o	t_a^o	t_c^o	V^o
1	1	21	1200	1	12	1200
2	5	11	700	5	11	700
3	7	30	1500	7	21	1500
4	12	21	1000	12	21	1000
5	17	24	800	17	24	800
6	19	30	1200	19	30	1200
7	22	36	1500	22	36	1500
8	24	33	1000	24	33	1000
9	27	38	1200	27	38	1200
10	30	45	1600	30	45	1600
11	34	60	1000	35	44	1000
12	35	44	1500	46	60	1500
	Total	14200		Total	14200	

From the analysis of the results reported in Figures 8 - 14, the following conclusions can be drawn:

- the competitive approach aims at assigning the entire order to the best performer (basing on the *index IP*) plant available reducing the number of plants involving in a single order (*p_div*); this strategy allows to save capacity in less performing plants to satisfy forthcoming orders. As a consequence this

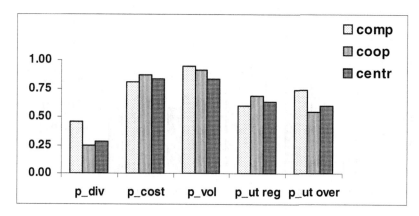

Fig. 8 Network with 4 plants and mark-up pricing strategy

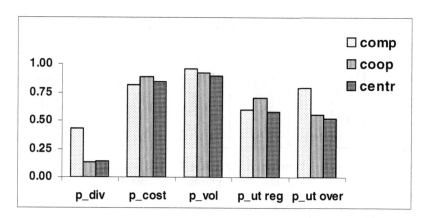

Fig. 9 Network with 8 plants and mark-up pricing strategy

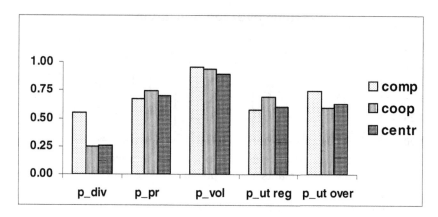

Fig. 10 Network with 4 plants and fixed price strategy

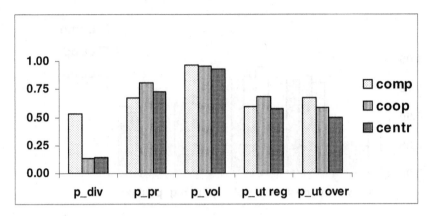

Fig. 11 Network with 8 plants and fixed price strategy

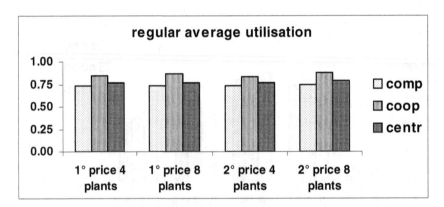

Fig. 12 Plants utilization rate in regular time

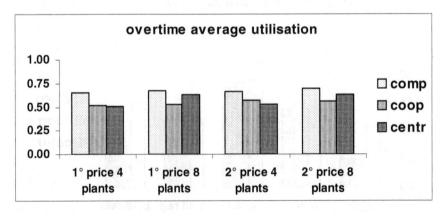

Fig. 13 Plants utilization rate in overtime

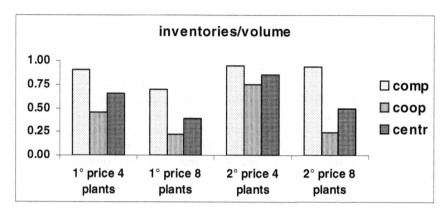

Fig. 14 Inventory comparison

approach allows to minimize production losses (*p_vol*), but, on the other hand this causes worse economic performance (*p_cost* or *p_pr*) because of higher utilization of overtime capacity and inventory ;

- the centralized approach optimizes the single order; that is for each order processed it allocates "the best" available resources belonging to all the plants; in order to do that, it tries to minimize the cost or maximize the profit for the single order, by planning it in regular time (to limit manpower costs) and near to the due date (to avoid inventory costs) limiting, in this way, the available capacity for the new ones. This causes the worst performance in term of *p_vol*.
- the cooperative approach represents a trade off between the centralized approach and the competitive one. Its explicit goal is to balance the workload among the plants (highest *p_ut reg*) because the order distribution is done considering a percentage of the best available resource (resulting from the local optimization). This allows to balance the regular plants utilization reducing drastically the production in over time. This distribution allows also to save capacity in all the plants, but it differs from the competitive approach because it does not allocate just the capacity of the best performer plant, but it distributed the order among all the plants; then, the residual available capacity in the future period is uniformly reduced. Moreover, each order is assigned to four/eight plants with lower volumes for each plant; this situation allows to the local genetic algorithm to find out a planning solution, which exploits the production capacity over the entire production horizon. This means higher lost in ordered volumes (respect to the competitive approach) but also higher economic performances, that are lower costs in the mark up strategy and higher profits in the price taking scenario.
- Finally, by observing results reported in Figure 14 it seems that the cooperation allows to minimize inventory while hard competition means high inventory. The approach robustness under different system utilization is confirmed by the results of the five mixes.

Of course, the choice among the three coordination strategies depends on what kind of performances is considered more important. The results also show how the centralized approach is not always the most performing one, especially when not all the information necessary to build up complex optimization algorithm is available. It is to be noticed, that the results here reported could be effected by the methodology of optimization used, a genetic algorithm. This will be the next path of the upcoming research, that is to test the three coordination policies under different optimization algorithms. Also, the testbed can effect the results, that are how order arrive, when they arrive, how they mix. Also this problem will be treated in the upcoming researches by building up proper experimental plans.

13 Negotiation Numerical Results

The negotiation model proposed cannot compare with the coordination approaches, because the operation environment is different; therefore, the performances are absolutely different. Table n. 5 reports the negotiation parameters value.

By using the developed simulation environment, a distributed network consisting of 4 plants has been simulated. The distributed production system is called to process 12 orders inputted by customers whose data are depicted in Table 6. The unit time is the week while the planning period is 60 weeks. Table 7 shows the results of the negotiated agreement that is the attributes characterizing the accepted order and the trade offs for NAP and NAi. Finally, Figure 15 shows the trade offs trend along the planning horizon.

In order to evaluate the negotiation process the following order data are collected:

- supplier utility is evaluates by the profit obtained for each order o pr_{r*}^{o} ;
- customer utility by the satisfaction for each order o U_{r*}^{o} that has reach an agreement (equation 35);
- the target profit of the supplier is the profit that it should obtain by the first counter-proposal submitted to the customer;
- the maximum satisfaction of the customer is $U_{r}^{o} = 3$, when the supplier proposal math all the requested parameters.

Table 5 Negotiation process parameters

rmax	ThU_min	ThU_max	e	mu_min	mu_max
5	1,5	3	0,7	1,1	1,8
	g	l	ThI_min	ThI_max	f
	0,65	1,4	0,2	0,5	0,7

Table 6 Order characteristics

Order	1	2	3	4	5	6
volume	1200	700	1500	1000	800	1200
R_V^o	2	0	1	1	0	1
Due date	21	15	30	25	28	35
R_{DD}^o	1	0	2	2	1	2
Price (x100)	1475	1625	1387,5	1512,5	1475	1637,5
R_P^o	1	1	1	1	2	1
Order	7	8	9	10	11	12
volume	1500	1000	1200	1600	1000	1500
R_V^o	0	1	2	1	0	1
Due date	45	39	45	55	54	60
R_{DD}^o	0	1	0	0	0	2
Price (x100)	1750	1725	1387,5	1700	1562,5	1500
R_P^o	0	0	2	2	0	1

Table 7 Negotiated accords and trade off

Order	1	2	3	4	5	6
Volume	1200	700	1500	1000	800	567
Due date	21	15	30	25	28	35
Price (x100)	1475	1625	1401	2121	1809	1637
NAi	1,00	1,00	0,94	0,67	0,67	0,67
NAp	0,98	1,00	0,81	1,04	0,93	0,61
Order	7	8	9	10	11	12
Volume	1500	536	1200	1448	1000	1500
Due date	45	39	45	55	54	60
Price (x100)	1750	1725	2531	1700	2004	1659
NAi	1	0,67	0,67	0,67	0,67	0,67
NAp	0,88	0,47	1,01	0,78	1,04	0,75

The performance of the negotiation process is evaluated by two tradeoff values, one for the customer and one for the supplier.

In particular, for the customer the tradeoff value is the ratio between the utility for the order that reach an agreement and the maximum value of the utility:

$$Trade\_off_{NA_i} = \frac{U_{r*}^o}{U_0^o} \tag{59}$$

For the supplier the tradeoff value is the ratio between the profit that it obtains and the profit of the first counter-proposal:

$$Trade\_off_{NA_P} = \frac{pr_{r*}^o}{pr_1^o} \tag{60}$$

$Trade\_off_{NA_i}$ is in the range [0,1], while $Trade\_off_{NA_P}$ can be major of 1, because the first counter-proposal is formulates by the cooperative approach that maximize the total profit of the network; for the subsequent round the counter-proposal is formulated by the competitive approach that can lead to a better profit value.

Fig. 15 Trade offs trend

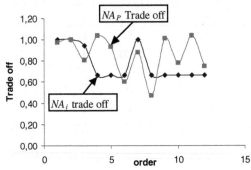

Fig. 16 Threshold vs. NAi utility

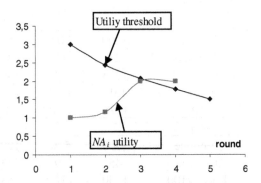

By examining figure 15 results the reader can notice that it is easier to find out a good compromise solution for early incoming orders; this is due because it has been assumed that the company mark up depends on the plants utilization; since plants utilization is lower at the beginning of the process it is easier a good trade off for both the actors. Afterwards, the trade off for the customer seems to stabilize at lower values: this is because NAi negotiates by comparing its utility, that does not decrease with the round, with the threshold value that decreases with the round as depicted in figure 16. On the contrary, NA_P trade off shows a fluctuation depending on the production plan and customer requirements.

It should be noticed that an agreement is obtained for all orders and negotiation never reaches the fifth round; this could imply that a lower compromise attitude of NAi would allow to reach a better performance for it.

Further researches within this field will deeply investigate several coordination and negotiation strategies aiming at building a framework to be used under different production planning conditions.

14 Conclusions and Discussion

This chapter investigates the distributed production planning at low level by the workload assignment to the plant of the network. A literature review is discussed, afterward introduces some very interesting innovative aspects in the multi-plan production planning problem. First a problem formalization is explained then, it has proposed three innovative approaches, based on the MAS concept, to face the multi-plant production planning problem; one approach is pure competitive, one is cooperative and the last one is based on a negotiation model. The proposed approaches are formalized by characterization of:

- Multi Agent Architecture to support the proposed approaches;
- the UML Activity diagram to formalize the interaction among the involved agents;
- the algorithm developed for each approach.

The proposed approaches are developed regard to two policy pricing: a mark-up policy; fixed pricing policy. This allows to test the approaches in different market condition and verify the robustness of the approaches proposed.
The originality aspects of the approaches proposed are the following:

- approaches based on MAS that can be support several applications: multi-plant production planning; in a Virtual Enterprise operation phase; Business to Business environment.
- the approaches operate when the demand is not known at priori;
- the approaches have been developed in order to operate in real case by cut down the computational time of the algorithm.

All the proposed approaches are suitable with an electronic network structure, and they act in a quasi real time fashion, and for such reasons, they could be implemented within a specific Business to Business application. In the next chapter is showed the model of benchmark and the simulation results to analyze the performance of the approaches.
Concerning the specific plant coordination problem here addressed, the following conclusion can be drawn:

- different coordination strategies leads to different performances for the networked enterprise;
- performance differences are quite neglectable when network congestion is high or when order overlapping is low, thus suggesting that coordination policy management it is not so important in those cases;
- in the other cases, plant coordination strategy management is very important for a networked enterprise; in particular, a cooperation oriented policy seems to lead, for the test case developed, to lowest production costs, while a

competitive oriented strategy to lowest production losses. Therefore, if the customer is considered highly strategic, an approach that puts plants in competition should be pursued to provide better customer satisfaction. Otherwise, if company's internal objective, such as cost objective, is considered strategic, an approach that allows cooperation among plants should be preferred. Finally, when company goals are not fixed a priori and they depend on orders under investigation, a mixed strategy should be pursued.

The proposed strategies can be also used to support decision maker in the quotation task giving information about the cost related to an order or/and in the negotiation procedure giving information on the reservation price.

The same approaches can be also used in MAS environment where agents belong to independent firms. In this case, a negotiation protocol could be useful because the competition nature of the resulting organization overcomes the collaborative one.

Further researches are investigating other directions: in particular, in case of a demand higher than the offer, it could be appropriate to allocate the entire order to the most efficient plant available (the plant with the lowest cost) and the remaining to the others; in this case, the price would be determined on the base of an accounting procedure that assigns higher costs as a consequence of piece of demand satisfied by less efficient plants. On the other hand, it could be also necessary to investigate the cooperative approach here presented allocating orders to plants with the lowest utilization. Moreover, other variables such as price, penalties for lateness, customer strategic importance and transportation costs are to be considered. Finally, revenue partition especially in the case of independent firms, should be also addressed. At more strategic level, the research shows that:

- Agent technology is an enabling technology to approach and solve complex distributed problems characterizing network enterprises;
- Discrete event simulation is a powerful tool for designing and testing distributed environment based on agent technology;
- Using open source technologies at the simulation package implementation phase allows code reusability for real platform development; in fact, the actual platform package can be built by utilizing the same code and architecture used for the simulation environment allowing time and investment cost saves. This reduces the risk related with investment in ICT and agent based technology in distributed production planning.

References

1. Alvarez, E.: Multi-plant production scheduling in SMEs. Robotics and Computer-Integrated Manufacturing 23, 608–613 (2007)
2. Amico, M., La Commare, U., Lo Nigro, G., Perrone, G., Renna, P.: Order Allocation strategies in distributed production planning environment. In: Proceedings of the V AITEM Conference, Bari, Italy, pp. 189–198 (2001)

3. Bhatnagar, R., Chandra, P.: Models for multi plant coordination. European Journal of Production Research 67, 141–160 (1993)
4. Brandolese, A., Brun, A., Portioli-Staudacher, A.: A multi-agent approach for the capacity allocation problem. International Journal of Production Economics 66, 269–285 (2000)
5. Cantamessa, M., Fichera, S., Grieco, A., La Commare, U., Perrone, G., Tolio, T.: Process and production planning in manufacturing enterprise networks. In: Proceedings of the 1st CIRP (UK) Seminar on Digital Enterprise Technology © 2002 School of Engineering, University of Durham (2002)
6. Dhaenens-Flipo, C., Finke, G.: An integrated model for industrial production-distribution problem. IIE Transactions 33, 705–715 (2001)
7. Ertogral, K., Wu, S.D.: Auction-theoretic coordination of production planning in the supply chain. IEE Transactions 32, 931–940 (2000)
8. Kanyalkar, A.P., Adil, G.K.: An integrated aggregate and detailed planning in a multi-site production environment using linear programming. International Journal of Production Research 43(20), 4431–4454 (2005)
9. Kutanoglu, E., Wu, S.D.: On combinatorial auction and Lagrangean relaxation for distributed resource scheduling. IEE Transactions 39(9), 813–826 (1999)
10. Kraus, S.: Negotiation and cooperation in multi-agent environments. Artificial Intelligence 94, 79–97 (1997)
11. Lin, J.T., Chen, Y.Y.: A multi-site supply network planning problem considering variable time buckets– A TFT-LCD industry case. Int. J. Adv. Manuf. Technol. 33, 1031–1044 (2007)
12. Lo Nigro, G., Noto La Diega, S., Perrone, G., Renna, P.: Coordination Policies to support decision making in distributed production planning. Robotics and Computer Integrated Manufacturing 19, 521–531 (2003)
13. Lo Nigro, G., La Commare, U., Perrrone, G., Renna, P.: Negotiation Strategies for Distributed Production Planning Problems. In: 12th International Working Seminar on Production Economics, Igls, Austria, February 2002, pp. 187–195 (2002)
14. Perrone, G., Renna, P., Cantamessa, M., Gualano, M., Bruccoleri, M., Lo Nigro, G.: An Agent Based Architecture for production planning and negotiation in catalogue based e-marketplace. In: 36th CIRP – International Seminar on Manufacturing Systems, Saarbruecken, Germany, June 03-05, 2003, pp. 47–54 (2003)
15. Raiffa: The Art and Science of Negotiation. Cambridge University Press, Cambridge (1982)
16. Tsiakis, P., Papageorgiou, L.G.: Optimal production allocation and distribution supply chain networks. Int. J. Production Economics 111, 468–483 (2008)
17. Timpe, C.H., Kallrath, J.: Optimal planning in large multi-site production networks. European Journal of Operational Research 126, 422–435 (2000)
18. Tharumarajah, A.: Survey of resource allocation methods for distributed manufacturing systems. Production Planning & Control 12(1), 58–68 (2001)
19. Wiendahl, H.-P., Lutz, S.: Production networks. Annals of CIRP 51(2), 5–22 (2002)

Part III
Knowledge Representation and
Acquisition

Extensions to Knowledge Acquisition and Effect of Multimodal Representation in Unsupervised Learning

Daswin De Silva, Damminda Alahakoon, and Shyamali Dharmage

The phenomenal behaviour and composition of human cognition is yet to be defined comprehensibly. Developing the same, artificially, is a foremost research area in artificial intelligence and related fields. In this chapter we look at advances made in the unsupervised learning paradigm (self organising methods) and its potential in realising artificial cognitive machines. The first section delineates intricacies of the process of learning in humans with an articulate discussion of the function of thought and the function of memory. The self organising method and the biological rationalisations that led to its development are explored in the second section. The next focus is the effect of structure restrictions on unsupervised learning and the enhancements resulting from a structure adapting learning algorithm. Generation of a hierarchy of knowledge using this algorithm will also be discussed. Section four looks at new means of knowledge acquisition through this adaptive unsupervised learning algorithm while the fifth examines the contribution of multimodal representation of inputs to unsupervised learning. The chapter concludes with a summary of the extensions outlined.

1 Introduction

At the highest level, learning in humans and most animals, is composed of two functions, the function of thought and the function of a commit or recall

Daswin De Silva
Cognitive and Connectionist Systems Lab, Monash University, Australia
e-mail: daswin.desilva@infotech.monash.edu.au

Damminda Alahakoon
Cognitive and Connectionist Systems Lab, Monash University, Australia
e-mail: damminda.alahakoon@infotech.monash.edu.au

Shyamali Dharmage
Centre for MEGA Epidemiology, University of Melbourne, Australia
e-mail: s.dharmage@unimelb.edu.au

A.-E. Hassanien et al. (Eds.): Foundations of Comput. Intel. Vol. 1, SCI 201, pp. 281–305.
springerlink.com

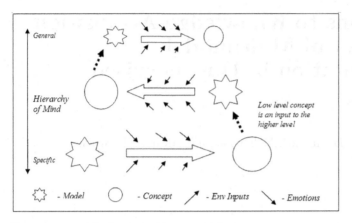

Fig. 1 Leonid Perlovsky's Theory of Mind

from memory. A comprehensive definition of thought will include references to consciousness, intelligence, concept formation and even conscience. However the process of thinking in learning can be defined in simpler terms. Aristotle [7] identified learning as the means by which hypothetical models of knowledge are narrowed down to logical concepts of understanding with the help of environmental inputs. Even though his definition was published as early as 4 BC, it has been enhanced more often than it has been challenged.

Leonid Perlovsky [15] proposed such an improvisation by incorporating two new aspects, that of knowledge instinct and hierarchical layering. Knowledge instinct is the connection of object models by emotional signals to instincts that they might satisfy and also to behavioral models that use them for instinct satisfaction. The second aspect is a hierarchy of layers where the object is understood in more general terms [14] . The general concept models of such higher layers accept as input signals, the resultant signals of object recognition from the first layer. These inputs are processed into correlations between objects, situations and knowledge. This process continues up the hierarchy towards more general models such as scientific theories, psychological concepts and the more transcendent theological concepts. Fig. 1 further illustrates this definition.

The second component of learning, the function and structure of memory is a well researched topic in the discipline of cognitive psychology. Endel Tulving [18] identified five major categories of human memory systems. These are listed in Table 1. The sequence of listing is believed to be in the order of the emergence of each, side by side with human evolution. Procedural memory is of cognitive action type, the type of memory that exists as skilled actions and not very well represented in symbolic form. The remaining four forms are cognitive representation systems, which can be represented in symbolic form and converted to verbal communication if required. Observations on the importance and practicality of each type of memory will be beneficial in

Table 1 Human Memory Systems

System	Subsystems	Retrieval
Procedural	Motor skills	Implicit
	Sensory reactions	
PRS (Priming)	Visual/ auditory priming	Implicit
Semantic	Spatial/ relational factual knowledge	Explicit
Primary	Visual/auditory short term memory	Explicit
Episodic	Events – spatial, temporal	Explicit

appreciating our focus on self organisation as the preferred means of artificial knowledge accumulation. Procedural memory is chiefly about learning to respond appropriately to sensory stimuli and naturally evolved first as it was essential for survival in the hostile environments of our very first ancestors. Priming or perceptual memory is a special form of perceptual learning, expressed in enhanced identification of objects. An initial encounter with an object primes this type of memory so that a subsequent encounter requires less stimulus information for proper identification.

With changes in surrounding environments, new memory systems developed. Semantic memory provides material for thought and contains factual knowledge of ones surroundings, factual knowledge required for cognitive operations on aspects of the world beyond immediate perception. Working memory became critical with increase in verbal communication. Similarly episodic and short term memory evolved to fulfil the requirement of knowledge accumulation by encoding and storing quantities of variable information, such as similar events at different times and counteracting associative interferences. Retrieval operations for these new types of memory are explicit compared to implicit recall of procedural memories [19]. This delineation of the categories of human memory systems provides a generous insight to the type of memory that would best fit as an artificial implementation. The problem spaces passed down to artificial systems for scrutiny and discovery of knowledge are essentially multimodal, carry a variety of attributes and contain patterns in similar as well as different contexts. Therefore best suited out of the five human memory systems is the type capable of processing different modals of information simultaneously to produce an integrated memory of the event or incident. This type is episodic memory.

Cognitive psychology [16] recognises the hippocampus part of the human brain as chiefly responsible for the creation of episodic memory. It is the meeting point for inputs from different cortical areas which are combined to form a snapshot event in episodic memory. This emergence of episodic memory is by means of unsupervised self organisation. Inputs reaching the hippocampus via the adjacent parahippocampal gyrus and entorhinal cortex contain inputs from

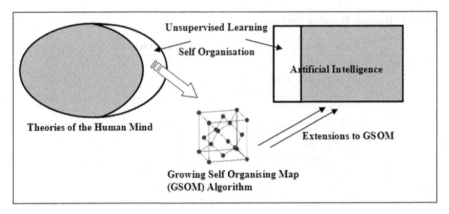

Fig. 2 Contribution of the GSOM Algorithm and its Extensions to Artificial Intelligence

virtually all association areas in the neocortex [17]. Thus it receives well detailed multimodal information, which has already been processed extensively along different and partially interconnected sensory pathways. These inputs are subjected to a process of self organisation in the hippocampus, resulting in episodic memory of events in time. A simple example of such an episodic event is a recollection of how one spent the previous evening, the location, faces and voices of the people he was with as well as emotional states. Inputs on rewards, punishments and other emotional states are also incorporated to an episodic event via the amygdala [8]. Research in cognitive psychology has underlined the desirable properties of episodic memory as multimodal knowledge representation and self organisation of such knowledge.

Fig. 2 positions the primary focus of this chapter, the Growing Self Organising Map (GSOM) algorithm and its extensions, within the research sphere of human cognition inspired artificial intelligence. Our discussion, so far, was on the human learning process and certain phenomena in this process have been understood well into the finer details. The next section outlines significant developments in machine intelligence and in the third section we look at how the self organising process of the human mind has been simplified and implemented as a machine algorithm. A major enhancement to this algorithm, the structure adapting functionality is also discussed. In this context, the extensions to the GSOM algorithm delineated in later sections of this chapter represent recent developments towards strengthening the biological foundations of unsupervised learning and self organisation in machines.

2 Related Work

Vernon et al [20] have identified two broad schools of thought in artificial intelligence, the cognitivist approach and the emergent approach. The cognitivist approach is based on symbolic information processing representational systems, which correspond to a rigid view of cognitive codes and computations. The emergent approach is based on principles of self organisation and embraces connectionist, dynamic and enactive paradigms. In cognitivist approaches, learning or the transformation of environmental inputs to internal information uses syntactic processing and is defined by a representational framework while emergent systems ground representations by autonomy preserving skill construction [20]. Adaptation implies acquisition of new knowledge for cognitivism whereas in emergent paradigms it entails re-organisation to reflect a new set of dynamics. Computational operation defines the core mode of action, cognitivism uses syntactic processing while emergent systems exploit self organisation through concurrent interactions in a network setting. The transformation of environmental inputs to internal information is defined by the representational framework.

The basis behind this internal information is semantic grounding, where in cognitivist systems it can be directly interpreted by humans while emergent systems ground representations by autonomy preserving skill construction, which cannot be understood by humans. Perception in cognitivism provides an interface between the external world and symbolic representation of the same, whereas in emergent concepts perception is changes in systems state in response to environmental factors. As highlighted in the foregoing discussion, cognitivist representational approaches bind the machine to an idealised description of cognition as defined by the designer [20]. Cognitivism makes the positivist assumption that the world we perceive is isomorphic with our perceptions of it as a geometric environment. In contrast, self organising precepts of the emergent variant provide an anticipatory and autonomous behaviour that resembles human cognition much directly.

Due to its autonomous and adaptive nature, self organising emergent theories of learning are the closest to biological observations of human learning. Hebbian learning is recognised as one of the most powerful computational implementation of biological learning. This learning rule adjusts synaptic efficacies based on coincident pre- and post synaptic activity. If two neurons are active simultaneously their connection is further strengthened. Hirsch and Spinelli demonstrated a link between the nature of an animals early visual environment and the subsequent response preference of cells in visual cortex [10]. Cortical neurons in the test animals showed a marked preference corresponding to the stimulus to which they were exposed. The cortex appears to be a continuously-adapting structure in dynamic equilibrium with both external and intrinsic input. This equilibrium is maintained by cooperative and competitive lateral interactions within the cortex mediated by lateral connections. Primary function of the lateral and afferent structures is to form a sparse, non-redundant encoding of the visual input.

Based on the Hebbian learning rule, Teuvo Kohonen developed computational models of these biological cortical maps, the self organising maps (SOM) [12]. SOMs have no global supervisor directing the learning process as cognitivist implementations do, and learning is usually coupled with normalisation so that synapses do not grow without bounds. The SOM has had wide multi disciplinary application due to its adaptive nature and unsupervised learning technique. The ability to apply a SOM algorithm without prior specialist knowledge of the problem space closely resembles the behaviour of human episodic memory as mentioned in the first section.

The SOM restricts growth of a feature map due to its pre-specified size and thus limits the quantity of learning by self organisation. It has been theoretically proven that the SOM in its original form does not provide complete topology preservation, and several researchers in the past have attempted to overcome this limitation. Human memory has no such restriction for learning. It is only remembering that can be affected by external factors such as a persons age and level of intelligence. The restricted growth of feature maps has been addressed by Alahakoon et al with an enhanced self organising algorithm. A growing variety of the SOM, the growing SOM (GSOM) addresses structural restrictions of the original SOM while also improvising on unsupervised learning [1]. The GSOM has had wide multi disciplinary application due to its hierarchical adaptive nature and unsupervised learning technique [2], [3], [4], [5], [6]. The intricacies of the GSOM are discussed next, along with an appreciation of its hierarchical self organising capability that exposes knowledge at different levels of understanding.

3   The Growing Self Organising Map (GSOM) Algorithm

Starting with four nodes, the GSOM algorithm makes use of error rate and an external parameter, the spread factor, for the generation of new nodes [1]. This allows for structural adaptation and the final map reflects the input dataset in both physical distribution and clustered knowledge. This reflection is useful both spatially and temporally. The GSOM algorithm is composed of three phases, initialization, growing and smoothing.

1. Initialization phase:

a) Initialize the weight vectors of the starting nodes (usually four) with random numbers between 0 and 1.

b) Calculate the growth threshold (GT) for the given data set of dimension D,

$$GT = -D \times ln(SF) \tag{1}$$

2. Growing Phase

a) Present input to the network.

b) Determine the weight vector that is closest to the input vector mapped to the current feature map (winner), using Euclidean distance. This step can be summarized as: find q' such that $|v - w_{q'}| \leq |v - w_q| \ \forall \in \mathbb{N}$ where v, w are the input and weight vectors respectively, q is the position vector for nodes and is the set of natural numbers.

c) The weight vector adaptation is applied only to the neighbourhood of the winner and the winner itself. The neighbourhood is a set of neurons around the winner, but in the GSOM the starting neighbourhood selected for weight adaptation is smaller compared to the SOM (localized weight adaptation). The amount of adaptation (learning rate) is also reduced exponentially over the iterations. Even within the neighbourhood, weights that are closer to the winner are adapted more than those further away. The weight adaptation can be described by

$$w_j(k+1) = \begin{cases} w_j(k) & \text{if } j \notin N_{k+1} \\ w_j(k) + LR(k) \times (x_k - w_j(k)) & \text{if } j \in N_{k+1} \end{cases}$$

where the learning Rate $LR(k)$, $k \in \mathbb{N}$ is a sequence of positive parameters converging to zero as $k \to \infty$. $wj(k)$, $wj(k + 1)$ are the weight vectors of the node j before and after the adaptation and N_{k+1} is the neighbourhood of the winning neuron at the $(k + 1)th$ iteration. The decreasing value of $LR(k)$ in the GSOM depends on the number of nodes existing in the map at time k.

d) Increase the error value of the winner (error value is the difference between the input vector and the weight vectors).

e) When $TE_i \geq GT$ (where TE_i is the total error of node i and GT is the growth threshold). Grow nodes if i is a boundary node. Distribute weights to neighbours if i is a non-boundary node.

f) Initialize the new node weight vectors to match the neighbouring node weights.
g) Initialize the learning rate (LR) to its starting value.

h) Repeat steps b) g) until all inputs have been presented and node growth is reduced to a minimum level.

3. Smoothing phase.

a) Reduce learning rate and fix a small starting neighbourhood.

b) Find winner and adapt the weights of the winner and neighbours in the same way as in growing phase.

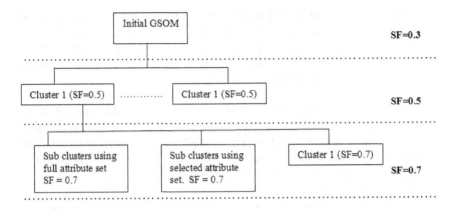

Fig. 3 Effect of the Spread Factor on Hierarchical Understanding of the GSOM

The GSOM adapts its weights and architecture to represent the input data. Therefore, in the GSOM, a node has a weight vector and two-dimensional coordinates that identify its position in the net, while in the SOM the weight vector is also the position vector [1]. The GSOM has two modes of activation: training mode or generating mode and testing mode or querying mode. The training mode consists of the three phases described above, and the testing mode is run to identify the positions of a set of inputs within an existing (trained) network. This can be regarded as a calibration phase if known data are used. For unclassified data the closeness of new inputs to the existing clusters in the network can be measured.

It is important to note the significance of the spread factor (SF) in facilitating levels of understanding (as the growth threshold). The spread factor takes values between zero and one and is independent of the number of dimensions in the data. It is possible to compare the results of different data sets with a different number of attributes by mapping them with the same spread factor [1]. With least spread, the algorithm obtains a very basic understanding of the dataset as only high level similarities are exposed by the self organisation process. As SF is increased further correlations are exposed leading to better understanding of the dataset (Fig. 3). A brief look at the derivation of the growth threshold (GT) is pertinent here.

The GT decides the amount of spread of the feature map to be generated. If only an abstract picture of the data is required, a large GT will result in a map with a fewer number of nodes [1]. Similarly, a smaller GT will result in the map spreading out more. Node growth in the GSOM is initiated when the error value of a node exceeds the GT. The total error value for node i is calculated as

$$TE_i = \sum_{H_i} \sum_{j=1}^{D} (x_{ij} - w_j)^2 \qquad (2)$$

where H_i is the number of hits to the node i and D is the dimension of the data. $X_{i,,j}$ and w_j are the input and weight vectors of the node i, respectively. For a boundary node to grow a new node, it is required that

$$TE_i \geq GT \qquad (3)$$

The GT value has to be experimentally decided depending on the requirement for the map growth. As can be seen from (1), the dimension of the data set will make a significant impact on the accumulated error *(TE)* value, and as such will have to be considered when deciding the GT for a given application. Since $0 ¡ Xi,j, Wj ¡ 1$, the maximum contribution to the error value by one attribute (dimension) of an input would be,

$$max|x_{i,j} - w_j| = 1. \qquad (4)$$

Therefore, from (1)

$$TE_{max} = D \times H_{max} \qquad (5)$$

where TE_{max} is the maximum error value and is the maximum possible number of hits. If $H(t)$ is considered to be the number of hits at time (iteration) t, the GT will have to be set such that,

$$0 \leq GT < D \times H(t) \qquad (6)$$

Therefore, GT has to be defined based on the requirement of the map spread. It can be seen from (4) that the GT value will depend on the dimensionality of the data set as well as the number of hits. Thus, it becomes necessary to identify a different GT value for data sets with different dimensionality. This becomes a difficult task, especially in applications such as data mining, since it is necessary to analyse data with different dimensionality as well as the same data under different attribute sets. It also becomes difficult to compare maps of several datasets since the GT cannot be compared over different datasets. Therefore, the user definable parameter is introduced. The SF can be used to control and calculate the GT for GSOM, without the data analyst having to worry about the different dimensions. The growth threshold is defined as,

$$GT = D \times f(SF) \qquad (7)$$

where $SF \epsilon \mathbb{R}, 0 \leq SF \leq 1$, and $f(SF)$ is a function of SF, which is identified as follows. The total error TE_i of a node i will take the values,

$$0 \leq TE_i \leq TE_{max} \qquad (8)$$

where TE_{max} is the maximum error value that can be accumulated. This can be written as

$$0 \le \sum_{H} \sum_{j=1}^{D} (x_{ij} - w_j)^2 \le \sum_{H_{max}} \sum_{j=1}^{D} (x_{ij} - w_j)^2 \qquad (9)$$

Since the purpose of the GT is to let the map grow new nodes by providing a threshold for the error value, and the minimum error value is zero, it can be argued that for growth of new nodes,

$$0 \le GT \le \sum_{H_{max}} \sum_{j=1}^{D} (x_{ij} - w_j)^2 \qquad (10)$$

Since the maximum number of hits (H_{max}) can theoretically be infinite, the above equation becomes $0 \le GT \le \infty$. According to the definition of spread factor, it is necessary to identify a function $f(SF)$ such that $0 \le Dtimes f(SF) \le \infty$ A function that takes the values zero to ∞, when x takes the values zero to one, is to be identified. A Napier logarithmic function of the type $y = -a \times ln(1-x)$ is one such equation that satisfies these requirements. If $\mu = 1 - SF$ and

$$GT = -D \times ln(1 - \eta) \qquad (11)$$

then,

$$GT = -D \times ln(SF) \qquad (12)$$

Therefore, instead of having to provide a GT, which would take different values for different data sets, the data analyst can now provide a value, SF, which will be used by the system to calculate the GT value depending on the dimensions of the data. This will allow GSOMs to be identified with their spread factors and can form a basis for comparison of different maps. An example of hierarchical clustering is shown in Fig. 4 and Fig. 5. The dataset contained a total of 18 attributes on animals, a sample of which is shown in Table 2. The type of animal is specified by a numeric value from 1 to 7. The name and the type are removed from those attributes, and the GSOM is generated and trained with the remaining 16 attributes.

Fig. 4 shows the GSOM of the zoo dataset with a 0.1 SF, while Fig. 5, an SF value of 0.85 is used. From Fig. 4 it can be seen that the different types of animals have been grouped together, and also similar subgroups have been mapped near each other. An SF around 0.3 would have given a better separated picture of the clusters. But it is still possible to see that there are three to four main groupings in the data set. One of the main advantages of

Table 2 A Sample Data Vector from the Zoo Dataset

Name	Hair	Feathers	Eggs	Milk	Airborne	Aquatic	Predator	Toothed
Lion	1	0	0	1	0	0	1	1

Backbone	Breathes	Venom	Fin	Legs	Tail	Domestic	Catsize	Type
1	1	0	0	4	1	0	1	1

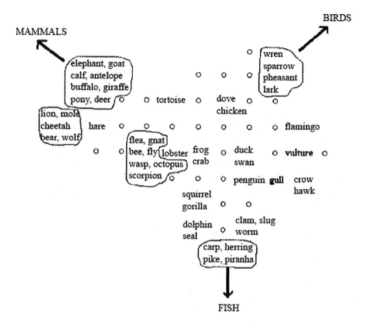

Fig. 4 The GSOM for the Zoo Dataset with a SF of 0.1

the GSOM over the SOM is thus highlighted; the GSOM indicates groupings in the data by its shape even when generated with a low SF. Figure 4 has branched out in three directions, indicating three main groups in the dataset (mammals, birds, and fish). The insects have been grouped together with some other animals but this group is not shown due to the low SF value.

Fig. 5 shows the same data set mapped with a higher (0.85) spread factor. It is possible to see the clusters clearly, as they are now spread out further. The clusters for birds, mammals, insects, and fish have been well separated. Since Fig. 5 is generated with a high SF value, even the sub-groupings have appeared in this map. The predatory birds have been separated into a separate sub-group from other birds. The other subgroups of birds can be identified as airborne and non-airborne (chicken, dove), and aquatic. The flamingo has been completely separated as it is the only large bird in the selected data. The mammals have been separated into predators and non-predators, and the non-predators have been separated into wild and domestic subgroups.

The process of hierarchical knowledge acquisition proposed in the GSOM algorithm bears close resemblance to Perlovskys theory on the hierarchy of mind (Fig. 1), where primitive knowledge lays the foundations for advanced understanding. For instance in the above example, the second level of clustering makes little sense without a basic understanding of the initial groupings, mammals, birds and fish.

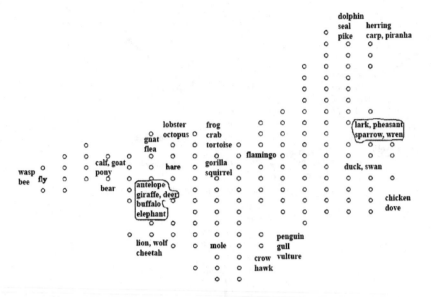

Fig. 5 The GSOM for the Zoo Dataset with a SF of 0.85

4 Inter-cluster Movements over Time

The previous section highlighted the importance of structure adaptation in unsupervised learning and detailed the workings of the GSOM algorithm. The benefits of hierarchical clustering were also explored and correlations drawn with models of incremental human understanding. This section looks at several other means deriving knowledge (or an understanding) of the dataset based on the composition of the clusters and their behaviour in the self organization process. The movement of data vectors along clusters groups, the effect of dynamic dimensions on cluster generation and cluster hierarchies expressed in terms of dominant dimensions are these derivations in brief. A continuous time dimension in the dataset as illustrated in Fig. 6 is a prerequisite for these extensions to be successfully applied.

4.1 Vertical Inter-cluster Movement

Vertical inter-cluster movement observes the behaviour of a single data vector and its movement across clusters from two different maps. The feature maps will be generated using the GSOM and will be of different spread factors, so that the drill down effect will be persistent. A graphical representation would be far more informative than a numerical representation given the dimensions involved. A sample graph is shown below Fig. 7.

The Y axis contains cluster IDs based on a predetermined order. Sorting can be either based on a stronger dimension /dimensions or on an aggregate of

Fig. 6 Disposition of
a Time Variant Data
Vector

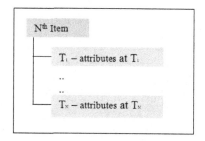

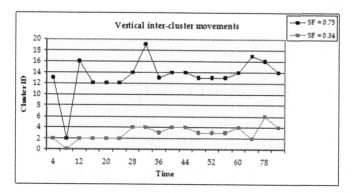

Fig. 7 Vertical Inter-Cluster Movement

the whole cluster. It is visible that the graph depicts movement at two distinct
levels, with one (SF=0.75) graph at higher cluster IDs and the other at lower
IDs. This difference corresponds to the effect of higher SF that produces a
better spread and lower SF which limits the spread and thus generating a
smaller number of nodes.

4.2 Horizontal Inter-Cluster Movement

Horizontal cluster movements are restricted to a single map. The objective
of this analysis is to compare movement of data vectors within the map,
over time. The data vector should be of the same disposition as before, with
each vector having several records at different points in time (Fig. 6). A
graphical representation will be similar to that of vertical analysis with time
and cluster ID axes. The clusters again, have to be sorted based on a predetermined criteria. For example in Fig. 8, A,B and C are three different data
items and the plot information displays the movement of time dependent data
vectors across the cluster generated. Data items A and B display a consistent
movement and thus a prediction of the next movement is possible, whereas
data item C exhibits no regularity.

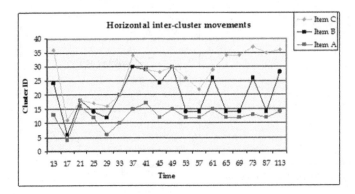

Fig. 8 Horizontal Inter-Cluster Movement

4.3   Inter-Cluster Dimensional Density

The clusters generated from datasets with continuous valued dimensions can be categorized based on the different densities of each of the dynamic dimension. Count of occurrences of each distinct value V for each dimension N of cluster C, this count as a percentage of the total cluster size will represent the dimensional density of dimension N for cluster C. The comparison of these densities on an inter-cluster basis can be used to determine patterns in the correlations that clustered the inputs together.

The comparisons were developed to be plotted in two dimensional graphs. Fig. 9 displays the density patterns in ten clusters for four dimensions. A noticeable pattern exists in the fluctuations of dimensions A and C. The same comparison can be done on clusters from different levels.

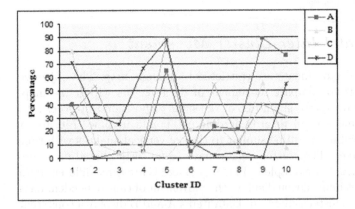

Fig. 9 Inter Cluster Dimensional Density

Intra-Cluster Dimensional Density

The above measurement can also be used for intra-cluster density comparisons, mainly when two dimensions are similar in context. Such a comparison will expose patterns not only in the two dimensions but also across levels of clustering. This would provide an added aspect of investigation, where the performance of the dimension can be examined, horizontally across the clusters and vertically through the cluster levels, using the SF value.

4.4 Experiments in Inter-cluster Movement and Dimensional Density

The extensions were experimented on an epidemiological dataset, a collection of information of 620 children born to families in which either parents or/and siblings have had asthma, eczema or hay fever. The eczema condition of each patient was observed over a period of 150 months, the information was recorded in the format specified in Fig. 10. Several observations of the rash spread are contained in each record (Fig. 10). Attributes SKINRASH to RASH6 list observations of the rash spread, while CORTISON and Non-cortico steroid cream (NON) are two types of treatment and the amounts of each administered. The ECZEMA attribute records overall spread of the rash.

The GSOM algorithm was applied to the full dataset for 500 epochs at a spread factor of 0.85. The learning rate was 0.25. Significant cluster groups were generated as per Fig. 11. The significance of each was further determined by performing an inter/intra dimensional density analysis of the generated clusters.

Fig. 12 and Fig. 13 display the results of dimensional density analysis on clusters A and B. For each attribute the first value within parentheses is the value of the attribute and the next is the percentage of clustered items with that value. A strong contrast is visible in the age distribution of the two clusters, with cluster A comprising of data vectors from infancy or the initial months while cluster B has grouped vectors mainly from 24 months and above. A second contrast is in the attributes SKINRASHFO to RASH6, where cluster A has high percentage of presence (binary value 1) while cluster B the opposite

NUMBER	AGE_OF_	SKIN_RAS	RASH	RASH6	CORTISO	NON	ECZEMA
1	4	1		1	0	7	0
1	78	1		0	0	0	0
1	104	0		0	0	0	0
2	4	0		0	0	0	0
2	78	0		0	0	0	0
2	104	0		0	0	0	0

Fig. 10 A Sample of the Ezcema Dataset

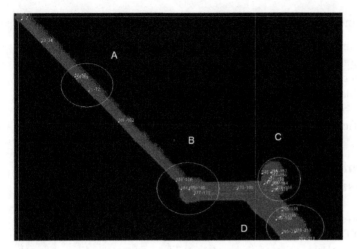

Fig. 11 Eczema Dataset: GSOM with Spread Factor 0.85

```
AGE, (4.0:1.649), (8.0:46.19), (16.0:1.142),(20.0:0.761), (24.0:0.634)
SKIN_RASH_FO, (0.0:94.16), (1.0:5.837)
RASH, (0.0:100.0)
RASH1, (0.0:99.74), (1.0:0.253)
RASH2, (0.0:93.52), (1.0:6.472)
RASH3, (0.0:99.87), (1.0:0.126)
RASH4, (0.0:99.61), (1.0:0.380)
RASH5, (0.0:99.74), (1.0:0.253)
RASH6, (0.0:100.0)
CORTISONE, (4.0:0.253), (10.0:0.761), (0.0:98.09)
NON, (4.0:0.126), (26.0:0.126), (10.0:0.126), (0.0:93.52), (14.0:0.126)
ECZEMA, (0.0:100.0)
```

Fig. 12 Eczema Dataset: Dimensional Density of Cluster A

```
AGE, (24.0:15.87), (36.0:12.69), (12.0:11.11), (60.0:20.63),
     (104.0:11.11), (4.0:9.523), (48.0:19.04)
SKIN_RASH_FO, (0.0:3.174), (1.0:96.82)
RASH, (0.0:38.09), (1.0:61.90)
RASH1, (0.0:26.98), (1.0:73.01)
RASH2, (1.0:100.0)
RASH3, (0.0:9.523), (1.0:90.47)
RASH4, (0.0:9.523), (1.0:90.47)
RASH5, (0.0:15.87), (1.0:84.12)
RASH6, (0.0:85.71), (1.0:14.28)
CORTISONE, (5.0:1.587), (7.0:1.587), (10.0:4.761), (0.0:52.38),
(6.0:1.587), (3.0:1.587), (28.0:23.80), (15.0:4.761), (20.0:1.587),
NON, (7.0:4.761), (10.0:14.28), (12.0:1.587), (0.0:39.68), (14.0:7.936),
(28.0:25.39), (15.0:1.587), (20.0:3.174)
ECZEMA, (0.0:100.0)
```

Fig. 13 Eczema Dataset: Dimensional Density of Cluster B

Dimensional density of Cluster D (Fig. 14) has a different disposition to both clusters A and B. AGE attribute is of normal distribution. SKIN-RASHFO, RASH2 and RASH3 attributes recorded 100% occurrence while RASH, RASH1, RASH4-6 recorded 100% absence. Another observation is

```
AGE, (36.0:16.66), (60.0:16.66), (94.0:50.0), (48.0:16.66)
SKIN_RASH_FO, (1.0:100.0)
RASH, (0.0:100.0)
RASH1, (0.0:100.0)
RASH2, (1.0:100.0)
RASH3, (1.0:100.0)
RASH4, (0.0:100.0)
RASH5, (0.0:100.0)
RASH6, (0.0:100.0)
CORTISONE, (1.0:100.0)
NON, (14.0:33.33), (28.0:33.33), (21.0:16.66), (20.0:16.66)
ECZEMA, (1.0:100.0)
```

Fig. 14 Eczema Dataset: Dimensional Density of Cluster D

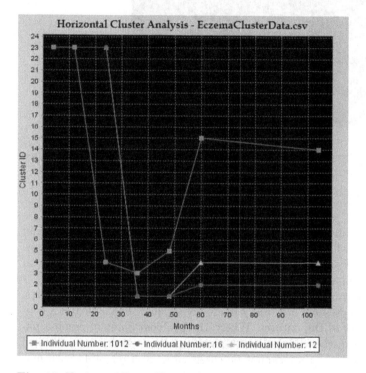

Fig. 15 Horizontal Inter-Cluster Analysis SF=0.85

the 100% occurrence of CORTISON and the ECZEMA attribute, both B and C had a total absence of these two attributes.

Investigations conducted on horizontal inter-cluster analysis produced the following results. Clusters from the previous map (Fig. 11) were used for this experiment as well.

Fig. 15 illustrates temporal movement across clusters of data vectors (or individual patients) 1012, 16 and 12. The clusters were sorted in ascending order of the summation of the attributes SKINRASHFO to RASH6. Individuals 16 and 12 show a very similar movement while 1012 has a similar

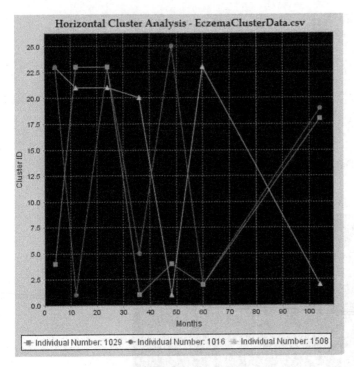

Fig. 16 Rapid Inter-Cluster Movements

pattern but at a different magnitude. In the original data vectors 1012 and 12 have high values for the CORTISON attribute while 16 comprises of zeros, all other attributes are similar across the vectors. This type of indication can be important for analytical diagnosis.

Fig. 16 depicts rapid movements of three individuals across the clusters. Such movement represents instability in clustering, and thus indicates a contrast with the rest of the dataset. Where the problem domain, eczema patients, is concerned, these irregularities signify patients with an exceptional record of the disease as they cannot be categorised even remotely with majority of the dataset.

For vertical inter-cluster analysis, a map from a different topology was necessary. Since the first map was at a wider spread with SF=0.85, the SF for the second was selected to be 0.45 and all other GSOM parameters the same. Spread was limited with three distinct cluster groups. The same cluster sorting criteria was used for this analysis. In an ideal case data vectors should exhibit similar behavior in cluster movement for any spread factor. That is a similar graph movement irrespective the magnitude of cluster sorting. Irregular movements resemble a peculiar trait that should be investigated. For example in the eczema dataset, individual 16 (Fig. 17) shows regularity in movement while individual 1258 displays rapid fluctuations.

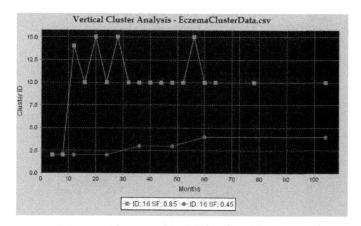

Fig. 17 Regular Movement: Individual 16

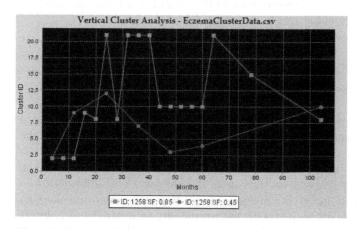

Fig. 18 Irregular Movement: Individual 1258

A rapid fluctuation from months 10-30 is noticeable for individual 16 at SF 0.85(Fig. 18). The reason for absence of similar behaviour in graph SF = 0.45 is the limited spread and hence the smaller number of clusters in the latter. This by itself is a pattern worth investigating as it demarcates the stronger clusters from the weaker ones for the respective data vector.

This section focused on extending the GSOM on to a completely new domain, that of cluster behaviour, where the possibility of knowledge acquisition from movement among clusters generated by the GSOM was explored. These extensions are resourceful when handling high dimensional, large and unstructured datasets. Inter-, Intra-cluster dimensional density, horizontal and vertical inter-cluster behaviour were the extensions proposed. Tests were conducted on a medical survey of eczema in infants from 4 to 104 months. The time variance of the data items was important for the experiments as

Fig. 19 Methodology for
Multi Modal Knowledge
Representation

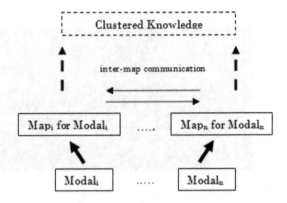

cluster movements are best examined over time. The outcome was encouraging with the exposure of significant patterns that could be used as preliminary information for the interpretation of the spread of eczema and effective medications at different stages in an infants life.

5 Effect of Multimodal Representation in Unsupervised Learning

This section focuses on the effect of expanded bases and increased variety of information on unsupervised learning with the objective of highlighting the importance of multimodal information for development of unconstrained artificial intelligence [13]. A simple example, humans perceive moments in time in terms of five (or six) senses, thus there is a magnitude of information to process, understand and create memories with. Remembering or recollecting such memories is made simple with the availability of a variety of pointers from the five different senses to the same event in time. However, with learning algorithms we have restricted input to a highly specific subset of the available and relevant domains of information.

As emphasised in the first sections of this chapter, the GSOM is successful in satisfying both ends of the cognitive spectrum and therefore was used for this investigation as well. With unsupervised learning, the GSOM attains autonomous and adaptive learning capability while hierarchical effects of the SF provide the psychological bearing for deeper levels of understanding. The methodology expands on the workings of the GSOM, improvisations have been made to facilitate different modals of information belonging to or describing the same entity. Each modal is allowed exclusive self organisation using the GSOM algorithm. The previous section discussed in great detail cognitive features of the GSOM that underline its position as a self organising algorithm. Fig. 19 further illustrates the methodology.

Although each modal self organises exclusively, correlations are maintained between the modals with inter-map communication for node growth and adjustments. Each modal i to n generates a separate feature map, each of these maps clusters knowledge specific to the respective modal. Inter-map communication on new node growth is required to maintain correspondence among the modals. Correlations are sustained amongst the shapes of the maps as well as the hit nodes and clusters created. Weight initialisation of these new nodes was retained as a local operation, as remote inputs would distort the consistency of local weight distribution.

The GSOM algorithm maintains consistency over local weight distribution or the magnification error (Kohonen, 1995) at non-boundary nodes with an error distribution function. The following two formulae are used for this distribution.

$$E_{t+1}^w = GT/2 \tag{13}$$

where is the error value of the winner node and GT, the growth threshold. The following formula increases the error value of immediate neighbours.

$$E_{t+1}^{n_i} = E_t^{n_i} + \gamma E_t^{n_i} \tag{14}$$

where $E_t^{n_i}$ (i can be 1-4) is the error value of the ith neighbour of the winner and γ a constant value - the factor of distribution, a value between 0 and 1. The effect of the two formulae is to spread the error outwards from the high error node and provide the non-boundary nodes an ability to initiate node growth.

The GSOM algorithm was altered to facilitate inter-map communication of remote node growth. The changes are highlighted below. In the growing phase of the algorithm, step (e) is expanded to notify other maps of new nodes grown. Communication occurs in a client-server context, with a server node managing exchange of information. If the case is weight distribution at a non-boundary node no communication occurs. Step (f) is executed as is, with remote node growth following local weight initialisation. Remote node growth does not involve remote weight adoption. A new step for communication with a central node to obtain information on node growth in remote maps is added to the algorithm soon after step (f).

The modified algorithm is as follows.

1. Initialization phase:

a) Create a new map or each of the modals m.

a) Initialize the weight vectors of the starting four nodes of each map with random numbers between 0 and 1.

b) Calculate the growth threshold (GT) for the given dimension D using,

$$GT = -D \times ln(SF) \tag{15}$$

where SF is spread factor.

2. Growing Phase

a) Present input to the network.

b) Determine the weight vector that is closest to the input vector mapped to the current feature map (winner), using Euclidean distance. This step can be summarized as: find q' such that $|v - w_{q'}| \leq |v - w_q|$ $\forall \epsilon \mathbb{N}$ where v, w are the input and weight vectors respectively, q is the position vector for nodes and is the set of natural numbers.

c) Apply weight vector adaptation to the neighbourhood of the winner and the winner itself. The neighbourhood is decided by the localized weight adaptation rule as used in the GSOM.

d) Increase the error value of the winner,

$$TE_i = \sum_{H_i} \sum_{j=1}^{D} (x_{ij} - w_j)^2 \tag{16}$$

e.1) When $TE_i \geq GT$ (where TE_i is the total error of node i and GT is the growth threshold). Grow nodes if i is a boundary node. Distribute weights to neighbours if i is a non-boundary node.

e.2) If new nodes grown in current map, m, inform server/controller of new node coordinates. These new nodes will be created in all m' maps.

f) Initialize the new node weight vectors to match the neighbouring node weights. Weight vectors of remote nodes are initialised by the local weight distribution rule.

g) Contact server/controller to obtain information on new node growth on other clients. Create these nodes on running map and update weights based on local weight distribution rule.

h) Initialize the learning rate (LR) to its starting value.

i) Repeat steps b) to g) until all inputs have been presented and node growth is reduced to a minimum level.

3. Smoothing phase.

a) Reduce learning rate and fix a small starting neighbourhood.

b) Find winner and adapt the weights of the winner and neighbours in the same way as in growing phase.

Fig. 20 Web categorisation dataset, sample modal on image information

ID	NO_JPG	NO_LINK	AVG_H	AVG_RES	AVG_W
edu_0	31.0	12.0	40.9	103.5	10.4
new_0	39.0	5.0	54.6	239.7	9.2
edu_1	11.0	11.0	97.6	271.9	13.2
mob_1	29.0	6.0	44.2	112.2	7.7
edu_2	10.0	8.0	79.8	113.1	13.6
ecom_4	4.0	5.0	73.3	222.4	20.6

The resulting clusters of knowledge from the self organisation process are unique to the respective modal. The modals have an underlying interconnection via data vectors. These interconnections can be used to examine and compare the composition of inter-modal clusters and thus substantiate the importance of multimodal knowledge representation for unsupervised learning.

A simple experiment was conducted on this extension using a dataset on the categorisation of websites. The basis of categorisation was different to any existing means, as we focused on the replication of modals much like those experienced by humans in the build up of cognition. Three such modals were identified, image information modal, statistics modal and the component information modal. The image information modal consisted of the following attributes, the number of GIF image types, number of JPG image types, average height of images in pixels, average width of images in pixels and average resolution in bits per inch. The statistics modal contained the count of images, count of hyperlinks, count of self-referral hyperlinks, count of words, count of repeating words. The component information model focused on the attributes, size of head tag (as a gauge on the level of client side scripting), number of form tags and two collectives on the complexity of form tags. Thus the dataset was composed of 14 attributes spanning across 3 modals. The data extraction application was allowed to randomly select 100 websites from a list and extract the mentioned attributes from the primary pages of each website. Fig. 20 shows a sample of the image information modal.

Running the modified GSOM with a spread factor of 0.85 for 2000 epochs generated interesting results. Websites containing news and current affairs were grouped together in both the image modal and the component modal.

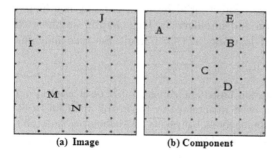

(a) Image   (b) Component

Fig. 21 Comparison of Maps from the Image and Component Modals

Output nodes A (Fig. 21-a) and I (Fig. 21-b) contained the data vectors representing news and current affairs sites. Distributed among nodes M and N from (Fig. 21-a), nodes C and D from (Fig. 21-b) were the data vectors representing content from educational institution websites. The separation was universities in one group and secondary schools in another. In another instance sites containing university information were found to be in the same clusters in all three modals of image, component and statistics. An inter-cluster comparison tool was used to calculate the percentage of similarity between the three modals. The results showed a success rate of 0.95 in matching clusters from the three modals.

6   Discussion and Conclusion

As outlined by Vernon et al [20], the emergent paradigm of artificial intelligence has the potential to realise a truly intelligent machine. Adaptive, anticipatory and autonomous behaviour that define an emergent architecture bears much similarity to what we perceive as human cognition and intelligence. It can be stated that unsupervised learning is a pragmatic approach to the development of an emergent architecture. The primary focus of this chapter was the growing self organising map algorithm (GSOM) [1], its benchmarks over the traditional SOM both from a structure adaptation and a hierarchical knowledge acquisition perspective. It was also noted how the hierarchical knowledge acquisition method of the GSOM closely follows the hierarchical theory of human thought, proposed by Perlovsky [15].

The next sections of the paper delineated extensions to the GSOM algorithm. The discussed extensions were able to expose correlations in the dataset in addition to the patterns identified by the initial self organising process. The extension of vertical/ horizontal inter-cluster movement over time, dimensional density of clusters were presented in the fourth section along with results from an experiment conducted on an epidemiological dataset. The effect of multimodality on unsupervised learning was investigated in the penultimate section. The original GSOM algorithm was modified to handle multiple modals of input, the results of which were outlined with tests on a novel web categorisation dataset. Substantial research is being conducted in the areas of artificial cognition and intelligence. This chapter emphasised the importance of unsupervised learning in this pursuit for true machine intelligence.

Acknowledgements. The population based epidemiological dataset used in our experiments was created and is currently maintained by David Hill and Cliff Hosking of the Centre for MEGA Epidemiology at University of Melbourne, Australia. We are grateful for their time, resources and support.

References

1. Alahakoon, D., Halgamuge, S., Srinivasan, B.: Dynamic self-organizing maps with controlled growth for knowledge discovery. IEEE Transactions On Neural Networks (2000)
2. Amarasiri, R., Alahakoon, D., Smith, K.: Applications of the growing self organizing map on high dimensional data. In: Proceedings of the Sixth International Conference on Intelligent Systems Design and Applications (2006)
3. Amarasiri, R., Wickremasinghe, L., Alahakoon, D.: Enhanced Cluster Visualization Using the Data Skeleton Model. In: Proceedings of the Third International Conference on Intelligent Systems Design and Application (2003)
4. Alahakoon, D.: Controlling the spread of dynamic self-organising maps. Neural Computing & Applications (2004)
5. De Silva, D., Alahakoon, D.: Analysis of Seismic Activity using the Growing SOM for the Identification of Time Dependent Patterns. In: Proceedings of the Second International Conference on Information and Automation for Sustainability (2006)
6. Hsu, A., Tang, S., Halgamuge, S.: An unsupervised hierarchical dynamic self-organizing approach to cancer class discovery and marker gene identification in microarray data. International Journal Of Bioinformatics (2003)
7. Barnes, J. (ed.): Aristotle (IV BC) Complete Works of Aristotle. Princeton University Press, Princeton (1992)
8. Field, D.J.: What is the goal of sensory encoding? Neural Computation 6 (1994)
9. Gilbert, C.D., Hirsch, J.A., Wiesel, T.N.: Lateral interactions in visual cortex. In: Cold Spring Harbour Symposia on Quantitative Biology. Cold Spring Harbour Press, LV (1990)
10. Hirsh, H., Spinelli, D.: Visual experience modifies distribution of horizontally and vertically oriented receptive fields in cats. Science 68, 869–871 (1970)
11. Kant, I.: Critique of Practical Reason, translated by J.H Bernard (1986) Hafner (1788)
12. Kohonen, T.: Self-Organising Maps. Springer, New York (1995)
13. Minsky, M.: Emotion Machine. Simon and Schuster, New York (2006)
14. Perlovsky, L.: Cognitive high level information fusion. Information Sciences (2007)
15. Perlovsky, L.: Towards physics of mind: Concepts, emotions, consciousness and symbols. Physics of Life (2006)
16. Rolls, E.T., Treves, A.: Neural Networks and Brain Function. Oxford University Press, Oxford (1999)
17. Rolls, E.T.: A model of the operation of the hippocampus and entorhinal cortex in memory. International Journal of Neural Systems (1995)
18. Tulving, E.: Organisation of memory: Quo Vardis. In: Memory: Systems, Process or Function? Oxford University Press, Oxford (1999)
19. Tulving, E.: Elements of Episodic Memory. Clarendon, Oxford (1983)
20. Vernon, D., Metta, G., Sandini, G.: A Survey of Artificial Cognitive Systems. IEEE Transactions on Evolutionary Computation, Special Issue on Autonomous Mental Development (2005)

A New Implementation for Neural Networks in Fourier-Space

Hazem M. El-Bakry and Mohamed Hamada

Abstract. Neural networks have shown good results for detecting a certain pattern in a given image. In this paper, fast neural networks for pattern detection are presented. Such processors are designed based on cross correlation in the frequency domain between the input image and the input weights of neural networks. This approach is developed to reduce the computation steps required by these fast neural networks for the searching process. The principle of divide and conquer strategy is applied through image decomposition. Each image is divided into small in size sub-images and then each one is tested separately by using a single fast neural processor. Furthermore, faster pattern detection is obtained by using parallel processing techniques to test the resulting sub-images at the same time using the same number of fast neural networks. In contrast to fast neural networks, the speed up ratio is increased with the size of the input image when using fast neural networks and image decomposition.

Keywords: Fast Pattern Detection, Neural Networks, Cross Correlation, Parallel Processin.

1 Introduction

Pattern detection is a fundamental step before pattern recognition. Its reliability and performance have a major   influence in a whole pattern recognition system. Nowadays, neural networks have shown very good results for detecting a certain pattern in a given image [2,4,6,8,9,10,12]. Among other techniques [3,5,7], neural networks are efficient pattern detectors [2,4,6,9]. But the problem with neural networks is that the computational complexity is very high because the networks have to process many small local windows in the images [5,7]. The main objective of this paper is to reduce the detection time using neural networks. Our idea is to

Hazem M. El-Bakry
Faculty of Computer Science & Information Systems,
Mansoura University, Egypt
e-mail: helbakry5@yahoo.com

Mohamed Hamada
University of Aizu, Aizu Wakamatsu, Japan
e-mail: Hamada@u-aizu.ac.jp

A.-E. Hassanien et al. (Eds.): Foundations of Comput. Intel. Vol. 1, SCI 201, pp. 307–330.
springerlink.com   © Springer-Verlag Berlin Heidelberg 2009

fast the operation of neural networks by performing the testing process in the frequency domain instead of spatial domain. Then, cross correlation between the input image and the weights of neural networks is performed in the frequency domain. This model is called fast neural networks. Compared to conventional neural networks, fast neural networks show a significant reduction in the number of computation steps required to detect a certain pattern in a given image under test. Furthermore, another idea to increase the speed of these fast neural networks through image decomposition is presented.

2 Fast Pattern Detection Using MLP and FFT

Here, we are interested only in increasing the speed of neural networks during the test phase. By the words "Fast Neural Networks" we mean reducing the number of computation steps required by neural networks in the detection phase. First neural networks are trained to classify face from non face examples and this is done in the spatial domain. In the test phase, each sub-image in the input image (under test) is tested for the presence or absence of the required face/object. At each pixel position in the input image each sub-image is multiplied by a window of weights, which has the same size as the sub-image. This multiplication is done in the spatial domain. The outputs of neurons in the hidden layer are multiplied by the weights of the output layer. When the final output is high this means that the sub-image under test contains the required face/object and vice versa. Thus, we may conclude that this searching problem is cross correlation in the spatial domain between the image under test and the input weights of neural networks.

In this section, a fast algorithm for face/object detection based on two dimensional cross correlations that take place between the tested image and the sliding window (20x20 pixels) is described. Such window is represented by the neural network weights situated between the input unit and the hidden layer. The cross correlation theorem in mathematical analysis says that a cross correlation between f and h is identical to the result of the following steps: let F and H be the results of the Fourier transformation of f and h in the frequency domain. Compute the conjugate of H (H^*). Multiply F and H^* in the frequency domain point by point and then transform this product into spatial domain via the inverse Fourier transform [1]. As a result, these cross correlations can be represented by a product in the frequency domain. Thus, by using cross correlation in the frequency domain a speed up in an order of magnitude can be achieved during the detection process [6,8,9,10,11,12,13,14,15,16].

In the detection phase, a subimage X of size mxn (sliding window) is extracted from the tested image, which has a size PxT, and fed to the neural network as shown in figure 1. Let W_i be the vector of weights between the input subimage and the hidden layer. This vector has a size of mxz and can be represented as mxn matrix. The output of hidden neurons $h(i)$ can be calculated as follows:

$$h_i = g\left(\sum_{j=1}^{m} \sum_{k=1}^{z} W_i(j,k)X(j,k) + b_i \right) \qquad (1)$$

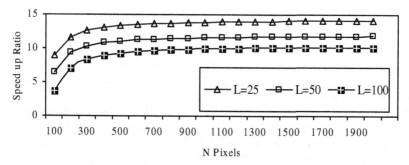

Fig. 1 The speed up ratio for images decomposed into different in size sub-images (L)

where g is the activation function and $b(i)$ is the bias of each hidden neuron (i). Eq.1 represents the output of each hidden neuron for a particular subimage I. It can be computed for the whole image Ψ as follows:

$$h_i(u,v)=g\left(\sum_{j=-m/2}^{m/2}\sum_{k=-z/2}^{z/2}W_i(j,k)\Psi(u+j,v+k)+b_i\right) \quad (2)$$

Eq. (2) represents a cross correlation operation. Given any two functions f and g, their cross correlation can be obtained by [1]:

$$g(x,y)\otimes f(x,y)=\left(\sum_{m=-\infty}^{\infty}\sum_{z=-\infty}^{\infty}f(x+m,y+z)g(m,z)\right) \quad (3)$$

Therefore, Eq. (2) can be written as follows [10-14]:

$$h_i = g\left(W_i \otimes \Psi + b_i\right) \quad (4)$$

where h_i is the output of the hidden neuron (i) and $h_i(u,v)$ is the activity of the hidden unit (i) when the sliding window is located at position (u,v) in the input image Ψ and $(u,v)\in[P-m+1,T-n+1]$.

Now, the above cross correlation can be expressed in terms of the Fourier Transform [1]:

$$W_i \otimes \Psi = F^{-1}\left(F(\psi)\bullet F^*\left(W_i\right)\right) \quad (5)$$

(*) means the conjugate of the *FFT* for the weight matrix. Hence, by evaluating this cross correlation, a speed up ratio can be obtained comparable to conventional neural networks. Also, the final output of the neural network can be evaluated as follows:

$$O(u,v)= g\left(\sum_{i=1}^{q}W_o(i)h_i(u,v)+b_o\right) \quad (6)$$

where q is the number of neurons in the hidden layer. $O(u,v)$ is the output of the neural network when the sliding window located at the position (u,v) in the input image Ψ. W_o is the weight matrix between hidden and output layer.

The complexity of cross correlation in the frequency domain can be analyzed as follows:

1. For a tested image of *NxN* pixels, the *2D-FFT* requires a number equal to $N^2 log_2 N^2$ of complex computation steps. Also, the same number of complex computation steps is required for computing the *2D-FFT* of the weight matrix for each neuron in the hidden layer.

2. At each neuron in the hidden layer, the inverse *2D-FFT* is computed. So, *q* backward and *(1+q)* forward transforms have to be computed. Therefore, for an image under test, the total number of the *2D-FFT* to compute is $(2q+1)N^2 log_2 N^2$.

3. The input image and the weights should be multiplied in the frequency domain. Therefore, a number of complex computation steps equal to qN^2 should be added.

4. The number of computation steps required by fast neural networks is complex and must be converted into a real version. It is known that the two dimensions Fast Fourier Transform requires $(N^2/2)log_2 N^2$ complex multiplications and $N^2 log_2 N^2$ complex additions [20,21]. Every complex multiplication is realized by six real floating point operations and every complex addition is implemented by two real floating point operations. So, the total number of computation steps required to obtain the *2D-FFT* of an *NxN* image is:

$$\rho = 6((N^2/2)log_2 N^2) + 2(N^2 log_2 N^2) \tag{7}$$

which may be simplified to:

$$\rho = 5N^2 log_2 N^2 \tag{8}$$

Performing complex dot product in the frequency domain also requires $6qN^2$ real operations.

5. In order to perform cross correlation in the frequency domain, the weight matrix must have the same size as the input image. Assume that the input object/face has a size of (nxn) dimensions. So, the search process will be done over subimages of (nxn) dimensions and the weight matrix will have the same size. Therefore, a number of zeros = (N^2-n^2) must be added to the weight matrix. This requires a total real number of computation steps = $q(N^2-n^2)$ for all neurons. Moreover, after computing the *2D-FFT* for the weight matrix, the conjugate of this matrix must be obtained. So, a real number of computation steps $=qN^2$ should be added in order to obtain the conjugate of the weight matrix for all neurons. Also, a number of real computation steps equal to *N* is required to create butterflies complex numbers $(e^{-jk(2\Pi n/N)})$, where $0<K<L$. These *(N/2)* complex numbers are multiplied by the elements of the input image or by previous complex numbers during the computation of the *2D-FFT*. To create a complex number requires two real floating point operations. So, the total number of computation steps required for fast neural networks becomes:

$$\sigma = (2q+1)(5N^2 log_2 N^2) + 6qN^2 + q(N^2-n^2) + qN^2 + N \tag{9}$$

which can be reformulated as:

$$\sigma = (2q+1)(5N^2 log_2 N^2) + q(8N^2-n^2) + N \tag{10}$$

6. Using a sliding window of size nxn for the same image of NxN pixels, $q(2n^2-1)(N-n+1)^2$ computation steps are required when using traditional neural networks for face/object detection process. The theoretical speed up factor η can be evaluated as follows:

$$\eta = \frac{q(2n^2-1)(N-n+1)^2}{(2q+1)(5N^2 \log_2 N^2)+q(8N^2-n^2)+N} \tag{11}$$

The theoretical speed up ratio (Eq. 11) with different sizes of the input image and different in size weight matrices is listed in Table 1. Practical speed up ratio for manipulating images of different sizes and different in size weight matrices is listed in Table 2 using 700 MHz processor and *MATLAB ver 5.3*. An interesting property with fast neural networks is that the number of computation steps does not depend on eith the size of the input subimage or the size of the weighth matrix (n). The effect of (n) on the number of computation steps is very small and can be ignored. This is incontrast to conventional networks networks in which the number of computation steps is increased with the size of both the input subimage and the weight matrix (n).

Table 1 The theoretical speed up ratio for images with different sizes.

Image size	Speed up ratio (n=20)	Speed up ratio (n=25)	Speed up ratio (n=30)
100x100	3.67	5.04	6.34
200x200	4.01	5.92	8.05
300x300	4.00	6.03	8.37
400x400	3.95	6.01	8.42
500x500	3.89	5.95	8.39
600x600	3.83	5.88	8.33
700x700	3.78	5.82	8.26
800x800	3.73	5.76	8.19
900x900	3.69	5.70	8.12
1000x1000	3.65	5.65	8.05
1100x1100	3.62	5.60	7.99
1200x1200	3.58	5.55	7.93
1300x1300	3.55	5.51	7.93
1400x1400	3.53	5.47	7.82
1500x1500	3.50	5.43	7.77
1600x1600	3.48	5.43	7.72
1700x1700	3.45	5.37	7.68
1800x1800	3.43	5.34	7.64
1900x1900	3.41	5.31	7.60
2000x2000	3.40	5.28	7.56

Table 2 Practical speed up ratio for images with different sizes Using MATLAB version 5.3

Image size	Speed up ratio (n=20)	Speed up ratio (n=25)	Speed up ratio (n=30)
100x100	7.88	10.75	14.69
200x200	6.21	9.19	13.17
300x300	5.54	8.43	12.21
400x400	4.78	7.45	11.41
500x500	4.68	7.13	10.79
600x600	4.46	6.97	10.28
700x700	4.34	6.83	9.81
800x800	4.27	6.68	9.60
900x900	4.31	6.79	9.72
1000x1000	4.19	6.59	9.46
1100x1100	4.24	6.66	9.62
1200x1200	4.20	6.62	9.57
1300x1300	4.17	6.57	9.53
1400x1400	4.13	6.53	9.49
1500x1500	4.10	6.49	9.45
1600x1600	4.07	6.45	9.41
1700x1700	4.03	6.41	9.37
1800x1800	4.00	6.38	9.32
1900x1900	3.97	6.35	9.28
2000x2000	3.94	6.31	9.25

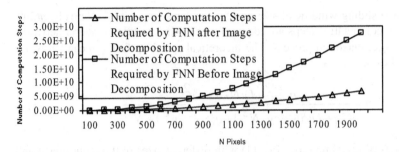

Fig. 2 A comparison between the number of computation steps required by FNN before and after Image decomposition.

The authors in [17-19] have proposed a multilayer perceptron (MLP) algorithm for fast face/object detection. The same authors claimed incorrect equation for cross correlation between the input image and the weights of the neural networks. They introduced formulas for the number of computation steps needed by conventional and fast neural networks. Then, they established an equation for the speed up ratio. Unfortunately, these formulas contain many errors which lead to invalid speed up ratio. Other authors developed their work based on these incorrect equations [48]. So, the fact that these equations are not valid must be cleared to all researchers. It is not only very important but also urgent to notify other researchers not to do research based on wrong equations. Some of these mistakes were corrected in [22-47]. In this paper, complete corrections are given.

The authors in [17-19] analyzed their proposed fast neural network as follows: For a tested image of NxN pixels, the *2D-FFT* requires $O(N^2(log_2N)^2)$ computation steps. For the weight matrix W_i, the *2D-FFT* can be computed off line since these are constant parameters of the network independent of the tested image. The *2D-FFT* of the tested image must be computed. As a result, q backward and one forward transforms have to be computed. Therefore, for a tested image, the total number of the *2D-FFT* to compute is $(q+1)N^2(log_2N)^2$ [17,19]. In addition, the input image and the weights should be multiplied in the frequency domain. Therefore, computation steps of (qN^2) should be added. This yields a total of $O((q+1)N^2(log_2N)^2+qN^2)$ computation steps for the fast neural network [17,18].

Using sliding window of size nxn, for the same image of NxN pixels, qN^2n^2 computation steps are required when using traditional neural networks for the face detection process. They evaluated theoretical speed up factor η as follows [17]:

$$\eta = \frac{qn^2}{(q+1)log^2N} \qquad (12)$$

The speed up factor introduced in [17] and given by Eq.14 is not correct for the following reasons:

a) The number of computation steps required for the *2D-FFT* is $O(N^2log_2N^2)$ and not $O(N^2log^2N)$ as presented in [17,18]. Also, this is not a typing error as the curve in Fig.2 in [17] realizes Eq.7, and the curves in Fig.15 in [18] realizes Eq.31 and Eq.32 in [18].

b) Also, the speed up ratio presented in [17] not only contains an error but also is not precise. This is because for fast neural networks, the term $(6qN^2)$ corresponds to complex dot product in the frequency domain must be added. Such term has a great effect on the speed up ratio. Adding only qN^2 as stated in [18] is not correct since a one complex multiplication requires six real computation steps.

c) For conventional neural networks, the number of operations is $(q(2n^2-1)(N-n+1)^2)$ and not (qN^2n^2). The term n^2 is required for multiplication of n^2 elements (in the input window) by n^2 weights which results in another new n^2 elements. Adding these n^2 elements, requires another (n^2-1) steps. So, the total computation steps needed for each window is $(2n^2-1)$. The search operation for a face in the input image uses a window with nxn weights. This operation is done at each pixel in the input image. Therefore, such process is repeated $(N-n+1)^2$ times and not N^2 as stated in [17,19].

d) Before applying cross correlation, the 2D-FFT of the weight matrix must be computed. Because of the dot product, which is done in the frequency domain, the size of weight matrix should be increased to be the same as the size of the input image. Computing the 2D-FFT of the weight matrix off line as stated in [17-19] is not practical. In this case, all of the input images must have the same size. As a result, the input image will have only a one fixed size. This means that, the testing time for an image of size 50x50 pixels will be the same as that image of size 1000x1000 pixels and of course, this is unreliable.

e) It is not valid to compare number of complex computation steps by another of real computation steps directly. The number of computation steps given by pervious authors [17-19] for conventional neural networks is for real operations while that is required by fast neural networks is for complex operations. To obtain the speed up ratio, the authors in [17-19] have divided the two formulas directly without converting the number of computation steps required by fast neural networks into a real version.

f) Furthermore, there is critical error in the activity of hidden neurons given in section 3.1 in [19] and also by Eq.(2) in [17]. Such activity given by those authors in [17,19] as follows:

$$h_i = g\left(\Psi \otimes W_i + b_i\right) \tag{13}$$

is not correct and should be written as Eq.(4) given here in this paper. This is because the fact that the operation of cross correlation is not commutative $(W \otimes \Psi \neq \Psi \otimes W)$. As a result, Eq. 13 (Eq.4 in their paper [17]) does not give the same correct results as conventional neural networks. This error leads the researchers who consider the references [17-19] to think about how to modify the operation of cross correlation so that Eq. 13 (Eq.4 in their paper [17]) can give the same correct results as conventional neural networks. Therefore, errors in these equations must be cleared to all the researchers. In [23-30], the authors proved that a symmetry condition must be found in input matrices (images and the weights of neural networks) so that fast neural networks can give the same results as conventional neural networks. In case of symmetry $W \otimes \Psi = \Psi \otimes W$, the cross correlation becomes commutative and this is a valuable achievement. In this case, the cross correlation is performed without any constrains on the arrangement of matrices. A practical proof for this achievement is explained by examples shown in appendix "A". As presented in [24-30], this symmetry condition is useful for reducing the number of patterns that neural networks will learn. This is because the image is converted into symmetric shape by rotating it down and then the up image and its rotated down version are tested together as one (symmetric) image. If a pattern is detected in the rotated down image, then, this means that this pattern is found at the relative position in the up image. So, if conventional neural networks are trained for up and rotated down examples of the pattern, fast neural networks will be trained only to up examples. As the number of trained examples is reduced, the number of neurons in the hidden layer will be reduced and the neural network will be faster in the test phase compared with conventional neural networks.

g) Moreover, the authors in [17-19] stated that the activity of each neuron in the hidden layer defined by Eq. 13 (Eq.4 in their paper [17]) can be expressed in terms of convolution between a bank of filter (weights) and the input image. This is not correct because the activity of the hidden neuron is a cross correlation between the input image and the weight matrix. It is known that the result of cross correlation between any two functions is different from their convolution. As we proved in [24-30] the two results will be the same, only when the two matrices are symmetric or at least the weight matrix is symmetric. A practical example which proves that for any two matrices the result of their cross correlation is different from their convolution unless that they are symmetric or at least the second matrix is symmetric as shown in appendix "B".

h) Images are tested for the presence of a face (object) at different scales by building a pyramid of the input image which generates a set of images at different resolutions. The face detector is then applied at each resolution and this process takes much more time as the number of processing steps will be increased. In [17-19], the authors stated that the Fourier transforms of the new scales do not need to be computed. This is due to a property of the Fourier transform. If $z(x,y)$ is the original and $a(x,y)$ is the sub-sampled by a factor of 2 in each direction image then:

$$a(x, y) = z(2x, 2y) \tag{14}$$

$$Z(u, v) = FT(z(x, y)) \tag{15}$$

$$FT(a(x, y)) = A(u, v) = \frac{1}{4} Z\left(\frac{u}{2}, \frac{v}{2}\right) \tag{16}$$

This implies that we do not need to recompute the Fourier transform of the sub-sampled images, as it can be directly obtained from the original Fourier transform. But experimental results have shown that Eq.16 is valid only for images in the following form:

$$\Psi = \begin{bmatrix} A\ A\ B\ B\ C\ C.................... \\ A\ A\ B\ B\ C\ C.................... \\ . \\ . \\ . \\ . \\ S\ S\ X\ X\ Y\ Y.................... \\ S\ S\ X\ X\ Y\ Y.................... \end{bmatrix} \qquad (17)$$

In [17], the author claimed that the processing needs $O((q+2)N^2log^2N)$ additional number of computation steps. Thus the speed up ratio will be [17]:

$$\eta = \frac{qn^2}{(q+2)log^2N} \qquad (18)$$

Of course this is not correct, because the inverse of the Fourier transform is required to be computed at each neuron in the hidden layer (for the resulted matrix from the dot product between the Fourier matrix in two dimensions of the input image and the Fourier matrix in two dimensions of the weights, the inverse of the Fourier transform must be computed). So, the term $(q+2)$ in Eq. 18 should be $(2q+1)$ because the inverse $2D$-FFT in two dimensions must be done at each neuron in the hidden layer. In this case, the number of computation steps required to perform $2D$-FFT for the fast neural networks will be:

$$\varphi = (2q+1)(5N^2log_2N^2) + (2q)5(N/2)^2log_2(N/2)^2 \qquad (19)$$

In addition, a number of computation steps equal to $6q(N/2)^2 + q((N/2)^2 - n^2) + q(N/2)^2$ must be added to the number of computation steps required by fast neural networks.

3   A New Faster Algorithm for Pattern Detection Based on Image Decomposition

In this section, a new faster algorithm for face/object detection is presented. The number of computation steps required for fast neural networks with different image sizes is listed in Tables 3 and 4. From these tables, we may notice that as the image size is increased, the number of computation steps required by fast neural networks is much increased. For example, the number of computation steps required for an image of size (50x50 pixels) is much less than that needed for an image of size (100x100 pixels). Also, the number of computation steps required for an image of size (500x500 pixels) is much less than that needed for an image of size (1000x1000 pixels). As a result, for example, if an image of size (100x100 pixels) is decomposed into 4 sub-images of size (50x50 pixels) and each sub-image is tested separately as shown in figure 2, then a speed up factor for face/object detection can be achieved. The number of computation steps required by fast neural networks to test an image after decomposition can be calculated as follows:

Image size	No. of computation steps in case of using FNN
25x25	1.9085e+006
50x50	9.1949e+006
100x100	4.2916e+007
150x150	1.0460e+008
200x200	1.9610e+008
250x250	3.1868e+008
300x300	4.7335e+008
350x350	6.6091e+008
400x400	8.8203e+008
450x450	1.1373e+009
500x500	1.4273e+009
550x550	1.7524e+009
600x600	2.1130e+009
650x650	2.5096e+009
700x700	2.9426e+009
750x750	3.4121e+009
800x800	3.9186e+009
850x850	4.4622e+009
900x900	5.0434e+009
950x950	5.6623e+009
1000x1000	6.3191e+009

Table 3 The number of computation steps required by fast neural networks (FNN) for images of sizes (25x25 - 1000x1000 pixels), q=30, n=20

Image size	No. of computation steps in case of using FNN
1050x1050	7.0142e+009
1100x1100	7.7476e+009
1150x1150	8.5197e+009
1200x1200	9.3306e+009
1250x1250	1.0180e+010
1300x1300	1.1070e+010
1350x1350	1.1998e+010
1400x1400	1.2966e+010
1450x1450	1.3973e+010
1500x1500	1.5021e+010
1550x1550	1.6108e+010
1600x1600	1.7236e+010
1650x1650	1.8404e+010
1700x1700	1.9612e+010
1750x1750	2.0861e+010
1800x1800	2.2150e+010
1850x1850	2.3480e+010
1900x1900	2.4851e+010
1950x1950	2.6263e+010
2000x2000	2.7716e+010
2050x2050	2.9211e+010

Table 4 The number of computation steps required by FNN for images of sizes (1050x1050 - 2000x2000 pixels), q=30, n=20

1. Assume that the size of the image under test is (NxN pixels).

2. Such image is decomposed into α (*LxL* pixels) sub-images. So, α can be computed as:

$$\alpha=(N/L)^2 \tag{20}$$

3. Assume that, the number of computation steps required for testing one (*LxL* pixels) sub-image is β. So, the total number of computation steps (T) required for testing these sub-images resulting after the decomposition process is:

$$T = \alpha\beta \tag{21}$$

The speed up ratio in this case (η_d) can be computed as follows:

$$\eta_d = \frac{q(2n^2 - 1)(N - n + 1)^2}{(q(\alpha + 1) + \alpha)(5N_s^2 \log_2 N_s^2) + \alpha q(8N_s^2 - n^2) + N_s^2 + \Delta} \tag{22}$$

Where,

Ns: is the size of each small sub-image.

Δ: is a small number of computation steps required to obtain the results at the boundaries between subimages and depends on the size of the subimage.

Table 5 The speed up ratio in case of using FNN and FNN after image decomposition into sub-images (25x25 pixels) for images of different sizes (from N=50 to N=1000, n=25, q=30)		
Image size	Speed up ratio in case of using FNN	Speed up ratio in case of using FNN after image decomposition
50x50	2.7568	5.0713
100x100	5.0439	12.4622
150x150	5.6873	15.6601
200x200	5.9190	17.3611
250x250	6.0055	18.4073
300x300	6.0301	19.1136
350x350	6.0254	19.6218
400x400	6.0059	20.0047
450x450	5.9790	20.3034
500x500	5.9483	20.5430
550x550	5.9160	20.7394
600x600	5.8833	20.9032
650x650	5.8509	21.0419
700x700	5.8191	21.1610
750x750	5.7881	21.2642
800x800	5.7581	21.3546
850x850	5.7292	21.4344
900x900	5.7013	21.5054
950x950	5.6744	21.5689
1000x1000	5.6484	21.6260

Table 6 The speed up ratio in case of using FNN and FNN after image decomposition into sub-images (25x25 pixels) for images of different sizes (from N=1050 to N=2000, n=25, q=30)		
Image size	Speed up ratio in case of using FNN	Speed up ratio in case of using FNN after image decomposition
1050x1050	5.6234	21.6778
1100x1100	5.5994	21.7248
1150x1150	5.5762	21.7678
1200x1200	5.5538	21.8072
1250x1250	5.5322	21.8434
1300x1300	5.5113	21.8769
1350x1350	5.4912	21.9079
1400x1400	5.4717	21.9366
1450x1450	5.4528	21.9634
1500x1500	5.4345	21.9884
1550x1550	5.4168	22.0118
1600x1600	5.3996	22.0338
1650x1650	5.3830	22.0544
1700x1700	5.3668	22.0738
1750x1750	5.3511	22.0921
1800x1800	5.3358	22.1094
1850x1850	5.3209	22.1257
1900x1900	5.3064	22.1412
1950x1950	5.2923	22.1559
2000x2000	5.2786	22.1699

The results of the detection before image decomposition (presented in the previous section) and after image decomposition are the same. A practical example which proves that the results of cross correlation before and after the decomposition are the same is listed in Appendix "C". To detect a face/object of size 20x20 pixels in an image of any size by using fast neural networks after image decomposition into sub-images, the optimal size of these sub-images must be computed. From Table 3, we may conclude that, the most suitable size for the sub-image which requires the smallest number of computation steps is 25x25 pixels. Also, the fastest speed up ratio can be achieved using this sub-image size (25x25) as shown in Figure 1. It is clear that the speed up ratio is reduced when the size of the sub-image (L) is increased. A comparison between the speed up ratio for fast neural networks and fast neural networks after image decomposition with different sizes of the tested images is listed in Tables 5 and 6. It is clear that the speed up ratio is increased with the size of the input image when using fast neural networks and image decomposition. This is in contrast to using only fast neural networks.

Table 7 The speed up ratio in case of using FNN and FNN after matrix decomposition into sub-matrices (25x25 elements) for very large matrices (from N=100000 to N=2000000, n=25, q=30)			Table 8 The practical speed up ratio in case of using FNN and FNN after image decomposition into sub-images (25x25 pixels) for images of different sizes (from N=100 to N=2000, n=25, q=30)		
Matrix size	Speed up ratio in case of using FNN	Speed up ratio in case of using FNN after matrix decomposition	Image size	Speed up ratio in case of using FNN	Speed up ratio in case of using FNN after image decomposition
			100x100	10.75	34.55
100000x100000	3.6109	22.7038	200x200	9.19	35.65
200000x200000	3.4112	22.7092	300x300	8.43	36.73
300000x300000	3.3041	22.7110	400x400	7.45	37.70
400000x400000	3.2320	22.7119	500x500	7.13	38.66
500000x500000	3.1783	22.7125	600x600	6.97	39.61
600000x600000	3.1357	22.7128	700x700	6.83	40.56
700000x700000	3.1005	22.7131	800x800	6.68	41.47
800000x800000	3.0707	22.7133	900x900	6.79	42.39
900000x900000	3.0448	22.7134	1000x1000	6.59	43.28
1000000x1000000	3.0221	22.7136	1100x1100	6.66	44.14
1100000x1100000	3.0018	22.7137	1200x1200	6.62	44.95
1200000x1200000	2.9835	22.7138	1300x1300	6.57	45.71
1300000x1300000	2.9668	22.7138	1400x1400	6.53	46.44
1400000x1400000	2.9516	22.7139	1500x1500	6.49	47.13
1500000x1500000	2.9376	22.7139	1600x1600	6.45	47.70
1600000x1600000	2.9245	22.7140	1700x1700	6.41	48.19
1700000x1700000	2.9124	22.7140	1800x1800	6.38	48.68
1800000x1800000	2.9011	22.7141	1900x1900	6.35	49.09
1900000x1900000	2.8904	22.7141	2000x2000	6.31	49.45
2000000x2000000	2.8804	22.7141			

As shown in Figure 2, the number of computation steps required by fast neural networks is increased rapidly with the size of the input image. Therefore the speed up ratio is decreased with the size of the input image. While in case of using fast neural networks and image decomposition, the number of computation steps required by fast neural networks is increased smoothly. Thus, the linearity of the computation steps required by fast neural networks in this case is better. As a result, the speed up ratio is increased. Increasing the speed up ratio with the size of the input image is considered an important achievement. Furthermore, for very large size matrices, while the speed up ratio for fast neural networks is decreased, the speed up ratio still increase in case of using fast neural networks and matrix decomposition as listed in Table 7. Moreover, as shown in Figure 3, the speed up ratio in case of fast neural networks and image decomposition is increased with the size of the weight matrix which has the same size (n) as the input window. For example, it is clear that the speed up ratio is for window size of 30x30 is larger than that of size 20x20. Simulation results for the speed up ratio in case of using

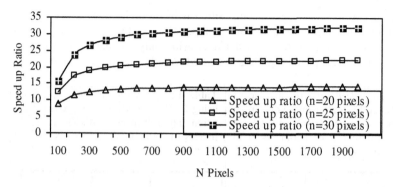

Fig. 3 The speed up ratio in case of image decomposition and different window size (n), (L=25x25)

| Table 9 The speed up ratio in case of using FNN and FNN after image decomposition into sub-images (5x5 pixels) for images of different sizes (from N=50 to N=1000, n=5, q=30) |
| Table 10 The speed up ratio in case of using FNN and FNN after image decomposition into sub-images (5x5 pixels) for images of different sizes (from N=50 to N=1000, n=5, q=30) |

Image size	Speed up ratio in case of using FNN	Speed up ratio in case of using FNN after image decomposition	Image size	Speed up ratio in case of using FNN	Speed up ratio in case of using FNN after image decomposition
50x50	0.3361	1.3282	1050x1050	0.2292	1.5689
100x100	0.3141	1.4543	1100x1100	0.2278	1.5695
150x150	0.2985	1.4965	1150x1150	0.2265	1.5700
200x200	0.2872	1.5177	1200x1200	0.2253	1.5704
250x250	0.2785	1.5303	1250x1250	0.2241	1.5709
300x300	0.2716	1.5388	1300x1300	0.2230	1.5713
350x350	0.2658	1.5448	1350x1350	0.2219	1.5716
400x400	0.2610	1.5493	1400x1400	0.2209	1.5720
450x450	0.2568	1.5529	1450x1450	0.2199	1.5723
500x500	0.2531	1.5557	1500x1500	0.2189	1.5726
550x550	0.2498	1.5580	1550x1550	0.2180	1.5728
600x600	0.2469	1.5599	1600x1600	0.2172	1.5731
650x650	0.2442	1.5615	1650x1650	0.2163	1.5733
700x700	0.2418	1.5629	1700x1700	0.2155	1.5735
750x750	0.2396	1.5641	1750x1750	0.2148	1.5738
800x800	0.2375	1.5652	1800x1800	0.2140	1.5740
850x850	0.2356	1.5661	1850x1850	0.2133	1.5742
900x900	0.2339	1.5669	1900x1900	0.2126	1.5743
950x950	0.2322	1.5677	1950x1950	0.2119	1.5745
1000x1000	0.2306	1.5683	2000x2000	0.2112	1.5747

fast neural networks and image decomposition is listed in Table 8. It is clear that simulation results confirm the theoretical computations and the practical speed up ratio after image decomposition is faster than using only fast neural networks. In addition, the practical speed up ratio is increased with the size of the input image.

Table 11 The speed up ratio in case of using FNN and FNN after image decomposition into sub-images (10x10 pixels) for images of different sizes (from N=50 to N=1000, n=10, q=30)

Image size	Speed up ratio in case of using FNN	Speed up ratio in case of using FNN after image decomposition
50x50	1.1202	3.1369
100x100	1.1503	3.9558
150x150	1.1303	4.2397
200x200	1.1063	4.3829
250x250	1.0842	4.4691
300x300	1.0647	4.5267
350x350	1.0474	4.5678
400x400	1.0321	4.5987
450x450	1.0185	4.6228
500x500	1.0063	4.6420
550x550	0.9952	4.6578
600x600	0.9851	4.6709
650x650	0.9758	4.6820
700x700	0.9672	4.6915
750x750	0.9593	4.6998
800x800	0.9519	4.7070
850x850	0.9451	4.7133
900x900	0.9386	4.7190
950x950	0.9325	4.7241
1000x1000	0.9268	4.7286

Table 12 The speed up ratio in case of using FNN and FNN after image decomposition into sub-images (10x10 pixels) for images of different sizes (from N=1050 to N=2000, n=10, q=30)

Image size	Speed up ratio in case of using FNN	Speed up ratio in case of using FNN after image decomposition
1050x1050	0.9214	4.7328
1100x1100	0.9163	4.7365
1150x1150	0.9114	4.7399
1200x1200	0.9068	4.7431
1250x1250	0.9023	4.7460
1300x1300	0.8981	4.7486
1350x1350	0.8941	4.7511
1400x1400	0.8902	4.7534
1450x1450	0.8865	4.7555
1500x1500	0.8829	4.7575
1550x1550	0.8795	4.7594
1600x1600	0.8762	4.7611
1650x1650	0.8730	4.7628
1700x1700	0.8699	4.7643
1750x1750	0.8669	4.7658
1800x1800	0.8640	4.7672
1850x1850	0.8613	4.7685
1900x1900	0.8586	4.7697
1950x1950	0.8559	4.7709
2000x2000	0.8534	4.7720

Also, to detect small in size matrices such as 5x5 or 10x10 using only fast neural networks, the speed ratio becomes less than one as shown in Tables 9,10,11, and 12. On the other hand, from the same tables it is clear that using fast neural networks and image decomposition, the speed up ratio becomes higher than one and increased with the dimensions of the input image. The dimensions of the new subimage after image decomposition *(L)* must not be less than the dimensions of the face/object which is required to be detected and has the same size as the weight matrix. Therefore, the following equation controls the relation between the subimage and the size of weight matrix (face/object to be detected) in order not to loss any information in the input image.

$$L \geq n \qquad (23)$$

For example, in case of detecting 5x5 pattern, the image must be decomposed into subimages of size no more 5x5.

To further reduce the running time as well as increase the speed up ratio of the detection process, a parallel processing technique is used. Each sub-image is tested using a fast neural network simulated on a single processor or a separated node in a clustered system. The number of operations (ω) performed by each processor / node (sub-images tested by one processor/node) =

$$\omega = \frac{The\ total\ number\ of\ sub-images}{Number\ of\ Processors/nodes} \tag{24}$$

$$\omega = \frac{\alpha}{Pr} \tag{25}$$

Where, Pr is the number of processors or nodes.

The total number of computation steps (γ) required to test an image by using this approach can be calculated as:

$$\gamma = \omega\beta \tag{26}$$

By using this algorithm, the speed up ratio in this case (η_{dp}) can be computed as follows:

$$\eta_{dp} = \frac{q(2n^2 - 1)(N - n + 1)^2}{ceil(((q(\alpha + 1) + \alpha)(5N_s^2 \log_2 N_s^2) + \alpha q(8N_s^2 - n^2) + N_s)/pr)} \tag{27}$$

Where, $ceil(x)$ is a $MATLAB$ function rounds the elements of x to the nearest integers towards infinity.

As shown in Tables 13 and 14, using a symmetric multiprocessing system with 16 parallel processors or 16 nodes in either a massively parallel processing system or a clustered system, the speed up ratio (with respect to conventional neural networks) for face/object detection is increased. A further reduction in the computation steps can be obtained by dividing each sub-image into groups. For each group, the neural operation (multiplication by weights and summation) is performed for each group by using a single processor. This operation is done for all of these groups as well as other groups in all of the sub-images at the same time. The best case is achieved when each group consists of only one element. In this case, one operation is needed for multiplication of the one element by its weight and also a small number of operations (ε) is required to obtain the over all summation for each sub-image. If the sub-image has n^2 elements, then the required number of processors will be n^2. As a result, the number of computation steps will be $\alpha q(1+\varepsilon)$, where ε is a small number depending on the value of n. For example, when $n=20$, then $\varepsilon=6$ and if $n=25$, then $\varepsilon=7$. The speed up ratio can be calculated as:

$$\eta = (2n^2 - 1)(N - n + 1)^2/\alpha(1 + \varepsilon) \tag{28}$$

Moreover, if the number of processors = αn^2, then the number of computation steps will be $q(1+\varepsilon)$, and the speed up ratio becomes:

$$\eta = (2n^2 - 1)(N - n + 1)^2/(1 + \varepsilon) \tag{29}$$

Image size	Speed up ratio
50x50	81.1403
100x100	199.3946
150x150	250.5611
200x200	277.7780
250x250	294.5171
300x300	305.8174
350x350	313.9482
400x400	320.0748
450x450	324.8552
500x500	328.6882
550x550	331.8296
600x600	334.4509
650x650	336.6712
700x700	338.5758
750x750	340.2276
800x800	341.6738
850x850	342.9504
900x900	344.0856
950x950	345.1017
1000x1000	346.0164

Table 13 The speed up ratio in case of using FNN after image decomposition into sub-images (25x25 pixels) for images of different sizes (from N=50 to N=1000, n=25, q=30) using 16 parallel processors or 16 nodes

Image size	Speed up ratio
1050x1050	346.8442
1100x1100	347.5970
1150x1150	348.2844
1200x1200	348.9147
1250x1250	349.4946
1300x1300	350.0300
1350x1350	350.5258
1400x1400	350.9862
1450x1450	351.4150
1500x1500	351.8152
1550x1550	352.1896
1600x1600	352.5406
1650x1650	352.8704
1700x1700	353.1808
1750x1750	353.4735
1800x1800	353.7500
1850x1850	354.0115
1900x1900	354.2593
1950x1950	354.4943
2000x2000	354.7177

Table 14 The speed up ratio in case of using FNN after image decomposition into sub-images (25x25 pixels) for images of different sizes (from N=1050 to N=2000, n=25, q=30) using 16 parallel processors or 16 nodes

Furthermore, if the number of processors = $q \alpha n^2$, then the number of computation steps will be $(1+\varepsilon)$, and the speed up ratio can be calculated as:

$$\eta = q(2n^2-1)(N-n+1)^2 / (1+\varepsilon) \tag{30}$$

In this case, as the length of each group is very small, then there is no need to apply cross correlation between the input image and the weights of the neural network in frequency domain.

4 Conclusion

Neural networks for fast pattern detection in a given image have been presented. It has been proved mathematically and practically that the speed of the detection process becomes faster than conventional neural networks. This has been accomplished by applying cross correlation in the frequency domain between the input image and the normalized input weights of the neural networks. New general formulas for fast cross correlation as well as the speed up ratio have been given. A faster neural network approach for pattern detection has been introduced. Such approach has decomposed the input image under test into many small in size sub-images. Furthermore, a simple algorithm for fast pattern detection based on cross correlations in the frequency domain between the sub-images and the weights of neural networks has been presented in order to speed up the execution time. Simulation results have shown that, using a parallel processing technique, large values of speed up ratio could be achieved. Moreover, by using fast neural networks and image decomposition, the speed up ratio has been increased with the size of the input image. Simulation results have confirmed theoretical computations by using MATLAB. The proposed approach can be applied to detect the presence/absence of any other object in an image.

Appendix "A"

An example proves that the cross correlation between any two matrices is not commutative

$$\text{Let} \quad X = \begin{bmatrix} 5 & 1 \\ 3 & 7 \end{bmatrix}, \quad \text{and} \quad W = \begin{bmatrix} 6 & 5 \\ 9 & 8 \end{bmatrix}$$

Then, the cross correlation between W and X can be obtained as follows:

$$W \otimes X = \begin{bmatrix} 6 & 5 \\ 9 & 8 \end{bmatrix} \otimes \begin{bmatrix} 5 & 1 \\ 3 & 7 \end{bmatrix} = \begin{bmatrix} 8 \times 5 & 8 \times 1 + 9 \times 5 & 9 \times 1 \\ 5 \times 5 + 8 \times 3 & 6 \times 5 + 5 \times 1 + 9 \times 3 + 8 \times 7 & 6 \times 1 + 9 \times 7 \\ 5 \times 3 & 6 \times 3 + 5 \times 7 & 6 \times 7 \end{bmatrix} = \begin{bmatrix} 40 & 53 & 9 \\ 49 & 118 & 63 \\ 15 & 53 & 42 \end{bmatrix}$$

On the other hand, the cross correlation the cross correlation between X and W can be computed as follows:

$$X \otimes W = \begin{bmatrix} 5 & 1 \\ 3 & 7 \end{bmatrix} \otimes \begin{bmatrix} 6 & 5 \\ 9 & 8 \end{bmatrix} = \begin{bmatrix} 7 \times 6 & 3 \times 6 + 7 \times 5 & 3 \times 5 \\ 1 \times 6 + 7 \times 9 & 5 \times 6 + 1 \times 5 + 3 \times 9 + 7 \times 8 & 5 \times 5 + 3 \times 8 \\ 1 \times 9 & 5 \times 9 + 1 \times 8 & 5 \times 8 \end{bmatrix} = \begin{bmatrix} 42 & 53 & 15 \\ 69 & 118 & 49 \\ 9 & 53 & 40 \end{bmatrix}$$

which proves that $X \otimes W \neq W \otimes X$.

Also, when one of the two matrices is symmetric the cross correlation between the two matrices is non-commutative as shown in the following example:

$$\text{Let} \quad X = \begin{bmatrix} 5 & 3 \\ 3 & 5 \end{bmatrix}, \quad \text{and} \quad W = \begin{bmatrix} 6 & 5 \\ 9 & 8 \end{bmatrix}$$

Then, the cross correlation between W and X can be obtained as follows:

$$W \otimes X = \begin{bmatrix} 6 & 5 \\ 9 & 8 \end{bmatrix} \otimes \begin{bmatrix} 5 & 3 \\ 3 & 5 \end{bmatrix} = \begin{bmatrix} 8 \times 5 & 8 \times 3 + 9 \times 5 & 9 \times 3 \\ 5 \times 5 + 8 \times 3 & 6 \times 5 + 5 \times 3 + 9 \times 3 + 8 \times 5 & 6 \times 3 + 9 \times 5 \\ 5 \times 3 & 6 \times 3 + 5 \times 5 & 6 \times 5 \end{bmatrix} = \begin{bmatrix} 40 & 69 & 27 \\ 49 & 112 & 63 \\ 15 & 43 & 30 \end{bmatrix}$$

On the other hand, the cross correlation the cross correlation between X and W can be computed as follows:

$$X \otimes W = \begin{bmatrix} 5 & 3 \\ 3 & 5 \end{bmatrix} \otimes \begin{bmatrix} 6 & 5 \\ 9 & 8 \end{bmatrix} = \begin{bmatrix} 5 \times 6 & 3 \times 6 + 5 \times 5 & 3 \times 5 \\ 3 \times 6 + 5 \times 9 & 5 \times 6 + 3 \times 5 + 3 \times 9 + 5 \times 8 & 5 \times 5 + 3 \times 8 \\ 3 \times 9 & 5 \times 9 + 3 \times 8 & 5 \times 8 \end{bmatrix} = \begin{bmatrix} 30 & 43 & 15 \\ 63 & 112 & 49 \\ 27 & 69 & 40 \end{bmatrix}$$

which proves that $X \otimes W \neq W \otimes X$.

The cross correlation between any two matrices is commutative only when the two matrices are symmetric as shown in the following example.

$$\text{Let} \quad X = \begin{bmatrix} 5 & 3 \\ 3 & 5 \end{bmatrix}, \quad \text{and} \quad W = \begin{bmatrix} 8 & 9 \\ 9 & 8 \end{bmatrix}$$

Then, the cross correlation between W and X can be obtained as follows:

$$W \otimes X = \begin{bmatrix} 8 & 9 \\ 9 & 8 \end{bmatrix} \otimes \begin{bmatrix} 5 & 3 \\ 3 & 5 \end{bmatrix} = \begin{bmatrix} 8 \times 5 & 9 \times 5 + 8 \times 3 & 9 \times 3 \\ 9 \times 5 + 8 \times 3 & 8 \times 5 + 9 \times 3 + 9 \times 3 + 8 \times 5 & 9 \times 5 + 8 \times 3 \\ 9 \times 3 & 5 \times 9 + 3 \times 8 & 8 \times 5 \end{bmatrix} = \begin{bmatrix} 40 & 69 & 27 \\ 69 & 122 & 69 \\ 27 & 69 & 40 \end{bmatrix}$$

On the other hand, the cross correlation between X and W can be computed as follows:

$$X \otimes W = \begin{bmatrix} 5 & 3 \\ 3 & 5 \end{bmatrix} \otimes \begin{bmatrix} 8 & 9 \\ 9 & 8 \end{bmatrix} = \begin{bmatrix} 5 \times 8 & 5 \times 9 + 3 \times 8 & 3 \times 9 \\ 5 \times 9 + 3 \times 8 & 5 \times 8 + 3 \times 9 + 3 \times 9 + 5 \times 8 & 5 \times 9 + 3 \times 8 \\ 3 \times 9 & 5 \times 9 + 3 \times 8 & 5 \times 8 \end{bmatrix} = \begin{bmatrix} 40 & 69 & 27 \\ 69 & 122 & 69 \\ 27 & 69 & 40 \end{bmatrix}$$

which proves that the cross correlation is commutative $(X \otimes W = W \otimes X)$ only under the condition when the two matrices X and W are symmetric.

Appendix "B"

An example proves that the cross correlation between any two matrices is different from their convolution

$$\text{Let} \quad X = \begin{bmatrix} 5 & 1 \\ 3 & 7 \end{bmatrix}, \quad \text{and} \quad W = \begin{bmatrix} 6 & 5 \\ 9 & 8 \end{bmatrix},$$

the result of their cross correlation can be computed as illustrated from the previous example (first result) in appendix "A". The convolution between X and W can be obtained as follows:

$$W \otimes X = \begin{bmatrix} 5 & 1 \\ 3 & 7 \end{bmatrix} \Diamond \begin{bmatrix} 8 & 9 \\ 5 & 6 \end{bmatrix} = \begin{bmatrix} 6 \times 5 & 5 \times 5 + 6 \times 1 & 5 \times 1 \\ 9 \times 5 + 6 \times 3 & 8 \times 5 + 9 \times 1 + 5 \times 3 + 6 \times 7 & 8 \times 1 + 5 \times 7 \\ 9 \times 3 & 8 \times 3 + 9 \times 7 & 8 \times 7 \end{bmatrix} = \begin{bmatrix} 30 & 31 & 5 \\ 63 & 106 & 43 \\ 27 & 87 & 56 \end{bmatrix}$$

which proves that $W \otimes X \neq W \Diamond X$.

When the second matrix W is symmetric, the cross correlation between W and X can be computed as follows:

$$W \otimes X = \begin{bmatrix} 8 & 9 \\ 9 & 8 \end{bmatrix} \otimes \begin{bmatrix} 5 & 1 \\ 3 & 7 \end{bmatrix}$$

$$= \begin{bmatrix} 8 \times 5 & 9 \times 5 + 8 \times 1 & 9 \times 1 \\ 9 \times 5 + 8 \times 3 & 8 \times 5 + 9 \times 3 + 9 \times 1 + 8 \times 7 & 8 \times 1 + 7 \times 9 \\ 9 \times 3 & 8 \times 3 + 9 \times 7 & 8 \times 7 \end{bmatrix}$$

$$= \begin{bmatrix} 40 & 87 & 9 \\ 79 & 106 & 71 \\ 45 & 53 & 56 \end{bmatrix}$$

while the convolution between X and W can be obtained as follows:

$$W \circledast X = \begin{bmatrix} 5 & 1 \\ 3 & 7 \end{bmatrix} \diamond \begin{bmatrix} 8 & 9 \\ 9 & 8 \end{bmatrix} = \begin{bmatrix} 8 \times 5 & 9 \times 5 + 8 \times 1 & 9 \times 1 \\ 9 \times 5 + 8 \times 3 & 8 \times 5 + 9 \times 3 + 9 \times 1 + 8 \times 7 & 8 \times 1 + 7 \times 9 \\ 9 \times 3 & 8 \times 3 + 9 \times 7 & 8 \times 7 \end{bmatrix} = \begin{bmatrix} 40 & 87 & 9 \\ 79 & 106 & 71 \\ 45 & 53 & 56 \end{bmatrix}$$

Which proves that under the condition that the second matrix is symmetric (or the two matrices are symmetric) the cross correlation between any the two matrices equals to their convolution.

Appendix "C"

An example for cross correlation with matrix decomposition

$$\text{Let } X = \begin{bmatrix} 5 & 1 & 8 & 6 \\ 3 & 7 & 3 & 4 \\ 1 & 2 & 9 & 5 \\ 6 & 5 & 4 & 2 \end{bmatrix}, \text{ and } W = \begin{bmatrix} 6 & 5 \\ 9 & 8 \end{bmatrix}$$

Then the cross correlation (CC) between W and X can be computed as follows:

$$CC = W \otimes X = \begin{bmatrix} 6 & 5 \\ 9 & 8 \end{bmatrix} \otimes \begin{bmatrix} 5 & 1 & 8 & 6 \\ 3 & 7 & 3 & 4 \\ 1 & 2 & 9 & 5 \\ 6 & 5 & 4 & 2 \end{bmatrix}$$

$$= \begin{bmatrix} 8x5 & 9x5 + 8x1 & 9x1 + 8x8 & 9x8 + 8x6 & 9x6 \\ 5x5 + 8x3 & 6x5 + 5x1 + 9x3 + 8x7 & 6x1 + 5x8 + 9x7 + 8x3 & 6x8 + 5x6 + 9x3 + 8x4 & 6x6 + 9x4 \\ 5x3 + 8x1 & 6x3 + 5x7 + 9x1 + 8x2 & 6x7 + 5x3 + 9x2 + 8x9 & 6x3 + 5x4 + 9x9 + 8x5 & 6x4 + 9x5 \\ 5x1 + 8x6 & 6x1 + 5x2 + 9x6 + 8x5 & 6x2 + 5x9 + 9x5 + 8x4 & 6x9 + 5x5 + 9x4 + 8x2 & 6x5 + 9x2 \\ 5x6 & 6x6 + 5x5 & 6x5 + 5x4 & 6x4 + 5x2 & 6x2 \end{bmatrix}$$

$$CC = \begin{bmatrix} 40 & 53 & 73 & 120 & 54 \\ 49 & 118 & 133 & 137 & 72 \\ 23 & 78 & 147 & 159 & 69 \\ 53 & 110 & 134 & 131 & 48 \\ 30 & 61 & 50 & 34 & 12 \end{bmatrix}$$

Suppose that X is decomposed into 4 smaller matrices x1, x2, x3, and x4 as follows:

$$x1 = \begin{bmatrix} 5 & 1 \\ 3 & 7 \end{bmatrix}, \ x2 = \begin{bmatrix} 8 & 6 \\ 3 & 4 \end{bmatrix}, \ x3 = \begin{bmatrix} 1 & 2 \\ 6 & 5 \end{bmatrix}, \ \text{and} \ x4 = \begin{bmatrix} 9 & 5 \\ 4 & 2 \end{bmatrix}$$

Then, the cross correlation between each resulting matrix and the matrix W can be computed as follows: $CC1 = W \otimes x1 = \begin{bmatrix} 6 & 5 \\ 9 & 8 \end{bmatrix} \otimes \begin{bmatrix} 5 & 1 \\ 3 & 7 \end{bmatrix}$

$$= \begin{bmatrix} 8x5 & 9x5+8x1 & 9x1 \\ 5x5+8x3 & 6x5+5x1+9x3+8x7 & 6x1+9x7 \\ 5x3 & 6x3+5x7 & 6x7 \end{bmatrix} = \begin{bmatrix} 40 & 53 & 9 \\ 49 & 118 & 69 \\ 15 & 53 & 42 \end{bmatrix}$$

$$CC2 = W \otimes x2 = \begin{bmatrix} 6 & 5 \\ 9 & 8 \end{bmatrix} \otimes \begin{bmatrix} 8 & 6 \\ 3 & 4 \end{bmatrix} = \begin{bmatrix} 8x8 & 9x8+8x6 & 9x6 \\ 5x8+8x3 & 6x8+5x6+9x3+8x4 & 6x6+9x4 \\ 5x3 & 6x3+5x4 & 6x4 \end{bmatrix}$$

$$= \begin{bmatrix} 64 & 120 & 54 \\ 64 & 137 & 72 \\ 15 & 38 & 24 \end{bmatrix}$$

$$CC3 = x3 \otimes W = \begin{bmatrix} 6 & 5 \\ 9 & 8 \end{bmatrix} \otimes \begin{bmatrix} 1 & 2 \\ 6 & 5 \end{bmatrix}$$

$$= \begin{bmatrix} 8x1 & 9x1+8x2 & 9x2 \\ 5x1+8x6 & 6x1+5x2+9x6+8x5 & 6x2+9x5 \\ 5x6 & 6x6+5x5 & 6x5 \end{bmatrix} = \begin{bmatrix} 8 & 25 & 18 \\ 53 & 110 & 57 \\ 30 & 61 & 30 \end{bmatrix}$$

$$CC4 = x4 \otimes W = \begin{bmatrix} 6 & 5 \\ 9 & 8 \end{bmatrix} \otimes \begin{bmatrix} 9 & 5 \\ 4 & 2 \end{bmatrix}$$

$$= \begin{bmatrix} 8x9 & 9x9+8x5 & 9x5 \\ 5x9+8x4 & 6x9+5x5+9x4+8x2 & 6x5+9x2 \\ 5x4 & 6x4+5x2 & 6x2 \end{bmatrix} = \begin{bmatrix} 72 & 121 & 45 \\ 77 & 131 & 48 \\ 20 & 34 & 12 \end{bmatrix}$$

The total result of cross correlating the resulting smaller matrices with the matrix W can be computed as:

$$CCT = CC1 + CC2 + CC3 + CC4$$

$$\mathbf{CCT} = \begin{bmatrix} CC1(1,1) & CC1(1,2) & CC1(1,3)+CC2(1,1) & CC2(1,2) & CC2(1,3) \\ CC1(2,1) & CC1(2,2) & CC2(2,3)+CC2(2,1) & CC2(2,2) & CC2(2,3) \\ CC1(3,1)+CC3(1,1) & CC1(3,2)+CC3(1,2) & CC1(3,3)+CC2(3,1)+CC3(1,3)+CC4(1,1) & CC2(3,2)+CC4(1,2) & CC2(3,3)+CC4(1,3) \\ CC3(2,1) & CC3(2,2) & CC3(2,3)+CC4(1,2) & CC4(2,2) & CC4(2,3) \\ CC3(3,1) & CC3(3,2) & CC3(3,3)+CC4(1,3) & CC4(3,2) & CC4(3,3) \end{bmatrix}$$

$$CCT = \begin{bmatrix} 40 & 53 & 9+64 & 120 & 54 \\ 49 & 118 & 69+64 & 137 & 72 \\ 15+8 & 53+25 & 42+15+18+72 & 38+121 & 24+45 \\ 53 & 110 & 57+77 & 131 & 48 \\ 30 & 61 & 30+20 & 34 & 12 \end{bmatrix}$$

$$CCT = \begin{bmatrix} 40 & 53 & 73 & 120 & 54 \\ 49 & 118 & 133 & 137 & 72 \\ 23 & 78 & 147 & 159 & 69 \\ 53 & 110 & 134 & 131 & 48 \\ 30 & 61 & 50 & 34 & 12 \end{bmatrix}$$

Which means that CC=CCT. This proves that the result of cross correlating a large matrix (X) with another matrix (W) equals to the results of cross correlating the resulting smaller sub matrices (X1,X2,X3, and X4) after decomposition.

References

[1] Klette, R., Zamperon: Handbook of image processing operators. John Wiley & Sonsltd, Chichester (1996)
[2] Rowley, H.A., Baluja, S., Kanade, T.: Neural Network - Based Face Detection. IEEE Trans. on Pattern Analysis and Machine Intelligence 20(1), 23–38 (1998)
[3] Schneiderman, H., Kanade, T.: Probabilistic modeling of local appearance and spatial relationships for object recognition. In: IEEE Conference on Computer Vision and Pattern Recognition (CVPR), SantaBarbara, CA, pp. 45–51 (1998)
[4] Feraud, R., Bernier, O., Viallet, J.E., Collobert, M.: A Fast and Accurate Face Detector for Indexation of Face Images. In: Proceedings of the Fourth IEEE International Conference on Automatic Face and Gesture Recognition, Grenoble, France, March 28–30 (2000)
[5] Zhu, Y., Schwartz, S., Orchard, M.: Fast Face Detection Using Subspace Discriminate Wavelet Features. In: Proc. of IEEE Computer Society International Conference on Computer Vision and Pattern Recognition (CVPR 2000), South Carolina, June 13-15, 2000, vol. 1, pp. 1636–1643 (2000)
[6] El-Bakry, H.M.: Automatic Human Face Recognition Using Modular Neural Networks. Machine Graphics & Vision Journal (MG&V) 10(1), 47–73 (2001)
[7] Srisuk, S., Kurutach, W.: A New Robust Face Detection in Color Images. In: Proc. of IEEE Computer Society International Conference on Automatic Face and Gesture Recognition, Washington D.C., USA, May 20-21, 2002, pp. 306–311 (2002)
[8] El-Bakry, H.M.: Face detection using fast neural networks and image decomposition. Neurocomputing Journal 48, 1039–1046 (2002)

[9] El-Bakry, H.M.: Human Iris Detection Using Fast Cooperative Modular Neural Networks and Image Decomposition. Machine Graphics & Vision Journal (MG&V) 11(4), 498–512 (2002)

[10] El-Bakry, H.M., Zhao, Q.: Fast Object/Face Detection Using Neural Networks and Fast Fourier Transform. International Journal of Signal Processing 1(3), 182–187 (2004)

[11] El-Bakry, H.M., Zhao, Q.: A Modified Cross Correlation in the Frequency Domain for Fast Pattern Detection Using Neural Networks. International Journal of Signal Processing 1(3), 188–194 (2004)

[12] El-Bakry, H.M., Zhao, Q.: Face Detection Using Fast Neural Processors and Image Decomposition. International Journal of Computational Intelligence 1(4), 313–316 (2004)

[13] El-Bakry, H.M., Zhao, Q.: Fast Complex Valued Time Delay Neural Networks. International Journal of Computational Intelligence 2(1), 16–26 (2005)

[14] El-Bakry, H.M., Zhao, Q.: A Fast Neural Algorithm for Serial Code Detection in a Stream of Sequential Data. International Journal of Information Technology 2(1), 71–90 (2005)

[15] El-Bakry, H.M., Zhao, Q.: A New Symmetric Form for Fast Sub-Matrix (Object/Face) Detection Using Neural Networks and FFT. International Journal of Signal Processing (to be published)

[16] El-Bakry, H.M., Zhao, Q.: Fast Pattern Detection Using Normalized Neural Networks and Cross Correlation in the Frequency Domain. EURASIP Journal on Applied Signal Processing (to be published)

[17] Ben-Yacoub, S., Fasel, B., Luettin, J.: Fast Face Detection using MLP and FFT. In: Proc. of the Second International Conference on Audio and Video-based Biometric Person Authentication (AVBPA 1999) (1999)

[18] Fasel, B.: Fast Multi-Scale Face Detection. IDIAP-Com 98-04 (1998)

[19] Ben-Yacoub, S.: Fast Object Detection using MLP and FFT. IDIAP-RR 11, IDIAP (1997)

[20] Cooley, J.W., Tukey, J.W.: An algorithm for the machine calculation of complex Fourier series. Math. Comput. 19, 297–301 (1965)

[21] Lewis, J.P.: Fast Normalized Cross Csorrelation,
http://www.idiom.com/~zilla/Papers/nvisionInterface/nip.html

[22] El-Bakry, H.M.: Face detection using fast neural networks and image decomposition. Neurocomputing Journal 48, 1039–1046 (2002)

[23] El-Bakry, H.M.: Comments on Using MLP and FFT for Fast Object/Face Detection. In: Proc. of IEEE IJCNN 2003, Portland, Oregon, pp. 1284–1288, July 20-24 (2003)

[24] El-Bakry, H.M., Stoyan, H.: Fast Neural Networks for Object/Face Detection. In: Proc. of the 30th Anniversary SOFSEM Conference on Current Trends in Theory and Practice of Computer Science, Hotel VZ MERIN, Czech Republic, January 24–30 (2004)

[25] El-Bakry, H.M., Stoyan, H.: Fast Neural Networks for Sub-Matrix (Object/Face) Detection. In: Proc. of IEEE International Symposium on Circuits and Systems, Vancouver, Canada, May 23-26 (2004)

[26] El-Bakry, H.M.: Fast Sub-Image Detection Using Neural Networks and Cross Correlation in Frequency Domain. In: Proc. of IS 2004: 14th Annual Canadian Conference on Intelligent Systems, Ottawa, Ontario, June 6-8 (2004)

[27] El-Bakry, H.M., Stoyan, H.: Fast Neural Networks for Code Detection in a Stream of Sequential Data. In: Proc. of CIC 2004 International Conference on Communications in Computing, Las Vegas, Nevada, USA, June 21-24 (2004)

[28] El-Bakry, H.M.: Fast Neural Networks for Object/Face Detection. In: Proc. of 5th International Symposium on Soft Computing for Industry with Applications of Financial Engineering, Sevilla, Andalucia, Spain, June 28 - July 4 (2004)

[29] El-Bakry, H.M., Stoyan, H.: A Fast Searching Algorithm for Sub-Image (Object/Face) Detection Using Neural Networks. In: Proc. of the 8th World Multi-Conference on Systemics, Cybernetics and Informatics, Orlando, USA, July 18-21 (2004)

[30] El-Bakry, H.M., Stoyan, H.: Fast Neural Networks for Code Detection in Sequential Data Using Neural Networks for Communication Applications. In: Proc. of the First International Conference on Cybernetics and Information Technologies, Systems and Applications: CITSA 2004, Orlando, Florida, USA, July 21-25, 2004, vol. IV, pp. 150–153 (2004)

[31] El-bakry, H.M., Abo-elsoud, M.A., Kamel, M.S.: Fast Modular Neural Networks for Human Face Detection. In: Proc. of IEEE-INNS-ENNS International Joint Conference on Neural Networks, Como, Italy, July 24-27, 2000, vol. III, pp. 320–324 (2000)

[32] El-bakry, H.M.: Fast Iris Detection using Cooperative Modular Neural Nets. In: Proc. of the 6th International Conference on Soft Computing, Japan, October 1-4 (2000)

[33] El-Bakry, H.M.: Automatic Human Face Recognition Using Modular Neural Networks. Machine Graphics & Vision Journal (MG&V) 10(1), 47–73 (2001)

[34] El-bakry, H.M.: Fast Iris Detection Using Cooperative Modular Neural Networks. In: Proc. of the 5th International Conference on Artificial Neural Nets and Genetic Algorithms, Sydney, Czech Republic, April 22-25, 2001, pp. 201–204 (2001)

[35] El-bakry, H.M.: Fast Iris Detection Using Neural Nets. In: Proc. of the 14th Canadian Conference on Electrical and Computer Engineering, Canada, May 13-16, 2001, pp. 1409–1415 (2001)

[36] El-bakry, H.M.: Human Iris Detection Using Fast Cooperative Modular Neural Nets. In: Proc. of INNS-IEEE International Joint Conference on Neural Networks, Washington, DC, USA, July 14-19, 2001, pp. 577–582 (2001)

[37] El-bakry, H.M.: Human Iris Detection for Information Security Using Fast Neural Nets. In: Proc. of the 5th World Multi-Conference on Systemics, Cybernetics and Informatics, Orlando, Florida, USA, July 22-25 (2001)

[38] El-bakry, H.M.: Human Iris Detection for Personal Identification Using Fast Modular Neural Nets. In: Proc. of the 2001 International Conference on Mathematics and Engineering Techniques in Medicine and Biological Sciences, Monte Carlo Resort, Las Vegas, Nevada, USA, July 25-28, 2001, pp. 112–118 (2001)

[39] El-bakry, H.M.: Human Face Detection Using Fast Neural Networks and Image Decomposition. In: Proc. the fifth International Conference on Knowledge-Based Intelligent Information & Engineering Systems, Osaka-kyoiku University, Kashiwara City, Japan, September 6-8, 2001, pp. 1330–1334 (2001)

[40] El-Bakry, H.M.: Fast Iris Detection for Personal Verification Using Modular Neural Networks. In: Proc. of the International Conference on Computational Intelligence, Dortmund, Germany, October 1-3, 2001, pp. 269–283 (2001)

[41] El-bakry, H.M.: Fast Cooperative Modular Neural Nets for Human Face Detection. In: Proc. of IEEE International Conference on Image Processing, Thessaloniki, Greece, October 7-10 (2001)

[42] El-Bakry, H.M.: Fast Face Detection Using Neural Networks and Image Decomposition. In: Proc. of the 6th International Computer Science Conference, Active Media Technology, Hong Kong – China, December 18-20, 2001, pp. 205–215 (2001)

[43] El-Bakry, H.M.: Face Detection Using Fast Neural Networks and Image Decomposition. In: Proc. of INNS-IEEE International Joint Conference on Neural Networks, Honolulu, Hawaii, USA, May 14-19 (2002)

[44] El-Bakry, H.M., Zhao, Q.: Fast Normalized Neural Processors For Pattern Detection Based on Cross Correlation Implemented in the Frequency Domain. Journal of Research and Practice in Information Technology 38(2), 151–170 (2006)

[45] El-Bakry, H.M., Zhao, Q.: A New High Speed Neural Model For Character Recognition Using Cross Correlation and Matrix Decomposition. International Journal of Signal Processing 2(3), 183–202 (2005)

[46] El-Bakry, H.M., Zhao, Q.: Speeding-up Normalized Neural Networks For Face/Object Detection. Machine Graphics & Vision Journal (MG&V) 14(1), 29–59 (2005)

[47] El-Bakry, H.M., Mastorakis, N.: New Fast Normalized Neural Networks for Pattern Detection. Journal of Image and Vision Computing (to be appeared)

[48] Ishak, K.A., Samad, S.A., Hussian, A., Majlis, B.Y.: A fast and robust face detection using neural networks. In: Proc. of the international Symposium on Information and Communication Technologies, Multimedia University, Putrajaya, Malaysia, October 7-8, vol. 2, pp. 5–8 (2004)

Part IV
Learning and Visualization

Dissimilarity Analysis and Application to Visual Comparisons

Sébastien Aupetit, Nicolas Monmarché, Pierre Liardet, and Mohamed Slimane

In this chapter, the embedding of a set of data into a vector space is studied when an unconditional pairwise dissimilarity w between data is given. The vector space is endowed with a suitable pseudo-euclidean structure and the data embedding is built by extending the classical kernel principal component analysis. This embedding is unique, up to an isomorphism, and injective if and only if w separates the data. This construction takes advantage of axis corresponding to negative eigenvalues to develop pseudo-euclidean scatterplot matrix representations. This new visual tool is applied to compare various dissimilarities between hidden Markov models built from person's faces.

1 Introduction

During the last decade, the amount of multivariate data generated by sensors or softwares has increased steadily. The analysis of these data has become a central issue that data mining can tackle. Data mining can be divided in two main areas: automatic and visual data mining. On one hand, automatic data mining techniques is able to treat a huge quantity of data but the results are often distorted by outliers and noise. Moreover, the models emerging from data analysis are often difficult to interpret. On the other hand, visual data mining methods can easily manage outliers and insure a better transmission

Sébastien Aupetit, Nicolas Monmarché, and Mohamed Slimane
Université François Rabelais de Tours, Laboraqtoire d'Informatique, EA 2101, Polytech'Tours, 64 Avenue J. Portalis, 37200 Tours, France
e-mail: `sebastien.aupetit,nicolas.monmarche,`
`mohamed.slimane@univ-tours.fr`

Pierre Liardet
Université de Provence, UMR-CNRS 6632, 39 rue F. Joliot-Curie, 13453 Marseille, cedex 13, France
e-mail: `liardet@cmi.univ-mrs.fr`

A.-E. Hassanien et al. (Eds.): Foundations of Comput. Intel. Vol. 1, SCI 201, pp. 333–361.
springerlink.com   © Springer-Verlag Berlin Heidelberg 2009

of knowledges to the user. Their main disadvantage is that only a few data can be analyzed. Nevertheless, interactions with the users may increase the size of set of data that can be processed. For instance, the user having ability to move a 3D scene is able to perform a better exploitation of the third dimension than a computational investigation. Certainly, the main interest of visual data mining is that it can take full advantage of the user cognition and expertise through interactions and representations.

Visual data mining methods are studied in the literature intensively but essentially depend on the kind of considered data. For multidimensional data, several representations are used like parallel coordinates [1], RadViz technique [2], scatterplot matrix [3], dimensional stacking [4], self organizing maps [5], multi-dimensional scalings [6, 7] or glyphs [8]. For trees and more general graphs, techniques like NicheWorks [9] or hyperbolic representations [10] are applied commonly. Finally, shaded similarity matrix [11], spring algorithms [12] or multi-dimensional scalings [6, 7] can use the key notion of dissimilarities between data.

In this chapter we develop a general approach of the dissimilarity problem which allows to embed distinct data into distinct points whereas they are not necessarily distinguished individually by their dissimilarities but they are possibly separated by considering them all together. The use of dissimilarity measures instead of distance is of a great interest in practice because they are very flexible: dissimilarities could be derived from various and complex algorithms (probability models, user judgments, ...). However, these algorithms usually never guaranty to furnish measurements that correspond to an euclidean distance nor even to any metric. Moreover, the usual requirement in a metrical meaning that "two similar data are closed, from each other, and two data with very different coordinates are very dissimilar" could be violated.

Recent works on kernel methods, and more precisely on indefinite kernel methods for classification tasks, have demonstrated that euclidean embeddings are not essential: pseudo-euclidean spaces can be used in place [13, 14]. Notice that, from our point of view (see *infra*), a dissimilarity is a real valued map, defined on all couples of data, which is characterized by only two properties: symmetry and null dissimilarity between identical objects. This is enough to embed data into a real vector space $\mathcal{W}$ equipped with a real quadratic form q such that, if ψ denote the embedding, for all data x and y, their dissimilarity is given by $q(\psi(x) - \psi(y))$. Fortunately, pseudo-euclidean spaces $(\mathcal{W}, q)$ have enough mathematical properties to permit a generalization of the scatterplot matrix in order to visualize the dominant part of principal axis components living in a pseudo-euclidean space.

This chapter is organized as follow. Section 2 develops the mathematical framework including new useful definitions. The notions of formal, minimal and standard embedding are introduced for any dissimilarity. The centered embedding is defined and the uniqueness of embeddings is discussed. Section 3 emphasizes some properties of pseudo-euclidean spaces that will be used in Section 4 where the classical scatterplot matrix representation is extended in order to manage also pseudo-euclidean embeddings. This new tool is applied in the next section to analyze dissimilarities between hidden Markov models obtained from learning persons' faces. The chapter ends with a conclusion section and an Appendix where technical proofs and remarks are reported.

2 Embeddings from a Dissimilarity

2.1 Dissimilarity and Its Representations from Quadratic Forms

Let $\mathcal{G}$ be a set of objects. These objects may take many forms, representing for examples patterns of recognition, features of faces, images, stimuli, sounds, brightness, times series, or points in a given multidimensional vector space as well. It is not required that a feature space for representing the objects or corresponding data is previously defined, but we usually suppose that a dissimilarity function is built from practical applications. To formalize such a function, we introduce the following classical definition.

Definition 1. *A map $w : \mathcal{G} \times \mathcal{G} \to \mathbb{R}$ will be said a dissimilarity or a dissimilarity function if the following two properties are satisfied:*

(i) $\forall x \in \mathcal{G}$, $w(x,x) = 0$,
(ii) $\forall (x,y) \in \mathcal{G} \times \mathcal{G}$, $w(x,y) = w(y,x)$.

One goal of this chapter is to represent in a standard manner any dissimilarity function by mapping the objects into a real quadratic space. In many real applications, $\mathcal{G}$ is already a set of data identified to a subset of an euclidean space H and $w(\cdot,\cdot)$ has the form

$$w(x,y) = ||x - y||_H^2 \tag{1}$$

where $||.||_H$ is the norm associated to the euclidean scalar product $(\cdot|\cdot)_H$ of H.

The definition of a dissimilarity function w given above is quite general. In particular $w(x,y)$ is not assumed to take only positive values if $x \neq y$. Our first aim is then, for a given finite subset E of $\mathcal{G}$, to built a real vector space $\mathcal{W}$ equipped with a bilinear symmetric form φ and a map $\psi : E \to \mathcal{W}$ such that for all couple (x,y) in $E \times E$,

$$w(x,y) = q(\psi(x) - \psi(y)) \tag{2}$$

where q denotes the quadratic form associated to φ ($q(x) := \varphi(x,x)$). Of course, such a solution (if any one exists) is not unique. For example, adding a constant to ψ does not change (2). Moreover, we ask for a minimal solution with respect to the dimension of $\mathcal{W}$ and we shall see that this solution is unique in the sense that if we built two solutions, up to a translation, they can be identified from an isomorphism of the underlying quadratic space structures (see Section 2.4).

We proceed in two steps. The first one is a formal construction, that exhibits an injective mapping ψ and a quadratic form satisfying (2). In the second step, using tools from general theory of symmetric bilinear forms, we built a more appropriated embedding map $\psi : E \to \mathcal{W}$ and a non singular quadratic form q on $\mathcal{W}$ such that the relation (2) holds. The case where the bilinear symmetric form associated to q is definite positive (euclidean case) was intensively exploited in the classical principal component analysis (PCA) and by the multidimensional scaling (MDS) method, introduced by W.S. Torgeson [15]. Multidimensional scaling approaches can tackle many other criteria using approximations (see for example [7, 14, 16, 17, 18, 19]). Notice that arbitrary values for the dissimilarity map w are accepted and the embedding exists without any approximation or modification of w. Such a construction has been proposed for pattern recognition in the seminal papers of Goldfarb [20, 21], and used in various recognition and machine learning problems (see [22, 23, 24] for more details and additional references).

2.2 Embedding in a Quadratic Space

For convenience of the reader and to fix notations and definitions we recall, in a concise manner, part of the theory of real symmetric bilinear forms specially used in our construction. We refer the reader to [25] for the fundamentals and to [24] for various applications. Recall that a real quadratic space is a couple $(\mathcal{W}, q)$ where $\mathcal{W}$ is a real vector space of finite dimension and q is a quadratic form, that is to say there exists a bilinear symmetric form $\varphi : \mathcal{W} \times \mathcal{W} \to \mathbb{R}$ such that $q(x) = \varphi(x, x)$ for all $x \in \mathcal{W}$. Recall that φ is constructed from q by the depolarization formula $\varphi(x, y) = \frac{1}{2}(q(x) + q(y) - q(x - y))$ which plays a crucial role in the sequel. A quadratic space is said to be pseudo-euclidean if the quadratic form is non-singular i.e., if its associated bilinear symmetric form is defined.

First Step: A Formal Embedding

Let $E = \{x_0, x_1, \ldots, x_T\}$ be a subset of $\mathcal{G}$ of $T+1$ elements with $T \geq 1$ and let c be any fixed element of E, called origin. Without loss of generality, we can assume that $c = x_0$. Let $\Psi : E \to \mathbb{R}^T$ be the map defined by $\Psi(x_0) = 0$ and $\Psi(x_k) = I_k$ $(1 \leq k \leq T)$ where I_k is the column vector whose k-th coordinate equal 1 and the other coordinates equal 0. The expected formula (2) and the above depolarization formula, suggest to set

$$^i m_j = \frac{1}{2}\big(w(c, x_i) + w(c, x_j) - w(x_i, x_j)\big) \tag{3}$$

for $0 \leq i, j \leq T$ and to consider the $T \times T$ matrix $M(w, c) = (^i m_j)_{1 \leq i, j \leq T}$, currently denoted by M for short. The matrix M is symmetric and defines a bilinear symmetric form $\Phi(X, Y) = {}^t Y M X$ on $\mathbb{R}^T$ (here ${}^t A$ denote the transpose of any matrix A). Notice that, according to (3), $^0m_j = {}^jm_0 = 0$ for all j, justifying to avoid the index 0 in the definition of M.

Let Q denote the quadratic form associated to Φ. By construction $\Phi(I_j, I_i) = {}^im_j$. In particular $Q(I_k) = w(c, x_k) = {}^km_k$ and

$$Q(I_i - I_j) = Q(I_i) + Q(I_j) - 2\Phi(I_i, I_j) = w(c, x_i) + w(c, x_j) - 2\,{}^im_j$$

so that $Q(\Psi(x_j) - \Psi(x_i)) = w(x_i, x_j)$ as expected from (2). By construction, $\Psi(E) + Z$ generates $\mathbb{R}^T$ for any $Z \in \mathbb{R}^T$. The triple (Ψ, T, Q) will be called a *formal embedding* with origin c. Of course, this embedding depends on the indexation of E. Obviously, the image $\Psi(E)$ does not figure out any geometrical structure of the data. This is a reason why we are looking for triple (ψ, r, q) where ψ is a map from E into a quadratic space $(\mathcal{W}, q)$ of dimension r as small as possible.

Real quadratic forms being classified by their Sylvester indices (p, n) (also called signature and denoted by $\sigma(q)$), we will choose for q the quadratic forms

$$Q(X) = \sum_{i=1}^{p}({}^iX)^2 - \sum_{j=p+1}^{p+n} ({}^jX)^2 \qquad (4)$$

where ^kX denotes the k-th coordinate of $X \in \mathbb{R}^T$ and $n+p \leq T$. If $n+p = T$ the quadratic form Q will be said *standard* and denoted by q_n, omitting the reference to the dimension T. The polar bilinear form of q_n will be denoted by $\langle \cdot | \cdot \rangle_n$, in other words

$$\langle X|Y\rangle_n = {}^tY J(p,n)X, \quad J(p,n) := \begin{pmatrix} Id_p & 0 \\ 0 & -Id_n \end{pmatrix} \qquad (5)$$

where Id_k denote the $k \times k$ identity matrix. With this notation, q_0 is the usual standard quadratic form $q_0(X) = ({}^1X)^2 + \cdots + ({}^TX)^2$ associated to the standard euclidean structure on $\mathbb{R}^T$ given by the scalar product $\langle X|Y\rangle_0 = {}^tYX$.

Definitions

In regard to the above construction, for a given finite subset E of objects (usually data) in $\mathcal{G}$, equipped with a dissimilarity function w, we introduce the following general definition.

Definition 2. *A triple (ψ, r, q) is said to be an embedding (solution) of the set E with a dissimilarity function w if q is a quadratic form on $\mathbb{R}^r$ and $\psi : E \to \mathbb{R}^r$ is a map such that:*

(i) $w(x, y) = q(\psi(x) - \psi(y))$.

The embedding is said to be full if in addition the property

(ii) the affine structure generated by $\psi(E)$ is supported by the full space (in other words, for all $Z \in \mathbb{R}^r$, the vector sets $\psi(E) + Z$ generates $\mathbb{R}^r$)

holds. The parameter r is called the dimension of the embedding, the Sylvester signature (p, n) of q will be called the signature of the embedding. Moreover, the embedding solution (ψ, r, q) will be said:

- minimal if its dimension is minimal among all possible dimensions of embeddings of E;
- proper if ψ is injective;
- regular if q is non-singular.

With this definition, $N + 1$ objects cannot have a full embedding into $\mathbb{R}^r$ if $r \geq N + 1$.

Proposition 1. *A minimal embedding is full.*

Proof. This is clear from the definitions.

Notice that the formal embedding is full, proper but in general it is not regular. A natural question is to identify dissimilarity w for which there exists a proper embedding in an euclidean. Such a dissimilarity will be said euclidean. The following characterization rephrase in our language a theorem of Schoenberg:

Theorem 1 (Schoenberg [26]). *There exists a proper embedding of the data set E with dissimilarity w into an euclidean space if and only if the following properties hold:*

(j) $(\forall (x, y) \in E \times E)(w(x, y) = 0 \Rightarrow x = y)$;
(jj) w is non negative;
(jjj) for all map $\lambda : E \to \mathbb{R}$, the equality $\sum_{x \in E} \lambda(x) = 0$ implies the inequality

$$\sum_{(x,y) \in E \times E} \lambda(x)\lambda(y)w(x, y) \leq 0.$$

It is easy to prove that a proper embedding in an euclidean space implies properties (j), (jj) and (jjj). For the converse we refer to [26]. For an interesting review on euclidean dissimilarity maps, see [22].

The above formal embedding of the data set E with $T = \text{card}(E) - 1$, defined in the first step, furnishes a key tool to build a full regular embedding (ψ, r, q) where r is the smallest possible integer, and $q = q_n$.

Definition 3. *Let E be a set of data equipped with a dissimilarity map w. An embedding (ψ, r, q) of E of signature (n, p) will be said standard if it is minimal and $q = q_n$.*

In this definition, we formally include the case where w is trivial (all dissimilarities are 0) by defining the trivial minimal embedding as the map sending

E into the null space $\mathbb{R}^0 = \{0\}$ endowed with the null quadratic form. For example, let $E = \{x_0, x_1, x_2\}$ be equipped with the trivial dissimilarity map and let $\psi : E \to \mathbb{R}^2$ be defined by $\psi(x_0) = \binom{0}{0}$, $\psi(x_1) = \binom{1}{1}$, $\psi(x_2) = \binom{-1}{-1}$. The triple $(\psi, 2, q_1)$ is a proper regular embedding of E in the hyperbolic plane $(\mathbb{R}^2, q_1)$ but it is not full, hence it is not minimal. The minimal embedding being in fact the trivial one.

Second Step: Standard Embedding

The aim of this subsection is to show that a standard embedding solution always exists. We use the above notations and hypothesis. In particular, recall that M is the $T \times T$ symmetric matrix with entries given by (3). From symmetry, there exists an orthonormal basis $V = (V_1, \ldots, V_T)$ (with respect to the standard euclidean structure on $\mathbb{R}^T$) of (column) eigenvectors V_i corresponding to eigenvalues λ_i of M, counting with their multiplicity. Let us recall that V is also an orthogonal basis with respect to Φ and $\Phi(V_i, V_i) = \lambda_i$. We order the basis such that positive eigenvalues verify $\lambda_1 \geq \ldots \geq \lambda_p$ and the negative ones verify $\lambda_{p+1} \leq \ldots \leq \lambda_{p+n}$. The remaining eigenvalues $\lambda_{p+n+1}, \ldots, \lambda_T$ are zero. Let $D(d_1, \ldots, d_T)$ denote the diagonal matrix D with diagonal entries $D_{ii} = d_i$. By construction, we have

$$M = VD(\lambda_1, \ldots, \lambda_T)^t V \qquad (6)$$

and the couple (p, n) is the signature of Q. The bilinear form Φ is singular if and only if $p + n < T$. It is worth noticing that the couple (p, n) does not depend on the choice of $c \in E$ (the origin of data) nor on the indexation of E. In fact, let $\{x'_0, \ldots, x'_T\}$ be a new indexation of E and let τ be the permutation of $\{0, \ldots, T\}$ defined by $x'_i = x_{\tau(i)}$ $(0 \leq i \leq T)$. If Φ' denotes the bilinear form associated to the matrix M' defined as above from the new indexation and if I_0 denote the null vector, then

$$\Phi'(I_j, I_i) = {}^i m'_j = \frac{1}{2}\left(w(x_{\tau(0)}, x_{\tau(i)}) + w(x_{\tau(0)}, x_{\tau(j)}) - w(x_{\tau(i)}, x_{\tau(j)})\right),$$

but $w(x_\ell, x_m) = Q(I_\ell - I_m) = Q((I_\ell - I_{\tau(0)}) - (I_m - I_{\tau(0)}))$ and using the polarization formula, $\Phi'(I_j, I_i) = \Phi(I_{\tau(j)} - I_{\tau(0)}, I_{\tau(i)} - I_{\tau(0)})$. Therefore, Φ' is the quadratic form Φ expressed into the new basis $\{I_{\tau(i)} - I_{\tau(0)} \,;\, 1 \leq i \leq T\}$.

Now, we introduce the following diagonal matrices

$$D^+ = D\left(\lambda_1^{1/2}, \ldots, \lambda_p^{1/2}\right), \quad D^- = D\left((-\lambda_{p+1})^{1/2}, \ldots, (-\lambda_{p+n})^{1/2}\right).$$

Let Δ^+ be the rectangular $p \times T$ matrix $\Delta^+ = (D^+ \ \mathbf{0} \ \mathbf{0})$ and let Δ^- be the rectangular $n \times T$ matrix $\Delta^- = (\mathbf{0} \ D^- \ \mathbf{0})$ (where the first null symbol represents the rectangular $n \times p$ matrix with null entries). Set $\Delta = \binom{\Delta^+}{\Delta^-}$, so that ${}^t \Delta J(p, n) \Delta = D(\lambda_1, \ldots, \lambda_T)$. Finally, define the $(p + n) \times T$ matrix

$$\Xi = \Delta^t V. \tag{7}$$

Using these constructions, the matrix M verifies

$$M = {}^t\Xi J(p,n)\Xi. \tag{8}$$

Let $\psi_c : E \to \mathbb{R}^{p+n}$ be the map defined by $\psi_c(x_j) := \Xi I_j$ for $j = 1, \ldots, p+n$. We equip the space $\mathbb{R}^{p+n}$ with the pseudo-euclidean structure given by the standard quadratic form q_n (see Formulas (4) and (5)) of signature (p,n). By construction of $\psi_c : E \to \mathbb{R}^{p+n}$ one has the formulas

$$\langle \psi_c(x_i)|\psi_c(x_j)\rangle_n = {}^jI\,{}^t\Xi J(p,n)\,\Xi\, I_i = {}^jm_i \tag{9}$$

which implies

$$w(x_i, x_j) = q_n(\psi_c(x_i) - \psi_c(x_j)). \tag{10}$$

Moreover, readily $\psi_c(E) + Z$ generates $\mathbb{R}^{p+n}$ for any $Z \in \mathbb{R}^{p+n}$. Therefore, the following theorem is proved:

Theorem 2. *The above construction* $(\psi_c, p+n, q_n)$ *realizes a standard embedding of* E.

Similarly to the classical euclidean case (see [16] for example), the *dual matrix* $\Xi^t\Xi$ is considered as a kernel. Therefore, by definition, principal axis are given for $i = 1, \ldots, p+n$ by

$$U_i = |\lambda_i|^{-1/2}\Xi\, V_i = |\lambda_i|^{-1/2}\sum_{j=1}^{T}{}^jV_i\psi_c(x_j) \tag{11}$$

The row vector $U = (U_1, \ldots, U_{p+n})$ is the canonical basis of $\mathbb{R}^{p+n}$ due to the definition (11) and Formulas (7) and (8) used to built the embedding $(\psi_c, p+n, q_n)$. In fact $\Xi V_i = \Delta^t V V_i = \Delta I_i = |\lambda|^{1/2}I_i$. Hence $U_i = I_i$. Set $\text{sign}(\lambda) = \lambda/|\lambda|$ if $\lambda \neq 0$ and $\text{sign}(0) = 0$. A straightforward computation shows that $\Xi^t\Xi = D(|\lambda_1|, \ldots, |\lambda_{p+n}|)$, and $\langle U_i|U_j\rangle_n = \text{sign}(\lambda_i)\,{}^i\delta_j$ for $1 \leq i, j \leq p+n$. Moreover, the linear expansion of $\psi_c(x_i)$ in the basis U is given by

$$\psi_c(x) = \sum_{i=1}^{p+n} \text{sign}(\lambda_i)\langle \psi_c(x)|U_i\rangle_n U_i$$

with

$$\langle \psi_c(x_j)|U_i\rangle_n = {}^tU_i\, J(p,n)\psi_c(x_j)$$

so that

$$\langle \psi_c(x_j)|U_i\rangle_n = |\lambda_i|^{-1/2}\sum_{k=1}^{T}{}^kV_i\langle \psi_c(x_j)|\psi_c(x_k)\rangle_n. \tag{12}$$

Any other standard embedding $(\psi', p+n, q_n)$ of E will replace U by a suitable orthonormal basis U' for $\langle \cdot | \cdot \rangle_n$. In fact, let Ξ' be the matrix corresponding to ψ' so that $M = {}^t\Xi' J(p, n)\Xi'$ and by construction $U'_i = |\lambda_i|^{-1/2} \Xi'V_i$. Then, $\langle U'_i | U'_j \rangle_n = |\lambda_i| {}^tV_j {}^t\Xi' J(p, n)\Xi' V_i = \text{sign}(\lambda_i)$ if $i = j$ and 0 otherwise.

Remark 1. The map ψ_c is injective on E if and only if all column vectors of the matrix Ξ are pairwise distinct and not null.

2.3 Centering

Assume we already have a standard embedding (ψ_0, r, q_n) of E_0 where $E_0 = \{x_1, \ldots, x_T\}$. We introduce the center μ of the "cloud" $\psi_0(E_0)$, namely $\mu = \frac{1}{T} \sum_{i=1}^{T} \psi_0(x_i)$ and replace ψ_0 by $\psi = \psi_0 - \mu$, so that $\sum_{i=1}^{T} \psi(x_i) = 0$. It is convenient to create an abstract object $c = x_0$ (distinct from all objects in E_0) called *center*, and then to extend ψ to c by $\psi(c) = 0$ (actually, 0 is the center of the cloud $\psi(E)$). The dissimilarity w is now extended to $E = E_0 \cup \{c\}$ by setting $w(c, x_i) = q_n(\mu - \psi_0(x_i))(= q_n(\psi(x_i)))$. Observe that the matrix M given by (3) with this new w verifies

$${}^im_j = \frac{1}{2}\big(q_n(\psi(x_i)) + q_n(\psi(x_j)) - q_n(\psi(x_i) - \psi(x_j))\big) = \langle \psi(x_i) | \psi(x_j) \rangle_n \,,$$

hence

$${}^im_j = \langle \psi_0(x_i) - \mu | \psi_0(x_j) - \mu \rangle_n \,. \tag{13}$$

Therefore, if Ξ' denotes the matrix which collects the vectors $\psi(x_i)$ $(1 \le i \le T)$ one has ${}^t\Xi' J(p, n)\Xi' = M$. Since the cloud $\psi(E_0)$ is centered, set $\tilde{w}_j = \frac{1}{T} \sum_{i=1}^{T} w_{ij}$, $\tilde{w} = \frac{1}{T^2} \sum_{i=1}^{T} \sum_{j=1}^{T} w_{ij}$. where $w_{i,j}$ stands for $w(x_i, x_j)$. By expanding the right member of Formula (13), one gets for $1 \le i, j \le T$:

$${}^im_j = \frac{1}{2}(\tilde{w}_i + \tilde{w}_j - w_{ij} - \tilde{w}) \tag{14}$$

which are the classical Torgeson relations [15] in case of euclidean spaces. It follows that the matrix M (from Formula (13)) is intrinsically defined from the dissimilarity only, *i.e.*, does not depend of any centered standard embedding in use.

2.4 The Problem of Uniqueness

Readily, translations have no effect on the dissimilarity formula (10), so that we can assume that all embeddings (ψ, s, q) of $E_0 = \{x_1, \ldots, x_T\}$ (endowed with dissimilarity w) are centered. We emphasize this fact by exhibiting, if necessary, a centered object $c = x_0$ as done above with $\psi(x_0) = 0 = \sum_{i=1}^{T} \psi(x_i)$. It is clear that from a standard embedding, we can construct another one using compositions of translations and automorphisms of the underlying pseudo-euclidean space. In fact, all standard embeddings are obtained in this way from any one of them. An old well-known fact says that in a given euclidean space $\mathcal{W}$, a finite subset E of $\mathcal{W}$ is determined, up the group of automorphisms of the affine euclidean structure of $\mathcal{W}$, by the mutual distance map $(x, y) \mapsto ||x - y||_{\mathcal{W}}$ $((x, y) \in E \times E)$. This result has been extended in the framework of pseudo-vector spaces over any field of characteristic not 2, and we refer to [27] for a proof. In our case we need an analogous result in order to identify two sets of $T + 1$ points which have, after a suitable indexation, the same mutual pseudo-distance map but without assuming in advance that their are subsets of the same standard pseudo-euclidean vector space. In fact, one has the following theorem (restricted to the field of real numbers for our purpose) which is proved in AppendixA.1:

Theorem 3. (uniqueness) *Let (ψ, r, q) and (ψ', r', q') be two standard embedding solutions of a given set E of data equipped with a dissimilarity w. Then $r = r'$ and the standard quadratic forms q and q' are identical. Moreover, there exists two translations τ, τ' on $\mathbb{R}^r$ and an automorphism A of the quadratic space $(\mathbb{R}^r, q)$ such that $\psi' = (\tau' \circ A \circ \tau) \circ \psi$.*

A standard embedding is not necessarily proper. The next result, proved in Appendix A.3, resolve this question but we first have to introduce the following natural notion:

Definition 4. *A dissimilarity function w on the set of data E is said to separate E if for all x and y in E, $x \neq y$, there exists $z \in E$ such that $w(x, z) \neq w(y, z)$.*

Notice that in the definition, the condition $w(x, z) \neq w(y, z)$ is obviously verified if $w(x, y) \neq 0$. In particular, if all pairwise dissimilarities $w(x, y)$ with $x \neq y$ are not zero, then the dissimilarity function separate the data. We are ready to set:

Theorem 4. *Let E be a data set equipped with a dissimilarity function w. A standard embedding of E is proper if and only if w separates E.*

3 Example

In this section, we consider a simple example to show the practical use of the preceding constructions to build standard embeddings. Let $\{o_0, o_1, o_2, o_3\}$ be a set of four distinct abstract objects equipped with the dissimilarity function depicted in Fig. 1(a) and given by the following dissimilarity matrix:

$$\omega = \begin{pmatrix} 0 & 25 & -9 & -16 \\ 25 & 0 & 16 & 9 \\ -9 & 16 & 0 & -49 \\ -16 & 9 & -49 & 0 \end{pmatrix}.$$

We consider five different origins to embed these objects and use the notations M_i, V_i, Δ_i and $\Xi_i = \Delta_i{}^t V_i$ to represent the matrices M, V, Δ and $\Xi = \Delta^t V$ with center o_i respectively.

Embedding with Origin o_0

In that case, o_0 plays the role of the origin $c = x_0$ in the construction of ψ_c given in Subsection 2.2 and so

$$M_0 = \begin{pmatrix} 25 & 0 & 0 \\ 0 & -9 & 12 \\ 0 & 12 & -16 \end{pmatrix}.$$

The eigenvalues of M_0 are 25, -25 and 0. To embed the objects, we have to compute V_0, Δ_0 and Ξ_0:

$$V_0 = \begin{pmatrix} 1 & 0 & 0 \\ 0 & 0.6 & -0.8 \\ 0 & -0.8 & 0.6 \end{pmatrix}, \Delta_0 = \begin{pmatrix} 5 & 0 & 0 \\ 0 & 5 & 0 \end{pmatrix}, \Xi_0 = \begin{pmatrix} 5 & 0 & 0 \\ 0 & 3 & -4 \end{pmatrix}.$$

Therefore, $\psi_0(o_0) = \begin{pmatrix} 0 \\ 0 \end{pmatrix}$, $\psi_0(o_1) = \begin{pmatrix} 5 \\ 0 \end{pmatrix}$, $\psi_0(o_2) = \begin{pmatrix} 0 \\ 3 \end{pmatrix}$ and $\psi_0(o_3) = \begin{pmatrix} 0 \\ -4 \end{pmatrix}$ (see Fig. 1(b)).

Embedding with Origin o_1

Suppose that o_1 plays the role of the origin. The corresponding matrix M_1 has eigenvalues $25 + 15\sqrt{11}$, $25 - 15\sqrt{11}$ and 0. Now, computations of the corresponding matrices V_1, Δ_1 and Ξ_1 lead to $\psi_1(o_0) \approx \begin{pmatrix} 5 \\ -0.25 \end{pmatrix}$, $\psi_1(o_1) = \begin{pmatrix} 0 \\ 0 \end{pmatrix}$, $\psi_1(o_2) \approx \begin{pmatrix} 5.15 \\ -3.25 \end{pmatrix}$ and $\psi_1(o_3) \approx \begin{pmatrix} 4.8 \\ 3.75 \end{pmatrix}$ (see Fig. 1(c)).

Embedding with Origin o_2

Suppose that o_2 plays the role of the origin. The matrix corresponding M_2 has eigenvalues $-21 + \sqrt{1891}$, $-21 - \sqrt{1891}$ and 0, and computations of V_2, Δ_3 and Ξ_3 lead to $\psi_2(o_0) \approx \begin{pmatrix} 0.51 \\ 3.04 \end{pmatrix}$, $\psi_2(o_1) \approx \begin{pmatrix} -4.56 \\ 2.19 \end{pmatrix}$, $\psi_2(o_2) = \begin{pmatrix} 0 \\ 0 \end{pmatrix}$ and $\psi_2(o_3) \approx \begin{pmatrix} 1.19 \\ 7.1 \end{pmatrix}$ (see Fig. 1(d)).

Embedding with Origin o_3

Suppose that o_3 plays the role of the origin. The corresponding matrix M_3 has eigenvalues $-28 + \sqrt{2409}$, $-28 - \sqrt{2409}$ and 0, and so we get the embedding $\psi_3(o_0) \approx \begin{pmatrix} 0.8 \\ 4.08 \end{pmatrix}$, $\psi_3(o_1) \approx \begin{pmatrix} -4.3 \\ 3.1 \end{pmatrix}$, $\psi_3(o_2) \approx \begin{pmatrix} 1.4 \\ 7.13 \end{pmatrix}$ and $\psi_3(o_3) = \begin{pmatrix} 0 \\ 0 \end{pmatrix}$ (see Fig. 1(e)).

Embedding with a Centered Origin

Now we suppose that the embedding is centered. To compute this embedding (see Subsection 2.3), we have to compute $\tilde{w}_j$ and $\tilde{w}$ from ω. We obtain $\tilde{w}_0 = 0$, $\tilde{w}_1 = 12.5$, $\tilde{w}_2 = -10.5$, $\tilde{w}_3 = -14$ and $\tilde{w} = -3$. The matrix M_c, which has eigenvalues $-3 + \sqrt{1886}/2$, $-3 - \sqrt{1886}/2$, 0 and 0, is given by

$$M_c = \begin{pmatrix} 3/2 & -19/4 & 3/4 & 5/2 \\ -19/4 & 14 & -11/2 & -15/4 \\ 3/4 & -11/2 & -9 & 55/4 \\ 5/2 & -15/4 & 554 & -25/2 \end{pmatrix}.$$

To embed the objects, we compute

$$V_c \approx \begin{pmatrix} 0.29 & -0.06 & -0.95 & -0.06 \\ -0.87 & -0.03 & -0.29 & 0.41 \\ 0.31 & -0.66 & 0.10 & 0.67 \\ 0.26 & 0.75 & 0 & 0.61 \end{pmatrix}, \quad \Delta_c = \begin{pmatrix} 4.33 & 0 & 0 & 0 \\ 0 & 4.97 & 0 & 0 \end{pmatrix},$$

$$\text{and } \Xi_c \approx \begin{pmatrix} 1.26 & -3.74 & 1.34 & 1.14 \\ -0.29 & -0.14 & -3.29 & 3.72 \end{pmatrix}.$$

Then, we have $\psi_c(o_0) \approx \left(\begin{smallmatrix} 1.26 \\ -0.29 \end{smallmatrix}\right)$, $\psi_c(o_1) \approx \left(\begin{smallmatrix} -3.74 \\ -0.14 \end{smallmatrix}\right)$, $\psi_c(o_2) \approx \left(\begin{smallmatrix} 1.34 \\ -3.29 \end{smallmatrix}\right)$ and $\psi_c(o_3) \approx \left(\begin{smallmatrix} 1.14 \\ 3.72 \end{smallmatrix}\right)$ (see Fig. 1 (f)). As expected $\sum_{i=0}^{3} \psi_c(o_i) = 0$.

Discussion about the Example

The five embeddings illustrate Theorem 3. In all cases, the signature of the quadratic form associated to the embedding is $(1,1)$ and, as expected, all embeddings are similar in their structure, in particular the alignment of objects o_0, o_2 and o_3 is preserved. All embeddings are equivalent but some of them are more interesting. Let motivate the reason why. The signification of a principal axis is proportional to the square root of the absolute eigenvalue associated with it. These values are given Table 1 for all embeddings. The table shows that for M_0 and M_c, both principal axis have the same signification in explaining the data cloud whereas for M_1, M_2 and M_3 one principal axis explains about 66% of the data. Consequently, if we eliminate the least significant principal axis for M_1, M_2 and M_3, then two opposed conclusions arise: (C1) o_1 is close to o_0 and (C2) o_1 is far from o_0. This means that the reduction of the dimensionality depends of the chosen origin for the embedding.

By considering the centered embedding, we know that the matrix M used to compute the embedding only depends on the dissimilarity function and not on the choice of an origin among the objects. Table 1 shows that in this case both axis are significant and that none of them could be eliminated without leading to a false conclusion, since each axis carries 50% of the variance. As

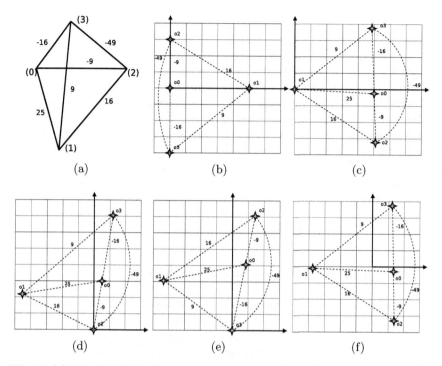

Fig. 1 (a) Pairwise dissimilarities between objects. Embedding with: (b) origin o_0, (c) origin o_1, (d) origin o_2, (e) origin o_3 and (f) where the embedding is centered

Table 1 Square roots of the absolute eigenvalues associated to the axis of the embeddings

Embedding	Horizontal Axis	Vertical Axis
The origin is at o_0	5	5
The origin is at o_1	8.64	4.97
The origin is at o_2	4.74	8.03
The origin is at o_3	4.59	8.78
Centered	4.33	4.97

expected and in accordance to the euclidean case, the centered embedding is the best candidate for visual analysis of objects with pairwise dissimilarities.

4   Pseudo-euclidean Scatterplot Matrix for the Analysis of Pairwise Dissimilarities

From now on we assume that a centered embedding of the objets is computed. The utilization of this embedding to analyze and to take decision is not a trivial task. If we consider classical data mining techniques, many quantities

may be computed on data. However, those quantities are based either on
pairwise dissimilarities or on the spatial configuration of the embedded data.

In the first case, pseudo-euclidean embedding is useless. In the second case,
an euclidean structure or at least a metric or near-metric structure is required
but such a structure does not exist if the embedding is neither euclidean nor
anti-euclidean. Another approach must be considered. To this aim we propose
to associate mathematical properties of the embedding with visual tools and
called this method: the pseudo-euclidean scatterplot matrix (PESM) method.
In order to present it, we proceed in three steps. First, we introduce a decomposition of the pseudo-euclidean spaces into an orthogonal sum of pseudo-
euclidean spaces of dimension 2, plus eventually one of dimension 1. After that,
we define criteria that can be used in selecting most significant principal axis to
reduce the dimensionality of the embedding. Finally, we present various visual
tools that constitute the PESM method in itself.

4.1   Spliting of Pseudo-euclidean Spaces

We start with five basic standard pseudo-euclidean spaces: the euclidean and
anti-euclidean spaces of dimension 1 and 2, namely $(\mathbb{R}, q_0)$, $(\mathbb{R}, -q_0)$, $(\mathbb{R}^2, q_0)$,
$(\mathbb{R}^2, -q_0)$, and the space $(\mathbb{R}^2, q_1)$. By construction, principal axes of a minimal
embedding into a pseudo-euclidean space $(\mathcal{W}, q)$, are orthogonal and can be
used to decompose $\mathcal{W}$ into an orthogonal sum

$$\mathcal{W} \simeq \begin{cases} \mathcal{W}_1 \oplus \cdots \oplus \mathcal{W}_t & \text{if } \mathcal{W} \text{ is of dimension } 2t \\ \mathcal{W}_1 \oplus \cdots \oplus \mathcal{W}_t \oplus \mathcal{W}_* & \text{if } \mathcal{W} \text{ is of dimension } 2t+1 \end{cases}$$

where t is a positive integer, each $\mathcal{W}_i$ being standard pseudo-euclidean space
of dimension 2 and $\mathcal{W}_*$ the standard euclidean or anti-euclidean space of
dimension 1.

Remark 2. The above decomposition is far from to be unique. As it will be
explained in following sections, we can exploit advantageously this diversity
(see section 4.3).

Suppose that $(\mathcal{W}, q)$ is a standard pseudo-euclidean space of dimension 2. For
$q = q_0$ (or $q = q_2 - q_0$) two points are close in the usual sense, that is to say,
their corresponding coordinates are also close each other and reciprocally.
The isoline (also called isovalues) $q(x) = c$ at level $c \geq 0$, corresponding to
the locus of constant dissimilarity c from the origin, is a circle.

The case $q = q_1$ corresponds to the classical hyperbolic plane. The isotropic
cone $C(q) = q^{-1}(0)$ is the union of the lines $\mathbb{R} \times \binom{1}{1}$ and $\mathbb{R} \times \binom{1}{-1}$, the isolines
$q(x) = c$ ($c \geq 0$ or $c \leq 0$) are equilateral hyperbola. In other words, objects
of identical dissimilarity with the centered object are sent on an hyperbola.
Notice that in this case, for any $\varepsilon > 0$, there are points x and y such that
$q_0(x - y) \geq 1$ and $q_1(x - y) \leq \varepsilon$. For our analysis, this fact should be fruitful.

In the case $q = q_2$ i.e., the anti-euclidean case, the isotropic cone is reduced
to $\{0\}$ and if we replace q by $-q$, the space becomes euclidean. Notice that

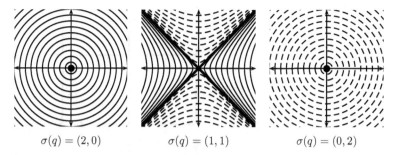

$$\sigma(q) = (2,0) \qquad\qquad \sigma(q) = (1,1) \qquad\qquad \sigma(q) = (0,2)$$

Fig. 2 Isolines of a 2-dimensional standard pseudo-euclidean space. Continuous lines are used for positive levels, dashed lines are used for negative levels and bold lines are set for the isotropic cone

here, all dissimilarities are negatives. The isolines are circle but depicted with a dashed line. Fig. 2 shows the isolines for each standard pseudo-euclidean space of dimension 2.

4.2 Criteria for Principal Axis Selection and Dimensionality Reduction

As explained in Section 2, the embedding of $T + 1$ objects leads to a pseudo-euclidean space of dimension r at most T. For large r, the embedding is untraceable and can not be used even after considering the direct sum decomposition. In practice, some approximation-reduction of the embedding must be used. One way to compute such a reduction is to eliminate the least significant axes as it is done in the classical principal component analysis [24]. Due to the structure of pseudo-euclidean spaces, we propose a selection of axis not only based on one criterion, but on four parameters denoted respectively η^+, η^-, η^* and η, and explained below. For a given standard embedding of signature (p, n), let $\lambda_1^+, \ldots, \lambda_p^+$ be the positive eigenvalues (counted with their multiplicity) sorted in decreasing order. Let $\lambda_1^-, \ldots, \lambda_n^-$ be the negative eigenvalues sorted in decreasing order of magnitude. We introduce the quantities $\lambda^+(k) = \sum_{i=1}^k \lambda_i^+$ ($k \leq p$) and $\lambda^-(k) = \sum_{i=1}^k |\lambda_i^-|$ ($k \leq n$). The selection of the first k^+ positive principal axis and the first k^- negative principal axis can be evaluated by the four ratios:

$$\eta^+ = \frac{\lambda^+(k^+)}{\lambda^+(p)}, \quad \eta^- = \frac{\lambda^-(k^-)}{\lambda^-(n)}, \quad \eta^* = \frac{\lambda^+(k^+) + \lambda^-(k^-)}{\lambda^+(p) + \lambda^-(n)}, \quad \eta = \frac{\lambda^-(n)}{\lambda^+(p)}.$$

These parameters quantify diverse selections and combinations of significant axises. More precisely, ratios η^+, η^-, η^* and η measure respectively the quantity of information carried by the selected positive principal axis, by the selected negative principal axis, by all selected principal axis, and by the

Fig. 3 Organization
of a scatterplot matrix
of dimension 3. v_1, v_2
and v_3 denote the three
dimensions of the data

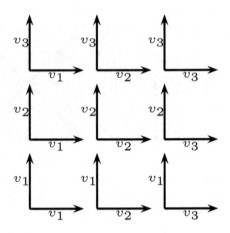

negative space against the positive space. Relatively large values of η (for example ≥ 0.3) mean that the negative part of the space is not negligible. When η is near to 0, the user might select only positive principal axis but if η is great, both positive and negative principal axis must be selected to explain the cloud. In that case, η^+, η^- and η^* must be sufficiently large to preserve the most important visual properties of the embedding.

4.3 Visual Tools for the Analysis

The Scatterplot Matrix

Scatterplot matrix is a classical visualization tool that allows to represent a cloud in $\mathbb{R}^r$ by several graphs. The matrix is composed of all two dimensional graphs corresponding to all possible couple of principal axes (see [3, 8]). On each 2D graph, data are plotted according to their coordinates in the corresponding couple of axes. Fig. 3 depicts the organization we have adopted for any scatterplot matrix. Being limited by the size of the visualization screen and the restricted number of axis that can be simultaneously exploited, we have to perform a dimensionality reduction. The method described in section 4.2 will be applied. This dimensionality reduction allows the user to preserve main properties of the embedding and at the same time to control the level of both approximation and abstraction.

On the Usefulness of Such Representation

Scatterplot matrices have many advantages: (A1) each principal axis plays the same visual role (see [28]); (A2) the analysis of high dimensional spaces is reduced to the analysis of many two dimensional spaces; (A3) all couples of principal axis are visualized at the same time. Advantage in (A1) implies that none of principal axes are favored by the representation, (A2) allows to simplify the analysis process and (A3) allows to eliminate the effect due to the multiple possible decomposition in orthogonal sums.

Scatterplot matrices are mainly used to detect three kinds of properties within data [3]: correlation between coordinates (i.e. dimensions), groups and outliers. By construction, the detection of correlation between dimensions is not interesting in our case because the embedding produces principal axes that are uncorrelated in the sense that the axes are orthogonal. However, detection of groups and detection of outliers can still be performed. In fact, spatial information carried by each scatterplot gives us enough information to detect groups and outliers. Moreover, the combination of the scatterplot matrix and of the embedding permit (see Section 5) to give meanings to principal axes and a better understanding of the dissimilarity.

Visual Tools on the Scatterplot Matrix

To help the interpretation of the representation by scatterplot matrices, many additional visual tools are needed. Most of then are proposed bellow.

– *Isolines:* In order to visualize the signature of a 2-dimensional pseudo-euclidean space we draw the isolines on the user's request: from a click on one scatterplot, the isolines, corresponding to the mouse position, are drawn. And when the user clicks on a embedded datum, the isolines allow to locate others points, both in the euclidean or anti-euclidean cases and in the hyperbolic case.

– *Reduction of visual marks for isolines:* in order to simplify the interpretation, it can be noticed that for groups and outliers detection, the sign of the level value associated with isolines does not have any importance. Indeed, only the scaling of the dissimilarity is useful to say if two data are similar or dissimilar. This allows to make two simplifications. Firstly, euclidean and anti-euclidean spaces can be considered in the same manner independently of the signature. Secondly, for hyperbolic spaces, it is not useful to differentiate the sign of the dissimilarity. This remark leads to plot isolines in the same manner independently.

– *Proportional scatterplots:* to facilitate the data analysis, scatterplots are not of the same size but proportional to the amplitude of associated eigenvalues. In a such configuration, the size of a plot emphasizes the contribution to this plan of the dissimilarity between objects. Moreover, this allows to keep the ratio on the drawing of the isolines.

– *Colored isolines:* we have added another trick to help the interpretation: isolines are colored according to the value of the associated levels. We color each isoline of level R with the color $(\max\{0, 1 - R\}, 0, 0)$ in the Red Green Blue color model. Colors facilitate the understanding of isolines in a pseudo-euclidean space by emphasizing the corresponding dissimilarity value.

– *Colored data according to criteria:* each point in the data is represented by a square that can be filled differently. At the beginning, all squares are white. The user has the possibility to map matters associated with data into each square. The mapping fill the square with a color depending on some properties of the data or fill the square with images. The color filling can be of two kinds: a specified color computed from a given attribute of the data, or a mixing of three specified colors computed by linear interpolation from the RGB components. The computed color values with are out of a preassigned range are minimized in consequence. This coloring could be used to determine the influence of attributes on the dissimilarity measure.

– *Groups selection:* Since we want to detect groups and outliers from dissimilarity, we have also added the possibility to mark data on the scatterplot. Marks could be used to show the link between all other scatterplots of the matrix. Similar techniques like the ones given in [29] are used. Marks can be of any color and are represented by empty squares around the marked square. Many points can be added to a selection at a time, using a rectangular selection. We do not impose that the selection corresponds to a rectangular one because this is of no utility for non euclidean spaces where far away points in term of coordinates can have small dissimilarity. So, to be useful, linking and brushing must be sufficiently flexible to permit the selection of non rectangular sections of the space. The easiest way to achieve that is to allow a selection composed of many rectangular parts of the space. Double click on a square representing a point creates a special selection that contains only one point. This point is selected in a tabular, viewing all attributes, that permit at request a specific analysis of the associated values.

In the next section, the PESM method is used to analyze, compare and choose various dissimilarity functions on pairwise hidden Markov models.

5 Comparison of Dissimilarities on Hidden Markov Models

Classification and learning tasks based on Hidden Markov Models (HMMs), may possibly depend on dissimilarity in between models, but various models are proposed in the literature and we need to choose a suitable one for our purpose.

5.1 Selection of an Hidden Markov Models

HMMs are widely used to model stochastic processes. They are applied in various research domains like speech recognition [30, 31], biology [32], tasks scheduling [33], information technology [34] or image recognition as well and a huge mount of papers are devoted to them, involving various kinds of HMM, like for instance the Hierarchical Hidden Markov Models[35]. We refer the interested reader to the overview [36]. We will focus our study to the basic model. Formally, this model is defined by a triple $\lambda = (A, B, \Pi)$ of stochastic matrices. In more details, if N denote the number of so-called *hidden* states and m the number of observed symbols, then A is the $N \times N$ transition matrix of the hidden states, $B = (b_j(k))_{1 \le j \le N, 1 \le k \le m}$ is the $N \times m$ matrix of distribution of the probabilities $b_j(k)$ to observe the k-th symbol at the j-th state, these probabilities being assumed to be time-independent. Finally, $\Pi = (\pi_i)_{1 \le i \le N}$ is the initial probability distribution on the space of hidden states. For any sequence, the training problem consists in finding a HMM λ^* which maximizes the probability $P(O|\lambda^*)$ to generate a given finite sequence O of observed symbols.

5.2 Dissimilarities between HMMs

Many dissimilarity measures between pairwise HMMs have been defined in the literature. See for example [36, 37, 38, 39]. We do not consider these dissimilarities measures for two reasons, one: their computation are very time-consuming; two: they are only able to compare HMMs having the same number of hidden states. The first reason is clearly justified. The second one is related to the fact that a lot of training criteria for HMMs do not impose a specific role to particular hidden states.

In fact, we consider the states as a set of random variables having (hidden) symbols for values and, for a given HMM λ, the probability distribution of these variables is denoted by $\gamma(\lambda)$. Two distributions are considered. The uniform distribution $\gamma_e(\lambda) = (1/N, \ldots, 1/N)$ and the limit distribution $\gamma_\infty(\lambda) = \lim_{t \to \infty} \Pi A^t$ where A is the matrix transition of the model λ and Π its initial distribution.

Under a given HMM λ, let $\mu(\gamma, \lambda)$ be the probability distribution of observed symbols where $\gamma = \gamma(\lambda)$ is the distribution of hidden states. By definition, for $k = 1, \ldots, m$, one has $\mu_k(\gamma, \lambda) = \sum_{i=1}^{N} \gamma_i b_i(k)$. Also, let $\rho(\gamma, \lambda)$ be the joint probability distribution of two consecutive observed symbols when the distribution of hidden states is γ. Then, for $k, \ell = 1, \ldots, m$ one has $\rho_{k,\ell}(\gamma, \lambda) = \sum_{i=1}^{N} \sum_{j=1}^{N} \gamma_i b_i(k) a_{i,j} b_j(\ell)$. The distribution μ can be considered as a first order distribution whereas ρ should be viewed as a second order distribution of observed symbols. Now, to compare two probability distributions $u = (u_i)_{1 \le i \le K}$ and $v = (v_i)_{1 \le i \le K}$, we consider the mean relative entropy given by

$$\mathrm{MRE}(u, v) = \tfrac{1}{2} \left(\sum_{i=1}^{K} u_i \ln \tfrac{u_i}{v_i} + \sum_{i=1}^{K} v_i \ln \tfrac{v_i}{u_i} \right).$$

Notice the the quantity $\sum_{i=1}^{K} u_i \ln \tfrac{u_i}{v_i}$ is the classical Kullback-Leibler divergence introduced in [40]. The MRE (up to the factor $1/2$) is introduced in [41] and has been exploited, for example, in [37] and [38]. Now we consider four dissimilarity functions given for each couple (x, y) of hidden Markov models having the same number of observed symbols but not necessarily the same number of hidden states:

$$w_e^{(1)}(x, y) = MRE(\mu(\gamma_e, x), \mu(\gamma_e, y)),$$

$$w_e^{(2)}(x,y) = MRE(\rho(\gamma_e, x), \rho(\gamma_e, y)),$$
$$w_\infty^{(1)}(x,y) = MRE(\mu(\gamma_\infty, x), \mu(\gamma_\infty, y)),$$
$$w_\infty^{(2)}(x,y) = MRE(\rho(\gamma_\infty, x), \rho(\gamma_\infty, y)).$$

For such dissimilarity functions we are able to compare HMMs with distinct numbering of the hidden states but also when HMMs have not the same number of hidden states.

5.3 Comparison of Dissimilarity Functions between HMMs

To study the above dissimilarities, we apply them on a set of HMMs. We can use either random HMMs or learned HMMs. The first case has the merit to be simple but since we can not compare with a "real" dissimilarity between such models, we can not determine the usefulness of our choice. The second case consists in using models obtained after learning from a dataset for which we already know that the learned models are good. So that, we are able to compare our dissimilarities and to verify if they are good approximations of the "real" dissimilarity and infer additional knowledge.

Protocol of Experimentation

For our experiments, the dataset in use is a subset of the ORL dataset [42] which is composed of 40 persons' faces. One has 10 photographs for each person. For our experiments, we only retain the first 5 persons in the database and for each one, we learn the 10 photographs. To build the observations we first re-sample grey levels from 256 to 64. To transform the bi-dimensional images into a one dimensional sequence of symbols, we divide each image in

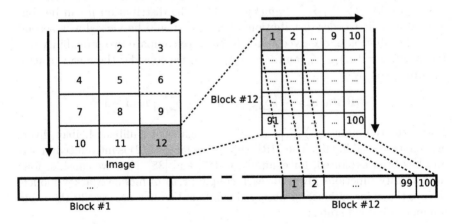

Fig. 4 Encoding of an image into a sequence

Table 2 Principal axis selected for Figures 5 and 6

Dissimilarity	η	η^+	η^-	η^*	Eigenvalue	
$w_e^{(1)}$	50%	57%			8	v_1
			55%		-4	v_2
		73%			2	v_3
		80%		71%	1	v_4
$w_e^{(2)}$	36%	59%			20	v_1
			55%		-7	v_2
		73%			5	v_3
		81%		74%	3	v_4
$w_\infty^{(1)}$	62%	63%			90	v_1
			62%		-55	v_2
		78%			21	v_3
			76%	78%	-13	v_4
$w_\infty^{(2)}$	55%	67%			170	v_1
			61%		-84	v_2
		78%			25	v_3
		85%		76%	18	v_4

Notes: all values have been rounded to the nearest integer value.

blocks of 10×10 pixels. Each block is linearized by appending rows end to end from the top to the bottom. The blocks are then concatenated, appending sequences end to end considering blocks from the top left to the bottom right, following the usual reading order (Fig. 4 shows this process).

Sequences are learned by the genetic hybrid algorithm described in [43]. Parameters setting of the genetic algorithm is standard and is not described here because we rely on the hypothesis that the algorithm finds the best possible models for any images, independently of the parameter settings.

Learned models, with the four dissimilarities, are then embedded in the pseudo-euclidean scatterplot matrix. Fig. 5 and 6 show the resulting representations when squares are filled by a color depending of the associated person. Principal axis that are selected for the representation are resumed in Table 2. The order of description of all principal axis is preserved in the figures (as in Fig. 3).

Analysis of the Results

The resulting embeddings lead to the following analysis:

– *Representativeness of the selected principal axis*: the first thing to be noticed is that only 4 axis are needed to accumulate more than 70% of the space structure; consequently, the PESMs are representative of the space structure.

– *Elimination of the second order dissimilarity measures*: the next thing that can be noticed is that indices η^+ and η^- of at least the first three

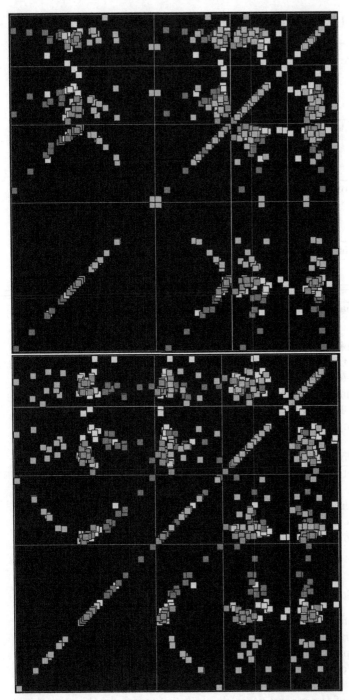

Fig. 5 The pseudo-euclidean scatterplot matrices for $w_e^{(1)}$ and $w_e^{(2)}$

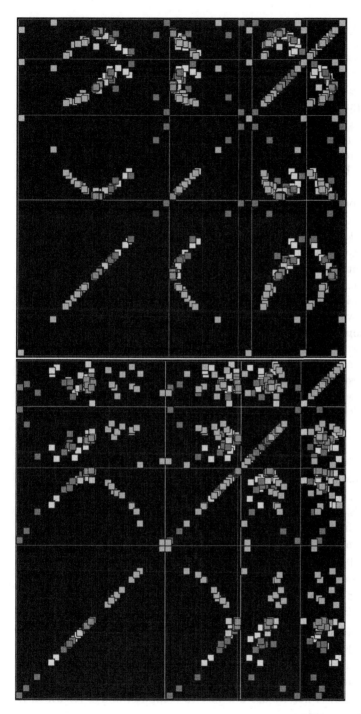

Fig. 6 The pseudo-euclidean scatterplot matrices for $w_\infty^{(1)}$ and $w_\infty^{(2)}$

eigenvalues are very similar for a given probability distribution of hidden states. Moreover, projections of data on these first three principal axis are very close if we take into account symmetries. These apparent properties of γ_e and γ_∞ show that the second order measures are useless because they imply much more computing time than the first order measures and they do not provide a significantly different structure in the space. From now on, we only consider the first order measures $w_e^{(1)}$ and $w_\infty^{(1)}$.

– *Meaning of the most significant principal axis:* when the squares that represent data are filled with the associated images (which are not shown here, due to the lack of place), we note that the first axis plays a singular role: it spans the images according to their global luminosity. Consequently, we can say that one important characteristic learned by HMMs is the global luminosity of the images.

– *Meaning of the second most significant principal axis:* moreover, we remark that the first plan (v_1,v_2), both for $w_e^{(1)}$ and $w_\infty^{(1)}$, is hyperbolic and the data send in that plan are mostly distributed on a curve corresponding to an isotropic cone. Consequently, the two first axis separate the images into two groups: light images and the others. The contribution of that plan to the dissimilarity between two data from one group is annihilated or quasi null. Therefore, the analysis can be divided into the analysis of the two independent groups for which only the third and fourth axis embed information.

– *Emerging property:* an important property emerges from these structures: data corresponding to a particular person are grouped by considering the coordinates only. Therefore, we can say that the two dissimilarity functions preserve natural grouping induced by trained models and so they can be efficiently used to manage dissimilarity between faces using classical methods.

– *Comparison of robustness of dissimilarity measures:* let us observe the class colored in green (lighter squares on black and white printing) taking into account all four axis. We can note that for $w_e^{(1)}$ one model is far from the green group whereas for $w_\infty^{(1)}$, that is not the case. Is this phenomenon results of the particular models obtained by the training phase? To answer this question, we did another training and built new PESM. And we have noticed that the phenomenon sometimes appears and sometimes it does not. This means that it is not due to an atypical HMM/image but to natural properties of the dissimilarity measure. Consequently, $w_\infty^{(1)}$ models groups better than the other one and so, is more robust against noises in data than $w_e^{(1)}$.

– *Impact of the dissimilarity measures on classical algorithms:* we have noticed that the parameter η is smaller with $w_e^{(1)}$ than with $w_\infty^{(1)}$. Therefore, $w_e^{(1)}$ is more subject to a better behavior with classical algorithm, such the k-means (preferring distance measures), than $w_\infty^{(1)}$.

Finally, if we need to choose a dissimilarity measure, we need to choose between $w_e^{(1)}$, which is not robust but fast to compute, and $w_\infty^{(1)}$, which is robust but time-consuming.

A Appendix

A.1 Uniqueness of the Standard Embeddings: Proof of Theorem 3

Let $E = \{x_0, x_1, \ldots, x_T\}$ be a set of $T + 1$ objets $(T \geq 1)$ equipped with a dissimilarity map w and let (ψ, r, q_n) and $(\psi', r', q_{n'})$ be two standard embeddings of E. To prove Theorem 3 we can assume without lost of generality that these embeddings are centered with the same origin, say x_0. Consequently the corresponding matrix M and M' given by (3) or equivalently by (14) are identical. By minimality we have $r = r'$.

Let $\{k_1, \ldots, k_r\} \subset E$ such that if $X_j = \psi(x_{k_j})$ then the row of vectors $\mathcal{X} = (X_1, \ldots, X_r)$ forms a basis of $\mathbb{R}^r$. Let $\mathcal{X}' = (X'_1, \ldots, X'_r)$ be defined for $1 \leq i \leq r$ by $X'_i = \psi'(x_{k_i})$. From the hypothesis, the Gram matrix $G_n(\mathcal{X}) = {}^t\mathcal{X} J(n, p) \mathcal{X}$ of $\mathcal{X}$ with respect to q_n is equal to the Gram matrix $G_{n'}(\mathcal{X}') = {}^t\mathcal{X}' J(n', p') \mathcal{X}'$ with respect to $q_{n'}$. This implies that $\mathcal{X}'$ is also a basis.

Let A be the square $r \times r$ matrix defined by $\mathcal{X}' = A\mathcal{X}$. By construction, for any couple of vectors (Y, Z) from $\mathbb{R}^r$, one has

$$\begin{aligned}
\langle Y | Z \rangle_n &= {}^t Z J(n, p) Y \\
&= {}^t Z \, {}^t\mathcal{X}^{-1} G_n(\mathcal{X}) \mathcal{X}^{-1} Y \\
&= {}^t Z \, {}^t\mathcal{X}^{-1} G'_n(\mathcal{X}') \mathcal{X}^{-1} Y \\
&= {}^t Z \, {}^t(\mathcal{X}' \mathcal{X}^{-1}) J(n', p')(\mathcal{X}' \mathcal{X}^{-1}) Y \\
&= \langle A Y | A Z \rangle_{n'} \ .
\end{aligned}$$

This means that A realizes an isomorphism between $(\mathbb{R}^r, q_n)$ and $(\mathbb{R}^r, q_{n'})$. Consequently, $n = n'$ (and $p = p'$). Since the quadratic form q_n is nonsingular, any vector $\psi(x_j)$ (resp. $\psi'(x_j)$) is determined by its dual coordinates $\langle \psi(x_j) | X_i \rangle_n$ (resp. $\langle \psi'(x_j) | X'_i \rangle_{n'}$) to the basis $\mathcal{X}$ (resp. $\mathcal{X}'$), so that $\psi' = A \circ \psi_c$.

A.2 From a Standard Embedding to Another One

Formula (11) suggests to introduce the operator $\Lambda : \mathbb{R}^T \to \mathbb{R}^{p+n}$ defined by

$$\Lambda = \Xi \left[\sum_{i=1}^{p+n} |\lambda_i|^{-1/2} \Pi_i \right] \tag{15}$$

where Π_i denote the usual orthogonal projection onto the vector space generated by V_i, i.e., $\Pi_i X = \langle X | V_i \rangle_0 V_i$. Consequently, $\Lambda(X) = \sum_{i=1}^{p+n} \langle X | V_i \rangle_0 I_i$ and, in particular,

$$\Lambda(V_i) = \begin{cases} U_i & i = 1, \ldots, p+n\,, \\ 0 & \text{otherwise,} \end{cases}$$

so that, by restriction, Λ realizes an isometry from $\mathrm{Ker}(Q)^{\perp}$ onto $\mathbb{R}^{p+n}$ both equipped with the standard euclidean structure. The following formulas

$${}^t\Lambda J(p,n)\Lambda = VD(\mathrm{sign}(\lambda_1), \ldots, \mathrm{sign}(\lambda_T))\,{}^tV \tag{16}$$

$$\Lambda\,{}^t\Lambda = Id_{p+n} \tag{17}$$

hold. Notice that ${}^t\Lambda\Lambda$ is the orthogonal projector (in $\mathbb{R}^T$) onto the subspace generated by $\{V_1, \ldots, V_{p+n}\}$. With the assumptions and notations of the above proof, let $\Xi = (\psi(x_1), \ldots, \psi(x_T))$ (resp. $\Xi' = (\psi'(x_1), \ldots, \psi'(x_T))$) denote the matrix associated to ψ (resp. ψ'). By assumption $M = {}^t\Xi J(p,n)\,\Xi = {}^t\Xi' J(p,n)\,\Xi'$. From Theorem 3, there exists a $r \times r$ matrix A such that $\Xi' = A\,\Xi$ and ${}^tA J(p,n)A = J(p,n)$. If we introduce the dual matrix $\Xi'\,{}^t\Xi'$ ($= A\,\Xi\,{}^t\Xi\,{}^tA$), its principal axis are given by the unitary eigenvectors $U_i' = AU_i$. The basis $U' = AU$ is orthonormal with respect to the pseudo-scalar product $\langle\cdot|\cdot\rangle_n$. In fact, we have $\langle U_i'|U_j'\rangle_n = \mathrm{sign}(\lambda_i)^i\delta_j$ and $\Xi'\,{}^t\Xi' = AD(|\lambda_1|, \ldots, |\lambda_{p+n}|)\,{}^tA$. The operator Λ' corresponding to Λ in the standard construction verifies $\Lambda' = A\,\Lambda$. Formula (16) remains unchanged but with Λ', and Formula (17) becomes $\Lambda'\,{}^t\Lambda' = A\,{}^tA$.

A.3 Proper Standard Embedding: Proof of Theorem 4

Notice that the Theorem 4 is obvious if the set of data has only two elements. Therefore let $E = \{x_0, x_1, \ldots, x_T\}$ be a set of $T+1$ objects, $T \geq 2$, equipped with a dissimilarity map w and let (ψ, r, q) be a given standard embedding of E associated to w. Assume that this embedding is proper. By Theorem 3, all standard embeddings are also proper, so that we can suppose that ψ is the standard embedding ψ_{x_0}, with origin x_0, constructed in Subsection 2.2.

Now let x_i and x_j be distinct elements such that $w(x_i, x_j) = w(x_i, x_k)$ for all $i \in \{0, 1, 2, \ldots, T\}$. We may always choose x_0 distinct from x_i and x_j. Let $M = ({}^im_j)_{1 \leq i,j \leq T}$ be the matrix given by formula (3) where $c = x_0$. By assumption on x_j and x_k one gets $MI_j = MI_k$ so that by (9) $\langle\psi_{x_0}(x_j)|\psi_{x_0}(x_i)\rangle_n = \langle\psi_{x_0}(x_k)|\psi_{x_0}(x_i)\rangle_n$ for all $i = 0, \ldots, T$. Since the embedding is proper one deduces the contradiction $\psi(x_j) = \psi(x_k)$. Hence w separates E.

Reciprocally, assume that w separates the set E but there exists a standard embedding (ψ, r, q) which is not proper. Therefore, there exists two distinct elements x_j and x_k in E such that $\psi(x_j) = \psi(x_k) = z$. By definition $w(x_j, x_i) = w(x_k, x_i) = q(z - \psi(x_i))$ for any i, proving that w does not separate E, in contraction with the hypothesis. Hence (ψ, r, q) is proper.

References

1. Inselberg, A., Dimsdale, B.: Parallel coordinates: a tool for visualizing multi-dimensional geometry. In: Kaufman, A., Rosenblum, L., Nielson, G.M. (eds.) Proceedings of the 1st conference on Visualization 1990, pp. 361–378. IEEE Computer Society Press, Los Alamitos (1990)

2. Brunsdon, C., Fotheringham, A., Charlton, M.: An Investigation of Methods for Visualising Highly Multivariate Datasets. In: Unwin, D., Fisher, P. (eds.) Technical report in Case Studies of Visualization in the Social Sciences, vol. 43, pp. 55–80 (1998)

3. Fayyad, U., Grinstein, G.G., Wierse, A. (eds.): Information visualization in data mining and knowledge discovery. Morgan Kaufmann, San Francisco (2001)

4. Ward, M.O., LeBlanc, J.T., Tipnis, R.: N-land: a graphical tool for exploring n-dimensional data. In: Proceedings of Computer Graphics International Conference 1994, Melbourne, Australia, p. 14 (1994),
davis.wpi.edu/~matt/docs/cgi94.ps

5. Vesanto, J.: Data mining techniques based on the self-organizing map, Master's thesis, Helsinki University of Technology, Espoo, Finland (1997)

6. Borg, I., Groenen, P.: Modern multidimensional scaling: theory and applications. Springer series in statistics. Springer, Heidelberg (1997)

7. Cox, T.F., Cox, M.A.A.: Multidimensional scaling, 2nd edn. Monographs on Statistics and Applied Probability. Chapman & Hall/CRC, Boca Raton (2000)

8. Wong, P.C., Bergeron, R.D.: 30 Years of Multidimensional Multivariate Visualization, Scientific Visualization, Overviews, Methodologies, and Techniques, pp. 3–33. IEEE Computer Society, Washington (1997)

9. Wills, G.J.: Nicheworks - interactive visualization of very large graphs. Journal of Computational and Graphical Statistics 8(2), 190–212 (1999)

10. Walter, J., Ritter, H.: On interactive visualization of high-dimensional data using the hyperbolic plane. In: Proceedings of the eighth ACM SIGKDD international conference on Knowledge discovery and data mining, Edmonton, Alberta, Canada, pp. 123–132 (2002)

11. Wang, J., Yu, B., Gasser, L.: Concept tree based clustering visualization with shaded similarity matrices. In: Kumar, V., Tsumoto, S., Zhong, N., Yu, P., Wu, X. (eds.) Proceedings of 2002 IEEE international conference on Data Mining, pp. 697–700. IEEE Computer Society, Maebashi (2002)

12. Eades, P., Lin, X.: Spring algorithms and symmetry. Theoretical Computer Science 240(2), 379–405 (2000)

13. Graepel, T., Herbrich, R., Bollmann-Sdorra, P., Obermayer, K.: Classification on pairwise proximity data. In: Proceedings of the 1998 conference on advances in neural information processing systems, pp. 438–444. MIT Press, Cambridge (1999)

14. Pekalska, E., Paclik, P., Duin, R.P.W.: A generalized kernel approach to dissimilarity-based classification. Journal of Machine Learning Research 2, 175–211 (2002)

15. Torgeson, W.S.: Multidimensional scaling of similarity. Psychometrika 30, 379–393 (1965)

16. Schölkopf, B., Smola, A.J., Müller, K.-R.: Kernel Principal Component Analysis. In: Advances in Kernel Methods – support vector learning, ch. 20, pp. 327–352. MIT Press, Cambridge (1999)

17. Camiz, S.: Contribution, à partir d'exemples d'application, à la méthodologie en analyse des données, Ph.D. thesis, Université Paris-IX Dauphine, Paris (2002)

18. Schnabel, R.B., Eskow, E.: A revised modified Cholesky factorization algorithm. SIAM Journal on Optimization 9(4), 1135–1148 (1999)

19. Cheng, S.H., Higham, N.J.: A modified Cholesky algorithm based on a symmetric indefinite factorization. SIAM Journal on Matrix Analysis and Applications 19(4), 1097–1110 (1998)

20. Goldfarb, L.: A unified approach to pattern recognition. Pattern Recognition 17, 575–582 (1984)

21. Goldfarb, L.: A new approach in pattern recognition. In: Kana, I.N., Rosenfeld, A. (eds.) Progress in Machine Intelligence and Pattern Recognition, vol. 2, pp. 241–402. Elsevier Sc. Publishers, Amsterdam (1985)

22. Pekalska, E.: Dissimilarity representations in pattern recognition, Concepts, theory and application, Thesis, Delft Univ. Tech, pp. 322 (2005)

23. Pekalska, E., Duin, R.P.W.: The Dissimilarity Representation for Pattern Recognition: Foundations And Applications (Machine Perception and Artificial Intelligence). World Scientific Publishing Company, Singapore (2005)

24. Harris, R.J.: A primer of multivariate statistics, 2nd edn. Academic Press, Inc., London (1985)

25. Kaye, R.W., Wilson, R.: Linear Algebra. Oxford University Press, Oxford (1998)

26. Schoenberg, I.J.: Metric spaces and positive definite functions. Trans. Amer. Math. Soc. 44, 522–536 (1938)

27. Boutin, M., Kemper, G.: On reconstructing n-point configurations from the distribution of distances or areas. Adv. Appl. Math. 32, 709–735 (2004)

28. van Wijk, J., van Liere, R.: Hyperslice - visualization of scalar functions of many variables. In: Press, I.C.S. (ed.) Proceedings visualization 1993, Los Alamitos, Canada (1993)

29. Becker, R.A., Cleveland, W.S.: Brushing Scatterplots. Technometrics 29, 127–142 (1987); reprinted in Cleveland, W.S., McGill, M.E.: Dynamic Graphics for Data Analysis. Chapman and Hall, New York (1988)

30. Baker, J.K.: The DRAGON system-An overview. IEEE Transactions on Acoustics, Speech, Signal Proceding 23(1), 24–29 (1975)

31. Jelinek, F., Bahl, L.R., Mercer, L.: Design of a linguistic statistical decoder for the recognition of continuous speech. IEEE Transactions on Information Theory 21(3), 250–256 (1975)

32. Brown, M.P., Hughey, R., Krogh, A., Mian, I.S., Sjolander, K., Haussler, D.: Using dirichlet mixture priors to derive hidden Markov models for protein families. In: A. Press (ed.) Proceedings of the 1st international conference on intelligent systems for molecular biology, pp. 47–55 (1993)

33. Soukhal, A., Kelarestaghi, M., Slimane, M., Martineau, P.: Hidden Markov Models and scheduling problem with transportation consideration. In: 15th Annual European Simulation Multi conference (ESM 2001), Prague, pp. 836–840 (2001)

34. Serradura, L., Vincent, N., Slimane, M.: Web pages indexing using hidden markov models. In: 6th International Conference on Document Analysis (ICDAR), Seattle, pp. 1094–1098 (2001)

35. Fine, S., Singer, Y., Tishby, N.: The hierarchical hidden markov model: analysis and applications. Machine Learning 32(1), 41–62 (1998)

36. Rabiner, L.R.: A tutorial on hidden Markov models and selected applications in speech recognition. Proceedings of the IEEE 77, 257–286 (1989)
37. Falkhausen, M., Reininger, H., Dietrich, W.: Calculation of distance measures between hidden Markov models. In: Proceedings of the Eurospeech 1995, Madrid, pp. 1487–1490 (1995)
38. Vihola, M., Harju, M., Salmela, P., Suontausta, J., Savela, J.: Two dissimilarity measures for HMMs and their application in phoneme model clustering. In: Proceedings ICASSP 2002, 2002 IEEE International Conference on Acoustics Speech and Signal Processing, Orlando, Florida, USA, pp. 933–936 (2002)
39. Do, M.N.: Fast approximation of Kullback-Leibler distance for dependence trees and hidden Markov models. IEEE Signal Processing Letters 10(8), 250–254 (2003)
40. Kullback, S., Leibler, R.: On information and sufficiency. Ann. Math. Statist. 22, 79–86 (1951)
41. Kullback, S.: Information theory and Statistics. Dover Publication, New York (1968)
42. Samaria, F., Harter, A.: Parameterisation of a stochastic model for human face identification. In: IEEE workshop on Applications of Computer Vision, Sarasota, Florida, pp. 138–142 (1994)
43. Slimane, M., Venturin, G., de Beauville, J.P.A., Brouard, T., Brandeau, A.: Optimizing hidden markov models with a genetic algorithm. In: Alliot, J.-M., Ronald, E., Lutton, E., Schoenauer, M., Snyers, D. (eds.) AE 1995. LNCS, vol. 1063, pp. 384–396. Springer, Heidelberg (1996)
44. Engelen, S., Hubert, M., Vanden Branden, K.: A comparison of three procedures for robust PCA in high dimensions. Austrian Journal of Statistics 34(2), 117–126 (2005)
45. Huber, M., Engelen, S.: Robust PCA and classification in biosciences. Bioinformatics 20, 1728–1736 (2004)

Dynamic Self-Organising Maps: Theory, Methods and Applications

Arthur L. Hsu, Isaam Saeed, and Saman K. Halgamuge

Summary. In an effort to counter the restrictions enforced by the fixed map size and aspect ratio of a Kohonen Self-Organising Map, many variants to the method have been proposed. As a recent development, the Dynamic Self-Organising Map, also known as the Growing Self-Organising Map (GSOM), provides a balanced performance in topology preservation, data visualisation and computational speed. In this book chapter, a comprehensive description and theory of GSOM is provided, which also includes recent theoretical developments. Methods of clustering and identifying clusters using GSOM are also introduced here together with their related applications and results.

1 Introduction

The classical, two-dimensional Self-Organising Map (SOM) is constructed based on a static, user-defined map size and aspect ratio. Such an imposition on feature map formation endangers accurate representation of the topology of an input space, especially when the predefined width and height of the map mismatch the input distribution [1]. Consequently, variants and extensions of SOM, categorised as incremental self-organising networks, have been proposed to overcome the limitation of a static network structure. A number of well-accepted incremental self-organising networks include the Growing Cell Structure (GCS) [2], Dynamic Cell Structures (DCS) [3], Growing Neural Gas (GNG) [4], Incremental Growing Grid (IGG) [5] and the Grid Growing Self-Organising Map (GGSOM) [6].

Isaam Saeed
Biomechanical Engineering Research Group
Department of Mechanical Engineering
Melbourne School of Engineering
e-mail: `i.saeed@pgrad.unimelb.edu.au`

A.-E. Hassanien et al. (Eds.): Foundations of Comput. Intel. Vol. 1, SCI 201, pp. 363–379.
springerlink.com   © Springer-Verlag Berlin Heidelberg 2009

As a recent development, the Dynamic Self-Organising Map, also known as the Growing Self-Organising Map (GSOM) [7], provides a balanced performance in topology preservation, data visualisation and computational speed. GSOM combines a number of merits from different incremental self-organising networks. Firstly, it maintains a two-dimensional grid (a three-dimensional grid is also possible) for ease of data visualisation and promotion of topology preservation. Secondly, it increments the network size parsimoniously: only a few new nodes are created per iteration, thereby reducing computational cost. Thirdly, the insertion of nodes occurs only when required, and at locations on the map that under-represent the input space under investigation.

Over the past decade, GSOM has evolved in several aspects. Fundamentally, it now provides the same alternatives in lattice structure as SOM and has become more consistent for data of largely different dimensionalities. Methods of clustering or identifying clusters using GSOM have been proposed, permitting its application to data with contrasting properties. Consequently, applications of GSOM are wide-ranging and span different industries and disciplines of research. In this book chapter, a comprehensive description and theory of GSOM is provided, which also includes recent theoretical developments. Methods of clustering and identifying clusters using GSOM are also introduced here together with their related applications and results.

2   Theoretical Foundation

The GSOM method consists of three progressive stages, following initialisation, that lead to the formation of a feature map. A growing phase that builds upon an initialised minimum node network aims at tackling under-representation of an input space by dynamically adding nodes to the map as per required. Two tuning phases, also referred to as smoothing phases, append map growth and serve to smooth the residual error over the fully grown map.

2.1   *Initialisation*

A foundation for map growth is a network that consists of a minimum number of nodes, dependent on the choice of lattice structure or topology. As illustrated in Fig. 1, four nodes are sufficient for a rectangular topology map, whereas seven nodes are necessary for a hexagonal topology.

Each node represents a weight vector (1),

$$w_i = [\eta_{i1}, \ldots, \eta_{iD}], \tag{1}$$

with the number of dimensions, D, equal to that of the input space; and has associated with it a neighbourhood encompassed, for example, by a Gaussian kernel.

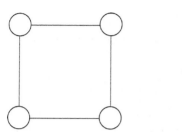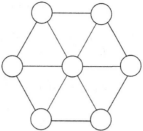

Fig. 1 Two examples of an initialised GSOM: **(a)** nodes connected in a rectangular topology **(b)** hexagonal topology

The components of these weight vectors are initialised, for instance, by the results of Principal Components Analysis on input data; random values from the domain of the input space; or simply set to constant values. However, during the early stages of map formation, only a few nodes are present in the map, consequently adaptation to input data will be comparable irrespective of the choice of initialisation technique.

2.2 Growth/Ordering

Dynamic insertion of nodes in response to input data is fundamental to GSOM. Underlying the concept of such growth is the concept of *Vector Quantisation* (VQ), a process by which a set of vectors in some D-dimensional space (input data) is approximated by a smaller set of reference vectors (weight vectors of each node), again in D-dimensional space. Now, underrepresentation, meaning that the weight vectors do not adequately quantise the input data presented thus far, in a particular region of a growing map induces a high error accumulation among nodes in that region and eventually forces the insertion of new nodes to alleviate the error.

To summarise this process, given a chosen distance metric d, each input vector is compared against every node (weight vector) and evaluated for similarity. The task that follows is to identify the node that minimises d, henceforth referred to as the Best Matching Unit (BMU). Once found, the accumulated error at the BMU is updated according to (2):

$$E_{BMU}(t+1) = E_{BMU}(t) + d(v, w_{BMU}).$$ (2)

A global indication of the input quantisation error at instant t is therefore calculated as the sum total of the error accumulation at each node (3):

$$QE = \sum_i E_i.$$ (3)

Repeated for each input vector, the BMU and its neighbours N_{BMU}, where $BMU \in N_{BMU}$, update their weight vectors according to (4):

$$w_j(t+1) = w_j(t) + LR(t) \times (x_i - w_j(t)), \tag{4}$$

where t is the current time instant, $j \in N_{BMU}$. The learning rate update $LR(t+1)$ is governed by (5):

$$LR(t+1) = \alpha \times \left(1 - \frac{R}{n(t)}\right) \times LR(t), \tag{5}$$

where R is an arbitrary constant, $n(t)$ is the number of nodes in the feature map at instant t, and $0 < \alpha < 1$ is required for convergence. It is permissible for a node to accumulate error until a predefined global *Growth Threshold* (GT) (6),

$$\frac{\delta QE}{\delta E_i} E_i > GT, \tag{6}$$

is reached. Once such a node has been identified, and if it is also a boundary node, additional nodes a grown in accordance with the chosen lattice structure, as shown in Fig 2 for a hexagonal lattice. Weight vectors of the newly generated nodes are initialised accordingly. If, on the other hand, a node is not located at the boundary of the map, half of its accumulated error is propagated in equal shares to each of its neighbours.

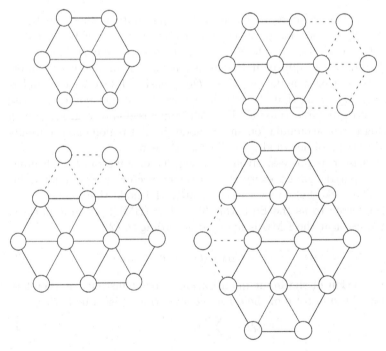

Fig. 2 Possible modes of node insertion for a hexagonal topolgy GSOM

Growth Threshold (GT)

Data analysts are able to control the size, and consequently the resolution, of feature maps produced by GSOM, with the adjustment of a dimension-independent parameter called the *Spread Factor* (SF). A lower SF will form relatively smaller maps, producing an overall visualisation of input data. The converse is similarly true, where a higher spread factor will produce larger, higher resolution maps. Here, resolution refers to the ratio of the number of nodes over the number of inputs in a fully grown map. To retain the afore-mentioned dimension-independent growth property, however, care should be taken to ensure consistency in implementation. For example, GT as it was initially proposed (7),

$$GT = -D_{mod} \times \ln(SF), \qquad (7)$$

is consistent with a squared Euclidean distance metric ($D_{mod} = D$), whereas the choice of a standard Euclidean distance measure results in a different derivation of GT ($D_{mod} = D^{\frac{1}{2}}$), and similar is the case for a general Minkowski distance metric ($D_{mod} = D^{\frac{1}{p}}$).

Though the derivation of GT is trivial, the emphasis on consistency is necessary, where a distance metric that is not reflected in GT will result in ill-sized feature maps that do not scale consistently with dimensionality. For example, the original GT equation (scales with D) when used with a standard Euclidean distance metric (scales with $D^{\frac{1}{2}}$) causes the error accumulation on each node to lag significantly, consequently impeding the natural growth of the map.

2.3 Smoothing

The quantisation error of a fully grown map is smoothed over two consecutive phases. Here, nodes are no longer added to the feature map. Instead, input vectors are presented to the feature map, as described previously, and weight adaptation occurs over a more localised neighbourhood. It is, however, desired that weight vectors do not undergo dramatic shifts in value. Therefore, the initial learning rate, $LR(0)$, is set to a lower value for the first of the two phases and lower still for the second of the two phases. Similarly, the learning rate update is chosen such that depreciation occurs at a reduced rate, achieved by selecting an appropriate value of α for each smoothing phase.

3 Evolution of GSOM

A number of extensions to GSOM have consequently brought forth improvements to feature map construction. Fundamental to a feature map, whether it be created by any given construction algorithm, is the ability to remain

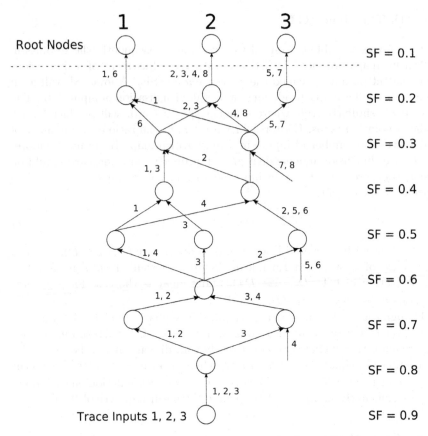

Fig. 3 A Dynamic SOM Tree model consisting of a hierarchy of varying spread factor GSOM

faithful to input space topology, to enhance visualisation of a particular set of data, and to be constructed in a fast manner. In response to such demand, improvements to each area have been unveiled.

3.1 Automated Cluster Identification

The SF parameter that is specific to GSOM can be exploited to automate cluster identification. As mentioned previously, SF can be thought of as a parameter with which to control the resolution of a feature map. By forming a hierarchy of increasing SF GSOMs (from upper-most layer, downward), interesting properties of the underlying input data become evident. This concept is embodied by the Dynamic SOM Tree Model (DSTM) [8]. Inputs that are mapped to the same neuron in one layer are traced back to nodes in the layer above. By keeping track of nodes that share mappings, a clustering of the

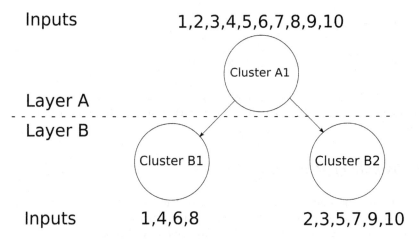

Fig. 4 Cluster separation and merging

data becomes apparent. Once root nodes have been identified (Fig. 3), they are considered *Abstract Clusters* for the reason that such clusters inherit the most abstract information about the input data. Neurons with overlapping root nodes are similarly children of the same cluster.

With such a setup, cluster relationships, such as separation and merging, can be explored. As a trivial example, consider two layers of a GSOM hierarchy, as shown in Fig. 4. When viewed relative to the top layer, the clusters that appear at the lower level have *merged*. Alternating the point of reference to the lower level, the clusters appear to have *separated* from the one cluster at the upper layer.

Attribute-Cluster Relationship (ACR) Model for a Hierarchical GSOM

The original ACR model was proposed with the capacity for a single layer GSOM. The following is an extension of the basic ACR model to handle a hierarchical GSOM [8]. ACR is essentially the addition of two layers to the DSTM. Directly above the DSTM is the *Cluster Summary Layer* (CSL) which connects itself to all nodes in each abstract cluster, thereby possessing the capability to extract, for instance, the mean, maximum, minimum and standard deviation of each attribute within each cluster. Above CSL is the *Attribute Layer* (AL), subsequently connected to all nodes in CSL, and is responsible for facilitating queries about cluster descriptions and/or queries regarding attributes. It is further possible for ACR to extract rules, in the guise of IF-THEN statements, that describe the input data.

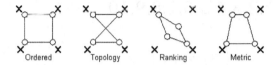

Fig. 5 Examples of Topographic Errors in feature maps. **(a)** correctly represented **(b)** error in map topology **(c)** ranking error **(d)** incorrect choice of a distance metric

3.2 Topology Preservation

To appreciate the concept of topology preservation, it is required that an objective measure be presented. There exist alternative choices for a definition of what is meant by an error in feature map topology. According to [9] there are three such modes for a topographic error, depicted in Fig. 5.

Nevertheless, for the following discussion, a simple measure used in Kohonen's SOM Pak [10] (8):

$$E_t = \frac{\sum_i^N f(r_{bi}, r_{sbi})}{N} \tag{8}$$

$$f(r_{bi}, r_{sbi}) = \begin{cases} 1, & \|r_{bi} - r_{sbi}\| > 1 \\ 0 & \end{cases} \tag{9}$$

will suffice, where, r_{bi}, r_{sbi} denote the indices of the BMU and the next-best BMU.

In light of this, a fundamental extension to the rectangular topology GSOM is the added support for a hexagonal topology [11], as it has been established that SOM with the capacity for both rectangular and hexagonal lattice structure is more faithful to input space topology when using the hexagonal variant [1]. For the following discussion, a hexagonal topology will be assumed.

Recursive Mean Directed Growing (RMDG)

It is during the growing phase of GSOM that topographic errors are introduced. The cause for this is nodes which are added incorrectly, and such errors cannot be rectified during the subsequent tuning phases. However, information *about* data can be used to preempt this. For instance, consider the centroid of the input space. If GSOM is able to adapt its growing procedure such that a feature map is grown in accordance with the extrinsic properties of input data, then topographic errors will be alleviated. In other words, by maintaining an updated location of the input space centroid, a map is grown such that it has a corresponding centroid. A technique that facilitates this is

Table 1 Comparison of feature map **quantisation error** using a default SOM, a SOM with optimal width and height (for the given dataset), standard GSOM and GSOM augmented with RMDG

Dataset	SOM (default)	SOM (optimal)	GSOM	GSOM with RMDG
Square	0.573	0.571	0.573	0.576
L-Shape	0.451	0.447	0.451	0.451
Iris	0.304	0.299	0.301	0.299

known as Recursive Means (RM) [1]. An adapted version for GSOM takes RM and tailors it specifically for map growth [12]. As each input vector is presented to the growing network, it contributes to a running average, essentially acting as a rudimentary sense of memory of all past inputs. Computing the value of the centroid of the input space (10),

$$RM_{data}(t+1) = \frac{RM(t) \times t + v_{t+1}}{t+1}, \tag{10}$$

and that of the growing map (11),

$$RM_{GSOM}(t+1) = RM_{GSOM}(t) + \frac{\sum_i \Delta w_i}{N}, \tag{11}$$

is straightforward. To promote, or direct, growth in a preferential direction, the error accumulation of the BMU is modified so that now (12)

$$E_{BMU}(t+1) = E_{BMU}(t) + (1-\rho) \times d(v,w) + \rho \times d(RM_{data}(t), RM_{GSOM}(t)), \tag{12}$$

where the parameter ρ governs the trade-off between direction of growth toward the input space centroid, and the natural unconstrained growth of the feature map. The optimal value has been empirically determined to be $\rho = 0.7$. Modification is also made to the initialisation of newly generated nodes: whereas previously node initialisation (13),

$$w_{new} = w_{BMU} + \beta \times (w_{BMU} - w_{opposite}), \tag{13}$$

was determined when $\beta = 1$, it has been empirically determined that in the optimal case considering topology preservation, $\beta = 0.7$.

These optima were determined by evaluating the results of GSOM on 2,400 setting variations, including various benchmark datasets, spread factors, learning rates, kernel widths and weight initialisations. As can be seen from Tables 1 and 2, GSOM with RMDG outperforms SOM as well as the standard GSOM algorithm in several benchmark datasets.

Fig. 6 Grid implementation of GSOM

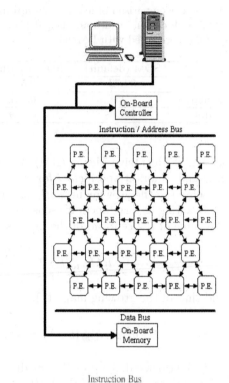

Fig. 7 Processing element (PE) within grid

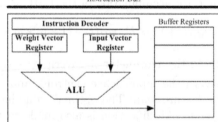

Table 2 Comparison of feature map **topographic error** using a default SOM, a SOM with optimal width and height (for the given dataset), standard GSOM and GSOM augmented with RMDG

Dataset	SOM (default)	SOM (optimal)	GSOM	GSOM with RMDG
Square	0.018	0.126	0.09	0.08
L-Shape	0.082	0.218	0.106	0.068
Iris	0.18	0.12	0.173	0.13

3.3 Speeding Up GSOM

Often times the slower speed of self-organising networks has been criticised as being impractical. However, there have been recent improvements that have specifically aimed at reducing the computational speed of GSOM. The first of these improvements relies on a hardware architecture that institutes a parallel implementation of GSOM [13], which maximally processes computations that are better suited for parallel execution. Similarly, an implementation of GSOM using Grid Computing (GC) has been reported [14] that also aims for parallel computation, but at a much lower cost than obtaining a dedicated hardware.

Hardware Architecture for GSOM

Hardware architecture for GSOM has been proposed with emphasis placed on maximising parallel computation while minimising hardware cost [13]. Since each neuron performs identical computations during each iteration, they are considered ideal candidates for parallel processing. Furthermore, floating-point arithmetic imposes additional hardware and computational cost, but both are circumvented by modifying the GSOM algorithm. The hardware architecture proposed in [13] consists of an array of processing elements (PE), an on-board memory (OBM) and a conventional (sequential) type of controller to manage the flow of parallel operations. Fig. 6 shows the schematic diagram of the proposed hardware module and Fig. 7 shows the elements of a single PE.

The size of the PE array is a significant design consideration of the architecture. It was reported that a 5x5 neighbourhood size is, in most cases, sufficient to map all neurons when performing weight adaptation.

The next consideration for hardware selection is the functional capacity of each processing element. The proposed architecture requires only Arithmetic Logic Units (ALUs) that support basic bitwise operations, such as additions, left-shift and right-shift operations, which can perform binary summation (subtraction is adding a negative number), division (divide by 2n) and multiplication (multiply by 2n) operations. Essentially, each PE consists of an instruction decoder, registers for storing temporary information and a simple ALU.

It is assumed that the OBM is large enough to handle most input datasets and the temporary variables needed for calculations. When the process begins, the main computer downloads the total input vector space onto the OBM and hands over control to the on-board controller, freeing the main computer for processing of other jobs. A comparison was conducted for a large number of data sets between the original GSOM algorithm and the modified GSOM algorithm, which uses only binary arithmetic operations. The modified GSOM algorithm produced comparable results to the original algorithm. The proposed hardware has a higher efficiency as the feature map grows to above 25 neurons. However, the efficiency of PE usage for weight adaptation drops as the neighbourhood kernel size for weight adaptation decreases. Although not all PEs are working at full load at all times, the overall throughput and speed will still be faster than sequential processing.

Grid Implementation of GSOM

The emergent technology of GC enables the sharing of distributed heterogeneous resources to solve computationally intensive and data intensive problems. Many projects have already been developed using GC in areas such as weather forecasting, financial modelling and earthquake simulation, to acquire various resources distributed at different locations, in order to satisfy their demands of huge computational exercises [15]. It is clear that GSOM can also benefit from GC to improve scalability, if we can divide the clustering task into independent sub-tasks and execute them separately.

A document clustering approach using a scalable two-level GSOM utilising grid resources to cluster massive text data sets was proposed in [14]. It attempts to achieve three goals [16]:

- Divide the original task into sub-tasks such that the processing of each sub-task is independent from the results of the others.
- Both the number of sub-tasks and the input data supplied to each sub-task should be great enough to compensate for the communication overheads.
- Data transfer between the central computer and each executing node should be minimised

The single growing phase in GSOM was divided into two separate growing phases. While the first growing phase is performed as a single process, the outcome is used to initiate the second growing phase running concurrently on different computer nodes, together with the remaining two tuning phases. This is intended to first obtain an abstract and rapid grouping of the entire input data during the first growing phase. It produces a feature map that only has a few neurons. The data are then segmented and sent to different computers on the computing grid and refined independently, and will be processed by another GSOM to achieve finer clustering. All the resulting outputs are then returned to the central computer and combined for further processing.

In our simulation, a grid consisting of 35 computers was used. Even though 35-fold performance increase should ideally be achieved, data is not evenly distributed to each computer node and the later task of producing the final map can not be realised without all the results returned. Therefore, the running time for the scalable GSOM is evaluated as the longest running time out of all 35 computer nodes, plus the time spent in generating the initial abstract map. It was noted that a 15-fold improvement in speed was achieved, but even so, the time improvement is very noticeable even for a relatively small dataset.

4 Applying GSOM: Unsupervised Data Mining

Self-organising networks are a special category of neural networks that is particularly suitable for unsupervised learning and can be easily extended to have classification capability. Unsupervised learners are used to analyse data, which do not have labels available, and generate descriptions of the groupings within the data. In light of this, GSOM has been applied in a wide range of applications to mine knowledge from datasets [17, 18, 19, 20, 21, 14, 22, 23]. Nevertheless, the form of how an algorithm is used to interpret data can also play an influential role in the knowledge extracted. As an illustration, we show two examples of different analysis methods in unsupervised GSOM. The first method, which is applied to the design of wireless sensor networks [24], uses the resource-competitive nature of GSOM to achieve excellent resource distribution within the sensor network. The second method uses a clustering quality measure and feature selection analysis to describe product quality and hence improving the production yield [25].

4.1 *Wireless Sensor Network Design*

The major challenge when designing wireless sensor networks is the amount of energy dissipated in the network. The energy used within the network may be a combination of processing energy and energy used in transmission and reception. A wireless sensor network typically consists of many sensors and a base station. The wireless sensors can have different modes of communication. They can either transmit directly to the base station, receive/transmit information to other sensors or forward information to the base station. The base station, which acts as a data-gathering unit, may be a powerful computer located in some remote location, distanced from the network. The base station does not have energy constraints as it is most likely connected to an external power source. Due to this advantage, the base station should be employed to perform computationally intensive tasks.

Consider the scenario where wireless sensors are used to monitor the tension in a set of mechanical bolts. A corresponding real-world application is the analysis of strain exerted within a structure that is fitted with a large number of tension measuring bolts [17]. By forming sensor clusters, tension readings can be monitored remotely, where necessary precautions can be taken when the need arises, rather than intermittent manual inspections. In this type of static sensor network, the communicating sensors are stationary [26]. Since the sensors are placed in pre-defined locations, the positions of the sensors are always known in advance and it is assumed that sensors are expected to be stationary after deployment. As the number of sensors used in this application is very large, the energy usage is made more economical by preventing each sensor from directly transmiting the measured information to the base station individually. Instead, a few sensors are delegated to gather the measured information from other sensors in their vicinity and transmit to the base station [27]. These delegate sensors are called 'cluster-heads'.

The choice of cluster-heads is often non-trivial, as it is often dynamically optimised for the amount of battery levels for all sensors. Clustering is a widely used method to improve the performance of wireless sensor networks. It is used for data fusion [28], routing and optimising energy consumption [27] to effectively increase the life-span of a sensor network. Formation of clusters and identifying cluster-heads will have a significant role in reducing the overall communication energy in sensor networks.

An extension to GSOM that takes energy consumption as part of the objective function being minimised is proposed in [24]. The energy consumption in the network is used as the criterion for growth in the growing phase of the GSOM algorithm. At the end of each epoch, the total energy consumption is calculated and compared with the growth threshold. If the total energy in the network is greater than the growth threshold, the cluster with highest energy consumption will grow.

The GSOM clustering method gave good results when compared to six other different clustering methods: Genetic Algorithm Clustering, Particle Swarm Optimisation, K-means, Fuzzy C-Means, Subtractive and GSOM without energy consumption criterion.

4.2 Manufacturing Process Improvement

Manufacturing datasets are complex due to the large number of processes, diverse equipment set and nonlinear process flows; often comprising a high volume of process control, process step and quality control data. It may not be straightforward to obtain a good clustering of this data. It is, therefore, necessary to monitor changes in the 'quality' of a cluster, and in some cases, perform a change of variables.

A *Clustering Quality* (CQ) measure has been proposed as a benchmark [29] to evaluate the results of running simulations with different parameter changes. A CQ of 1 would mean a perfect separation of good and faulty products, and a CQ of 0 would mean no separation at all. The proposed CQ takes the complete cluster map into account and can be generated automatically as an objective quantifier of the programme to separate good and faulty products from the data provided.

Furthermore, due to the large size and the high complexity of manufacturing datasets, the data may posses a substantial amount of noise. Noise reduction from the data is one of the primary conditions for obtaining good clustering. Manufacturing data may contain many categorical variables, which are comprised of letters or a combination of numbers and letters. These categorical data should be transformed to numerical data to allow their use in data mining programmes. One of the ways to transform categorical data to numerical data is to use binary values for each of the categories, such that a new attribute is created for each categorical variable and a value of 1 or 0 is provided for the existence and non-existence of that variable in the respective data. In this process each column with categorical data will be expanded to many columns depending on the number of different categorical values in that column. This makes the dataset very large and noisy.

Investigation of sample manufacturing datasets has revealed that there are many categorical variables that do not have any effect on separating good and faulty products. For example, a categorical variable may have an equal distribution among good and faulty products. These variables, when expanded, only contribute to the addition of noise to the dataset. To filter out these unnecessary categorical variables, a *Filtration Index* (FI) function is proposed to identify and remove noisy variables. If FI for a variable is close to zero, the variable is to be considered equally distributed among good and faulty products and hence should not be considered for analysis. For example, if a dataset has 50 good products and 25 faulty products and among them 10 good products and 5 faulty products have the same variable, then FI will be $\left[\left(\frac{10}{50}\right) - \left(\frac{5}{25}\right)\right] = 0$. As the variable is similarly distributed among the good and faulty products, it will not have any effect on clustering. However, it is not expected that a variable would have FI value exactly zero. In practice, the user has to define the value (or range of values) for FI depending on the characteristics of the dataset. Although FI has been proposed for categorical variables, it can similarly be used for numerical variables.

5 Conclusions

GSOM improves on the limitations of a static SOM through dynamic map growth, and enhancements thereof have furthered its rank in terms of the fundamental requirements of a data visualisation tool: topology preservation, visualisation and computational speed. Interestingly, the nature of data itself has changed thanks to the evolution in the technology that produces it. Data collected nowadays grow in size as well as complexity, and also vary in the extent of completeness. Unsupervised learning has been one of the most applied methods of analysis for gaining understanding of the collected data, and it will remain as such. The inherent efficacy that characterises GSOM, promotes its application in a variety of fields and problem domains. By no means an encompassing domain for GSOM, applications that have been explored in this book chapter range from bioinformatics to sensor networks to manufacturing processes. Ultimately, GSOM has played a key role in the provision of data analysis.

References

1. Kohonen, T.: Self-Organizing Maps, 3rd edn. Springer, Heidelberg (2007)
2. Fritzke, B.: Growing cell structures - a self-organising network for unsupervised and supervised learning. Neural Networks 7, 1441–1460 (1994)

3. Bruske, J., Sommer, G.: Dynamic cell structures. In: Advances in Neural Information Processing Systems (NIPS 1994), pp. 497–504 (1994)
4. Fritzke, B.: A growing neural gas network learns topologies. In: Advances in Neural Information Processing Systems, vol. 7 (1995)
5. Blackmore, J., Miikkulainen, R.: Visualizing high-dimensional structurewith the incremental grid growing neural network. In: Proceedings of the 12th International Conference on Machine Learning (1995)
6. Bauer, H., Vilmann, T.: Growing a hypercubical output space in a self-organising feature map. Technical report, Berkerly (1995)
7. Alahakoon, D., Halgamuge, S.K., Srinivasan, B.: Dynamic self-organizing maps with controlled growth for knowledge discovery. IEEE Transactions on Neural Networks 11(3), 601–614 (2000)
8. Hsu, A., Alahakoon, D., Halgamuge, S.K., Srinivasan, B.: Automatic clustering and rule extraction using a dynamic som tree. In: Proceedings of the 6th International Conference on Automation, Robotics, Control and Vision (2000)
9. Villmann, T., Herrmann, M., Der, R., Martinetz, M.: Topology preservation in self-organising feature maps: Exact definition and measurement. IEEE Transactions on Neural Networks 18 (1997)
10. Kohonen, T., Hynninen, J., Kangas, J., Laaksonen, J., Torkkola, K.: University of technology, laboratory of computer and information sciencecomputer and information science, http://www.cis.hut.fi/research/som_lvq_pak.shtml
11. Hsu, L.A., Tang, S.-L., Halgamuge, S.K.: An unsupervised hierarchical dynamic self-organizing approach to cancer class discovery and marker gene identification in microarray data. Bioinformatics 19, 2131–2140 (2003)
12. Hsu, L.A., Halgamuge, S.K.: Enhancement of topology preservation and hierarchical dynamic self-organising maps for data visualisation (2003)
13. Preethichandra, D.M.G., Hsu, A., Alahakoon, D., Halgamuge, S.K.: A modified dynamic self-organizing map algorithm for efficient hardware implementation. In: Proceedings of the 1st International Conference on Fuzzy Systems and Knowledge Discovery (2002)
14. Zhai, Y.Z., Hsu, A., Halgamuge, S.K.: Scalable dynamic self-organising maps for mining massive textual data. In: King, I., Wang, J., Chan, L.-W., Wang, D. (eds.) ICONIP 2006. LNCS, vol. 4234, pp. 260–267. Springer, Heidelberg (2006)
15. Foster, I., Kesselman, C.: The grid: blueprint for a new computing infrastructure. Elsevier, Amsterdam (2004)
16. Depoutovitch, A., Wainstein, A.: Building grid enabled data-mining applications, http://www.ddj.com/184406345
17. Guru, S.M., Hsu, A., Halgamuge, S., Fernando, S.: Clustering sensor networks using growing self-organising map. In: Proceedings of the 2004 Intelligent Sensors, Sensor Networks and Information Processing Conference, pp. 91–96, December 14-17 (2004)
18. Wickramasinghe, L.K., Alahakoon, L.D.: Discovery and sharing of knowledge with self-organized agents. In: IEEE/WIC International Conference on Intelligent Agent Technology, 2003. IAT 2003, pp. 126–132 (2003)
19. Wickramasinghe, L.K., Alahakoon, L.D.: Adaptive agent architecture inspired by human behavior. In: Proceedings of IEEE/WIC/ACM International Conference on Intelligent Agent Technology, 2004 (IAT 2004), pp. 450–453 (2004)

20. De Silva, L.P.D.P., Alahakoon, D.: Analysis of seismic activity using the growing som for the identification of time dependent patterns. In: International Conference on Information and Automation, 2006. ICIA 2006, pp. 155–159 (2006)
21. Amarasiri, R., Alahakoon, D.: Applying dynamic self organizing maps for identifying changes in data sequences, pp. 682–691 (2003)
22. Cho, J., Lan, J., Thampi, G.K., Principe, J.C., Motter, M.A.: Identification of aircraft dynamics using a som and local linear models. In: The 2002 45th Midwest Symposium on Circuits and Systems, 2002. MWSCAS 2002, vol. 2, pp. 148–151 (2002)
23. Wang, H., Azuaje, F., Black, N.: Biomedical pattern discovery and visualisation based on self-adaptive neural networks. In: Proc. of the 4th Annual IEEE Conf. on Information Technology Applications in Biomedicine, pp. 306–309 (2003)
24. Guru, S.M., Hsu, A., Halgamuge, S.K., Fernando, S.: An extended growing self-organising map for selection of clustering in sensor networks. Journal of Distributed Sensor Networks 1 (2005)
25. Karim, M.A., Halgamuge, S.K., Smith, A.J.R., Hsu, A.: Manufacturing yield improvement by clustering. In: King, I., Wang, J., Chan, L.-W., Wang, D. (eds.) ICONIP 2006. LNCS, vol. 4234, pp. 526–534. Springer, Heidelberg (2006)
26. Tilak, S., Abu-Ghazaleh, N., Heinzelman, W.: A taxonomy of wireless microsensor network models. ACM SIGMOBILE Mobile Computing and Communications Review 6, 28–36 (2002)
27. Heinzelman, W.B., Chandralasan, A.P., Balakrishan, H.: An application-specific protocol architecture for wireless microsensor networks. IEEE Transactions on Wireless Communications 1, 660–670 (2002)
28. Wang, A., Heinzelman, W.R., Chandrakasan, A.P.: Energy-scalable protocols for battery-operated microsensor networks. In: Signal Processing Systems (1999)
29. Russ, G., Karin, M.A., Islam, A., Hsu, A., Halgamuge, S.K., Smith, A.J.R., Kruse, R.: Detection of faulty semiconductor wafers using dynamic growing self-organising maps. In: IEEE Tencon (2005)

Hybrid Learning Enhancement of RBF Network with Particle Swarm Optimization

Sultan Noman, Siti Mariyam Shamsuddin, and Aboul Ella Hassanien

Abstract. This study proposes RBF Network hybrid learning with Particle Swarm Optimization (PSO) for better convergence, error rates and classification results. In conventional RBF Network structure, different layers perform different tasks. Hence, it is useful to split the optimization process of hidden layer and output layer of the network accordingly. RBF Network hybrid learning involves two phases. The first phase is a structure identification, in which unsupervised learning is exploited to determine the RBF centers and widths. This is done by executing different algorithms such as k-mean clustering and standard derivation respectively. The second phase is parameters estimation, in which supervised learning is implemented to establish the connections weights between the hidden layer and the output layer. This is done by performing different algorithms such as Least Mean Squares (LMS) and gradient based methods. The incorporation of PSO in RBF Network hybrid learning is accomplished by optimizing the centers, the widths and the weights of RBF Network. The results for training, testing and validation of five datasets (XOR, Balloon, Cancer, Iris and Ionosphere) illustrates the effectiveness of PSO in enhancing RBF Network learning compared to conventional Backpropogation.

Keywords: Hybrid learning, Radial basis function network, K-means, Least mean squares, Backpropogation, Particle swarm optimization, Unsupervised and supervised learning.

1 Introduction

Radial Basis Function (RBF) Networks form a class of Artificial Neural Networks (ANNs), which has certain advantages over other types of ANNs, such as better

Sultan Noman and Siti Mariyam Shamsuddin
Soft Computing Group,
Faculty of Computer Science and Information System, UTM, Malaysia
e-mail: sultannoman@yahoo.com, mariyam@utm.my

Aboul Ella Hassanien
Information Technology Department,
Faculty of Computer Science and Information System, Cairo University
e-mail: aboitcairo@gmail.com

A.-E. Hassanien et al. (Eds.): Foundations of Comput. Intel. Vol. 1, SCI 201, pp. 381–397.
springerlink.com   © Springer-Verlag Berlin Heidelberg 2009

approximation capabilities, simpler network structures and faster learning algorithms. The RBF Network is a three layer feed forward fully connected network, which uses RBFs as the only nonlinearity in the hidden layer neurons. The output layer has no nonlinearity and the connections of the output layer are only weighted, the connections from the input to the hidden layer are not weighted [1].

Due to their better approximation capabilities, simpler network structures and faster learning algorithms, RBF Networks have been widely applied in many science and engineering fields. It is three layers feedback network, where each hidden unit implements a radial activation function and each output unit implements a weighted sum of hidden units' outputs. Its training procedure is usually divided into two stages. The first stage includes determination of centers and widths of the hidden layer which are obtained from clustering algorithms such as K-means, vector quantization, decision trees, and self-organizing feature maps. The second stage involves weights establishment by connecting the hidden layer with the output layer. This is determined by Singular Value Decomposition (SVD) or Least Mean Squares (LMS) algorithms [2]. Clustering algorithms have been successfully used in training RBF Networks such as Optimal Partition Algorithm (OPA) to determine the centers and widths of RBFs. In most traditional algorithms, such as the K-means, the number of cluster centers need to be predetermined, which restricts the real applications of the algorithms. In addition, Genetic Algorithm (GA), Particle Swarm Optimization (PSO) and Self-Organizing Maps (SOM) are also been considered in clustering process [4].

In this study, PSO is explored to enhance RBF learning mechanism. The paper is structured as follows. Section 2, related work about RBF Network training is introduced. Section 3 presents RBF Network model and parameter selection problem. In section 4 describes PSO algorithm. BP-RBF Network model is given in Section 5. Section 6 describes our proposed approach. Sections 7 and 8 give the experiments setup, results and validation results of the proposed model on datasets respectively. The comparison between PSO-RBF Network and BP-RBF Network is presented in section 9 and finally, the paper is concluded in Section 10.

2   Related Work

Although there are many studies in RBF Network training, but research on training of RBF Network with PSO is still fresh. This section presents some existing work of training RBF Network based on Evolutionary Algorithms (EAs) such as PSO especially based on unsupervised learning only (Clustering).

In [11], they have proposed a PSO learning algorithm to automate the design of RBF Networks, to solve pattern classification problems. Thus, PSO-RBF finds the size of the network and the parameters that configure each neuron: center and width of its basis function. Supervised mean subtractive clustering algorithm has been proposed [13] to evolve RBF Networks and the evolved RBF acts as fitness evaluation function of PSO algorithm for feature selection. The method performs feature selection and RBF training simultaneously. PSO algorithm has been introduced [12] to train RBF Network related to automatic configuration of network architecture related to centers of RBF. Two training algorithm were compared. One was PSO algorithm. The other was newrb routine that was included in Matlab neural networks toolbox as standard training algorithm for RBF network.

A hybrid PSO (HPSO) was proposed [15] with simulated annealing and Chaos search technique to train RBF Network. The HPSO algorithm combined the strong ability of PSO, SA, and Chaos. An innovative Hybrid Recursive Particle Swarm Optimization (HRPSO) learning algorithm with normalized fuzzy c-mean (NFCM) clustering, PSO and Recursive Least Squares (RLS) has been presented [16] to generate RBF networks modeling system with small numbers of descriptive RBFs for fast approximating two complex and nonlinear functions. On other hand, a newly evolutionary search technique called Quantum-Behaved Particle Swarm Optimization, in training RBF Network has been used [17]. The proposed QPSO-Trained RBF Network was test on nonlinear system identification problem.

Unlike previous studies, this research shares consideration of parameters of RBF (unsupervised learning) which are centers and length of width or spread of RBFs with different algorithms such as K-means and K-nearest neighbors or standard deviations algorithms respectively. However, training of RBF Network need to enhance with PSO to optimize the centers and widths values which are obtained from the clustering algorithms and PSO also used to optimize the weights which connect between hidden layer and output layer (supervised learning). Also this paper has been presented to train, test and validate the PSO-RBF Network on the datasets.

3   Architecture of RBF Network

RBF Network is structured by embedding radial basis function a two-layer feed-forward neural network. Such a network is characterized by a set of inputs and a set of outputs. It is used in function approximation, time series prediction, and control, and the network architecture is constructed with three layers: input layer, hidden layer, and output layer. The input layer is made up of source nodes that connect the network to its environment. The second layer, the only hidden layer of the network, applies a non-linear transformation from the input space to a hidden space. The nodes in the hidden layer are associated with centers that determine the behavior structure of network. The response from the hidden unit is activated through RBF using Gaussian function or other functions. The output layer provides the response of the network to the activation pattern of the input layer that serves as a summation unit.

In a RBF model, the layer from input nodes to hidden neurons is unsupervised and the layer from hidden neurons to output nodes is supervised. The model is given by the following equation for the jth output $y_j(i)$:

$$y_j(i) = \sum_{k=1}^{K} w_{jk} \Phi_k[x(i), c_k, \sigma_k]$$

$$j = 1, 2, \ldots, n \qquad i = 1, 2, \ldots, N,$$

(1)

Where K is the number of RBFs used, and $c_k \in R^m$, $\sigma_k \in R^m$, are the center value vector and the width value vector of RBF, respectively. These vectors are defined as:

$$c_k = [c_{k1} \quad c_{k2} \quad \cdots \quad c_{km}]^T \in R^m \ , \quad k = 1, \ldots, K$$

$$\sigma_k = [\sigma_{k1} \quad \sigma_{k2} \quad \cdots \quad \sigma_{km}]^T \in R^m \ , \quad k = 1, \ldots, K, \tag{2}$$

Where, $\{w_{jk} \mid k = 1, 2, \ldots, K\}$ are the weights of RBFs connected with the jth Output. Fig. 1 shows the structure of RBF Network.

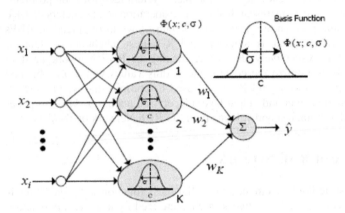

Fig. 1 Structure of RBF network

RBF Network can be implemented in a two-layered network. For a given set of centers, the first layer performs a fixed nonlinear transformation which maps the input space onto a new space. Each term $\Phi_k(.)$ forms the activation function in a unit of the hidden layer. The output layer then implements a linear combination of this new space.

$$\Phi_k(x, c_k, \sigma_k) = \prod_{i=1}^{m} \phi_{ki}(x_i, c_{ki}, \sigma_{ki}) \tag{3}$$

Moreover, the most popular choice for $\varphi(.)$ is the Gaussian form defined as

$$\phi_{ki}(x_i, c_{ki}, \sigma_{ki}) = \exp[-(x_i - c_{ki})^2 / 2\sigma_{ki}^2] \tag{4}$$

In this case, the jth output in equation (1) becomes:

$$y_j(i) = \sum_{k=1}^{K} w_{jk} \exp\{-\sum_{i=1}^{m} [(x_i - c_{ki})^2 / 2\sigma_{ki}^2]]\} \tag{5}$$

In equation (5), σ_k indicates the width of the kth Gaussian RBF functions. One of the σ_k selection methods is shown as follows:

$$\sigma_k^2 = \frac{1}{mk} \sum_{x \in \theta k} \left\| x - ck \right\| \tag{6}$$

Where θ_k is the kth cluster of training set and M_k is the number of sample data in the kth cluster.

The neuron number of the hidden layer, i.e., the cluster number of training set, must be determined before the parameter selection of RBF Network. If the neuron numbers of hidden layer has been decided, the performance of RBF depends on the selection of the network parameters. There are three types of parameters in RBF model with Gaussian basis functions: RBF centers (hidden layer neurons); Widths of RBFs (standard deviations in the case of a Gaussian RBF); and Output layer weights. There are two categories of training algorithms in RBF models: supervised and unsupervised learning.

4 Particle Swarm Optimization

Particle Swarm Optimization (PSO) algorithm, originally introduced by Kennedy and Eberhart in 1995 [5], simulates the knowledge evolvement of a social organism, in which each individual is treated as an infinitesimal particle in the n-dimensional space, with the position vector and velocity vector of particle i being represented as $X_i(t) = (X_{i1}(t), X_{i2}(t),...., X_{in}(t))$ and $V_i(t) = (V_{i1}(t), V_{i2}(t),...., V_{in}(t))$. The particles move according to the following equations:

$$V_{id}(t+1) = W \times V_{id}(t) + c_1 r_1 (P_{id}(t) - X_{id}(t)) + c_2 r_2 (P_{gd}(t) - X_{id}(t)) \tag{7}$$

$$X_{id}(t+1) = X_{id}(t) + V_{id}(t+1)$$

$$i = 1,2,...,M ; d = 1,2,...,n \tag{8}$$

Where c_1 and c_2 are the acceleration coefficients, Vector $P_i = (P_{i1}, P_{i2},..., P_{in})$ is the best previous position (the position giving the best fitness value) of particle i known as the personal best position (pbest); Vector $P_g = (P_{g1}, P_{g2},..., P_{gn})$ is the position of the best particle among all the particles in the population and is known as the global best position (gbest). The parameters r_1 and r_2 are two random numbers distributed uniformly in (0, 1). Generally, the value of V_{id} is restricted in the interval $[-V_{max}, V_{max}]$. Inertia weight w was first introduced by Shi and Eberhart in order to accelerate the convergence speed of the algorithm [6].

5 BP-RBF Network

In our paper, the standard BP is selected as the simplest and most widely used algorithm to train feed-forward RBF Networks and considered for the full-training paradigm; customizing it for half-training is straightforward and can be done simply by eliminating gradient calculations and weight-updating corresponding to the appropriate parameters.

The following is the procedure of BP- RBF Network algorithm:
1. Initialize network.
2. Forward pass: Insert the input and the desired output; compute the network outputs by proceeding forward through the network, layer by layer.
3. Backward pass: Calculate the error gradients versus the parameters, layer by layer, starting from the output layer and proceeding backwards: $\partial E/\partial w, \partial E/\partial c, \partial E/\partial \sigma^2$.
4. Update parameters: (weight, center and width of the RBF Network respectively)

$$\text{a.} \quad w_{kj}(t+1) = w_{kj}(t) + \Delta w_{kj}(t+1) \tag{9}$$

$$\Delta w_{kj}(t+1) = \eta_3 \delta_k O_j \tag{10}$$

With

$$\delta_k = O_k(1-O_k)(t_k - O_k) \tag{11}$$

$$O_j = \exp(-(x-c_j)^2/2\sigma_j^2) \tag{12}$$

Where $w_{kj}(t)$ is the weight from node k to node j at time t, Δw_{kj} is the weight adjustment, η_3 is the learning rate, δ_k is error at node k, O_j is the actual network output at node j, O_k is the actual network output at node k and t_k is the target output value at node k.

$$\text{b.} \quad c_{ji}(t+1) = c_{ji}(t) + \Delta c_{ji}(t+1) \tag{13}$$

$$\Delta c_{ji}(t+1) = \eta_2 \delta_k w_{kj} O_j / \sigma_j^2 (x_{ji} - c_{ji}) \tag{14}$$

Where $c_{ji}(t)$ is the centre from node j to node i at time t, Δc_{ji} is the center adjustment, η_2 is the learning rate, δ_k is error at node k, O_j is the actual network output at node j, σ_j is the width at node j, w_{kj} is the weight connected between node j and k and x_{ji} is the input node j to node i.

$$\text{c.} \quad \sigma_j(t+1) = \sigma_j(t) + \Delta \sigma_j(t+1) \tag{15}$$

$$\Delta \sigma_j(t+1) = \eta_1 \delta_k w_{kj} O_j((x-c_j)^2/2\sigma_j^4) \tag{16}$$

Where $\sigma_j(t)$ is the width of node j at time t, $\Delta\sigma_j$ is the width adjustment, η_1 is the learning rate, δ_k is error at node k, O_j is the actual network output at node j, σ_j is the width at node j, w_{kj} is the weight connected between node j and k, x_{ji} is the input at node i and η_3, η_2, η_1 are learning rate factors in the range [0; 1].

5. Repeat the algorithm for all training inputs. If one epoch of training is finished, repeat the training for another epoch.

BP-RBF Network doesn't need the momentum term as it is common for the MLP. It does not help in training of the RBF Network [14].

6 PSO-RBF Network

PSO has been applied to improve RBF Network in various aspects such as network connections (centers, weights), network architecture and learning algorithm. The main process in this study is to employ PSO-based training algorithm on center, width and weight of RBF network, and investigate the efficiency of PSO in enhancing RBF training. Every single solution of PSO called (a particle) flies over the solution space in search for the optimal solution. The particles are evaluated

```
For each particle do
        initialize particle position and velocity
End for

While stopping criteria are not fulfilled do
        For each particle do
            Calculate fitness value (MSE in RBF Network)
            If fitness value is better than best fitness value pBest in particle
            history   then
                Set current position as pBest
            End if
        End for
        Choose as gBest the particle with best fitness value among all particles
in
        current iteration
        For each particle do
            Calculate particle velocity based on eq. (7)
            Update particle position(center, width and weight) based on eq. (8)
        End for
End while
```

Fig. 2 PSO-RBF Network Algorithm

using a fitness function to seek the optimal solution. Particles (center, width) values are then initialized with values which are obtained from the k-means algorithm while particles (weight, bias) values are initialized randomly or from LMS algorithm. The particles are updated accordingly using the equation eq. (7) and eq. (8).

The procedure for implementing PSO global version (gbest) is shown in Figure 2. Optimization of RBF network parameters (the center and width of RBF and the weight, the bias) with PSO, the fitness value of each particle (member) is the value of the error function evaluated at the current position of the particle and position vector of the particle corresponds to the (center, width, weight and bias) matrix of the network. The pseudo code of the procedure is as follows:

7 Experiments

7.1 Experimental Setup

The experiments of this work included the standard PSO and BP for RBF Network training. For evaluating all of these algorithms we used five benchmark classification problems obtained from the machine learning repository [10].

Table 1 Execution parameters for PSO

Parameter	Value
Population Size	20
Iterations	10000
W	[0.9,0.4]
C_1	2.0
C_2	2.0

The parameters of the PSO algorithm were set as: weight w decreasing linearly between 0.9 and 0.4, learning rate $c_1 = c_2 = 2$ for all cases. The population size used by PSO was constant. The algorithm stopped when a predefined number of iterations have been reached. Values selected for parameters are shown in table 1.

Table 2 Parameters of the experiments

Parameter	Dataset				
	XOR	Balloon	Cancer	Iris	Ionosphere
Train data	5	12	349	120	251
Test data	3	4	175	30	100
Validation data	8	16	175	150	351
Input dimension	3	4	9	4	34
Output neuron	1	1	1	3	1
Network Structure	3-2-1	4-2-1	9-2-1	4-3-3	34-2-1

The XOR data set (3 features and 8 examples) is a logical operation on three operands that results in a logical value of true if and only if one of the operands but not both have a value of true. The Balloon data set (4 features and 16 examples) is used in cognitive psychology experiment. There are four data sets representing different conditions of an experiment. All have the same attributes. The cancer dataset (9 features and 699 examples) is related to the diagnosis of breast cancer in benign or malignant. The Iris dataset (4 features and 150 examples) is used for classifying all the information into three classes. Finally, the Ionosphere dataset (34 features and 351 examples) is radar data was collected by a system in Goose Bay; Labrador is used for classifying all the information into "Good" or "Bad" results.

The number of maximum iterations is set differently to bound the number of forward propagations to 4×10^4 and for comparison purposed. The maximum iterations in BP-RBFN is set to 2×10^4 (number of forward propagations = $2 \times$ maximum number of iterations), while the maximum number of iterations in PSO-RBFN is set to 10000 (number of forward propagations = swarm size × maximum number of iterations) [9]. The stopping criteria are the maximum number of iterations that the algorithm has been reached or the minimum error.

The architecture of the RBF Network was fixed in one hidden layer (number of inputs of the problem - 2 hidden units - 1 output units) in XOR, Balloon, Cancer, Ionosphere and (number of inputs of the problem - 3 hidden units - 3 output units) in Iris dataset. The parameters of the experiments are described in Table 2.

7.2 Experimental Results

This section presents the results of the study on PSO-trained RBF Network and BP-trained RBF Network. The experiments are conducted by using five datasets: XOR, Balloon, Cancer, Iris and Ionosphere. The results for each dataset are compared and analysed based on the convergence, error and classification performance.

7.2.1 XOR Dataset

A connective in logic known as the "exclusive or" or exclusive disjunction is a logical operation on three operands. Two algorithms used to train and test of RBF Network. The stopping conditions of PSO-RBFN are set as minimum error of 0.005 or maximum iteration of 10000. On the other hand, the stopping conditions for BP-RBFN are set as the minimum error of 0.005 or the iterations have reached to 20000. The results for PSO-based RBFN and BP-based RBFN are illustrated in Table 3 and Figure 3. From Table 3, PSO-RBFN converges at 93 iterations compared to BP-RBFN with 5250 iterations for the whole learning process. Both algorithms are converged with given minimum error. For the classification, it shows that BP-RBFN is better than PSO-RBFN with 93.88% compared to 93.51%. However, PSO-RBFN converges faster compared to BP-RBFN. The classification rate for testing of XOR problem is not good due to smaller amount of data to be learned by the network.

Table 3 Result of BP-RBFN and PSO-RBFN on XOR dataset

	BP-RBFN		PSO-RBFN	
	Train	Test	Train	Test
Learning Iteration	5250	1	93	1
Error Convergence	0.00500	0.32972	0.004998	0.28693
Classification (%)	93.88	64.72	93.51	64.03

Figure 3 illustrates that PSO-RBFN significantly reduces the error with minimum iterations compared to BP-RBFN.

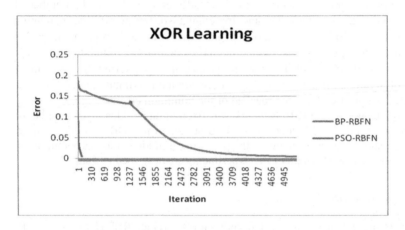

Fig. 3 Convergence rate of XOR dataset

7.2.2 Balloon Dataset

This data is used in cognitive psychology experiment. There are four data sets representing different conditions of an experiment. All have the same attributes.

It contains 4 attributes and 16 instances. The stopping conditions of PSO-RBFN are set to a minimum error of 0.005 or maximum iteration of 10000. Conversely, the stopping conditions for BP-RBFN are set to the minimum error of 0.005 or the iterations have reached to 20000. From Table 4, we conclude that PSO-RBFN converges faster compared to BP-RBFN for the whole learning process. However, both algorithms have converged to the given minimum error. For the classification, it shows that PSO-RBFN is better than BP-RBFN with 95.05% compared to 91.27%.

Figure 4 illustrates the learning process for both algorithms. In PSO-RBFN, 20 particles work together to find the lowest error (gbest) at each iteration and consistently reducing the error. While in BP-RBFN, it seems that the error is decreasing at each iteration, and the learning is discontinued at a specified condition.

Table 4 Result of BP-RBFN and PSO-RBFN on Balloon dataset

	BP-RBFN		PSO-RBFN	
	Train	Test	Train	Test
Learning Iteration	20000	1	3161	1
Error Convergence	0.01212	0.23767	0.0049934	0.16599
Classification (%)	91.27	75.41	95.05	78.95

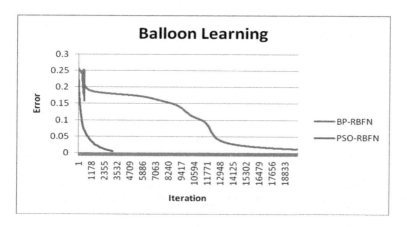

Fig. 4 Convergence of Balloon dataset

7.2.3 Cancer Dataset

The purpose of the breast cancer data set is to classify a tumour as either benign or malignant based on cell descriptions gathered by microscopic examination. It contains 9 attributes and 699 examples of which 485 are benign examples and 241 are malignant examples. The first 349 examples of the whole data set were used for training, the following 175 examples for validation, and the final 175 examples for testing [8]. The ending conditions of PSO-RBFN are set to minimum error of 0.005 or maximum iteration of 10000. Alternatively, the stopping conditions for BP-RBFN are set to a minimum error of 0.005 or maximum iteration of 20000 has been achieved.

Table 5 Result of BP-RBFN and PSO-RBFN on Cancer dataset

	BP-RBFN		PSO-RBFN	
	Train	Test	Train	Test
Learning Iteration	20000	1	10000	1
Error Convergence	0.03417	0.27333	0.0181167	0.27464
Classification (%)	92.80	70.37	97.65	71.77

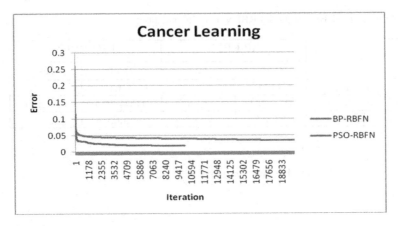

Fig. 5 Convergence of Cancer dataset

In Cancer learning process, from Table 5 shows PSO-RBFN takes 10000 iterations compared to 20000 iterations in BP-RBFN to converge. In this experiment, PSO-RBFN is managed to converge at iteration 10000, while BP-RBFN converges at a maximum iteration of 20000. Table 5 illustrates that PSO-RBFN is better than BP-RBFN with an accuracy of 97.65% and 92.80%. Figure 5 shows PSO-RBFN significantly reduce the error with small number of iterations compared to BP-RBFN.

7.2.4 Iris Dataset

The Iris dataset is used for classifying all the information into three classes which are iris setosa, iris versicolor, and iris virginica. The classification is based on its four input patterns which are sepal length, sepal width, petal length and petal width. Each class refers to type of iris plant contain 50 instances. For Iris dataset, the minimum error of PSO-RBFN is set to 0.05 or maximum iteration of 10000. While, the minimum error for BP-RBFN is set to 0.05 or the network has reached maximum iteration of 20000. Table 6 shows that BP-RBFN is better than PSO-RBFN with an accuracy of 95.66% compared to 95.48%. However, PSO-RBFN converges faster at 3774 iterations compared to 10162 iterations in BP-RBFN.

For Iris learning, both algorithms converge using the maximum number of prespecified iteration. PSO-RBFN takes 3774 iterations to converge at a minimum error of 0.0499949 while minimum error for BP-RBFN is 0.05000 with 10162

Table 6 Result of BP-RBFN and PSO-RBFN on Iris dataset

	BP-RBFN		PSO-RBFN	
	Train	Test	Train	Test
Learning Iteration	10162	1	3774	1
Error Convergence	0.05000	0.04205	0.0499949	0.03999
Classification (%)	95.66	95.78	95.48	95.64

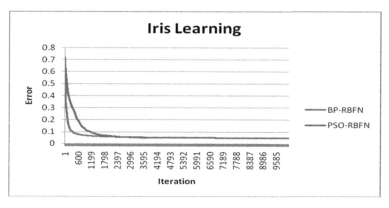

Fig. 6 Convergence of Iris dataset

iterations. Figure 6 shows that PSO-RBFN reduces the error with minimum iterations compared to BP-RBFN.

7.2.5 Ionosphere Dataset

This radar data was collected by a system in Goose Bay, Labrador. This system consists of a phased array of 16 high-frequency antennas with a total transmitted power on the order of 6.4 kilowatts. The targets were free electrons in the ionosphere. "Good" radar returns are those showing evidence of some type of structure in the ionosphere. "Bad" returns are those that do not; their signals pass through the ionosphere. For Ionosphere problems, the stopping conditions for BP-RBFN is minimum error of 0.05 or maximum iteration of 20000. The minimum error of PSO-RBFN is 0.05 or maximum iteration of 10000. The experimental results for PSO-based RBFN and BP-based RBFN are shown in Table 7 and Figure 7.

Table 7 Result of BP-RBFN and PSO-RBFN on Ionosphere dataset

	BP-RBFN		PSO-RBFN	
	Train	Test	Train	Test
Learning Iteration	20000	1	5888	1
Error Convergence	0.18884	0.23633	0.0499999	0.01592
Classification (%)	62.27	62.71	87.24	90.70

In Ionosphere learning process, Table 7 shows PSO-RBFN takes 5888 iterations compared to 20000 iterations in BP-RBFN to converge. In this experiment, PSO-RBFN is managed to converge using minimum error at iteration of 5888, while BP-RBFN trapped at the local minima and converges at a maximum iteration of 20000. For the correct classification percentage, it shows that PSO-RBFN result is better than BP-RBFN with 87.24% compared to 62.27%. Figure 7 shows PSO-RBFN significantly reduce the error with small number of iterations compared to BP-RBFN. The results for this data are not promising for BP-RBFN since

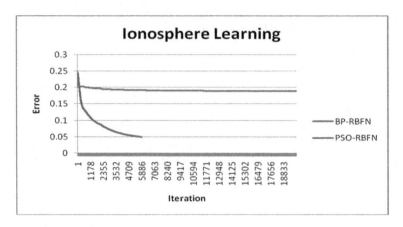

Fig. 7 Convergence of Ionosphere dataset

it depends on the data and repeatedly traps in local minima. The local minima problem in BP-RBFN algorithm is usually caused by disharmony adjustments between centers and weights of RBF Network. To solve this problem, the error function has been modified as suggested [17].

8 Validation Results

In artificial neural network methodology, data samples are divided into three sets; training, validation and testing in order to obtain a network which is capable of generalizing and performing well with new cases. There is no precise rule on the optimum size of the three sets of data, although authors agree that the training set must be the largest. Validations are motivated by two fundamental problems either in model selection or in performance estimation.

Table 8 Validation Result of BP-RBFN and PSO-RBFN on all dataset

Dataset	BP-RBFN		PSO-RBFN	
	Train	Test	Train	Test
XOR	0.11332	0.32864	0.00494071	0.47354
Balloon	0.06004	0.33155	0.00499450	0.27348
Cancer	0.03046	0.04233	0.00541208	0.02733
Iris	0.05000	0.06227	0.0499760	0.05792
Ionosphere	0.20743	0.22588	0.0499953	0.06325

To create a N-fold partition of the dataset we simplifies that for each of N experiments, use N-1 folds for training and the remaining one for testing and the true error is estimated as the average error rate. The results demonstrate the evaluation of our algorithms with respect to the convergence rate on the training and testing dataset. These results for BP-RBFN and PSO-RBFN on all dataset are shown in Table 8.

On the whole data, the experiments showed that PSO gives high performance for RBF Network training. PSO-RBFN reduces the error with minimum iterations compared to BP-RBFN.

9 Comparison between PSO-RBF Network and BP-RBF Network

This analysis is carried out to compare the results between PSO-RBFN and BP-RBFN. The learning patterns for both algorithms in both experiments are compared using a five datasets. The classification results for all datasets are shown in Figure 8.

For Balloon, Cancer and Ionosphere dataset, the results show that PSO-RBFN is better in terms of convergence rate and correct classification. PSO-RBFN converges in a short time with high classification rates. For XOR and Iris dataset, both algorithms converge to the solution within specified minimum error; it shows that at this time, BP-RBFN classifications are better than PSO-RBFN. But in terms of convergence rate, it shows that PSO-RBFN is better than BP-RBFN, and PSO-RBFN significantly reduces the error with minimum iterations.

For overall performance, the experiments show that PSO-RBFN produces feasible results in terms of convergence rate and classification accuracy.

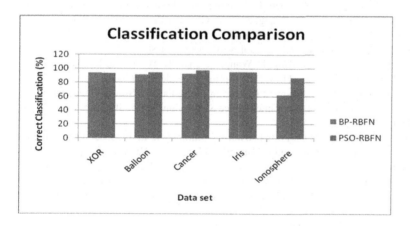

Fig. 8 Comparison of Classification Accuracy of PSO-RBFN and BP-RBFN

10 Conclusion

This paper proposes PSO based Hybrid Learning of RBF Network to optimize the centers, widths and weights of network. Based on the results, it is clear that PSO-RBFN is better than BP-RBFN in term of convergence and error rate and PSO-RBFN reached optimum because it reduces the error with minimum iteration and obtains the optimal parameters of RBF Network. In PSO-RBFN, network architecture and selection of network parameters for the dataset influence the convergence and the performance of network learning.

In this paper, both the algorithms need to be used the same network architecture. Choosing PSO-RBFN parameters also depend on the problem and dataset to be optimized. These parameters can be adjusted accordingly to achieve better optimization. However, to have better comparison, the same parameters for all five datasets have been used. For BP-RBFN, the learning rate which is critical for standard BP network is provided with a set of weight. This is to ensure the convergence time is faster with better results. Although Standard BP learning becomes faster based on those parameters, the overall process including parameters selection in BP-RBFN takes lengthy time compared to the process in PSO-RBFN.

Acknowledgment. This work is supported by Ministry of Higher Education (MOHE) under Fundamental Research Grant Scheme. Authors would like to thank Research Management Centre (RMC) Universiti Teknologi Malaysia, for the research activities and Soft Computing Research Group (SCRG) for the support and incisive comments in making this study a success.

References

1. Leonard, J.A., Kramer, M.A.: Radial basis function networks for classifying process faults. Control Systems Magazine 11(3), 31–38 (1991)
2. Liu, Y., Zheng, Q., Shi, Z., Chen, J.: Training radial basis function networks with particle swarms. In: Yin, F.-L., Wang, J., Guo, C. (eds.) ISNN 2004. LNCS, vol. 3173, pp. 317–322. Springer, Heidelberg (2004)
3. Chen, J., Qin, Z.: Training RBF Neural Networks with PSO and Improved Subtractive Clustering Algorithms. In: King, I., Wang, J., Chan, L.-W., Wang, D. (eds.) ICONIP 2006. LNCS, vol. 4233, pp. 1148–1155. Springer, Heidelberg (2006)
4. Cui, X.H., Polok, T.E.: Document Clustering Using Particle Swarm Optimization. In: Proceedings 2005 IEEE Swarm Intelligence Symposium, 2005. SIS 2005, pp. 185–191 (2005)
5. Kennedy, J., Eberhart, R.C.: Particle Swarm Optimization. In: Proceedings of IEEE International Conference on Neural Networks, Piscataway, NJ, vol. IV, pp. 1942–1948 (1995)
6. Shi, Y., Eberhart, R.C.: A Modified Particle Swarm. In: Proceedings of 1998 IEEE International Conference on Evolutionary Computation, Piscataway, NJ, pp. 1945–1950 (1998)
7. Wang, X.G., Tang, Z., Tamura, H., Ishii, M.: A modified error function for the backpropagation algorithm. Neurocomputing 57, 477–488 (2004)
8. Zhang, C., Shao, H., Li, Y.: Particle Swarm Optimization for Evolving Artificial Neural Network. In: 2000 IEEE International Conference on Systems, Man, and Cybernetics, vol. 4, pp. 2487–2490 (2000)
9. Al-kazemi, B., Mohan, C.K.: Training Feedforward Neural Network Using Multiphase Particle Swarm Optimization. In: Proceeding of the 9th International Conference on Neural Information Processing, vol. 5, pp. 2615–2619 (2002)
10. Blake, C., Merz, C.J.: UCI Repository of Machine Learning Databases (1998), http://www.ics.uci.edu/~mlearn/MLRepository.html

11. Qin, Z., Chen, J., Liu, Y., Lu, J.: Evolving RBF Neural Networks for Pattern Classification. In: Hao, Y., Liu, J., Wang, Y.-P., Cheung, Y.-m., Yin, H., Jiao, L., Ma, J., Jiao, Y.-C. (eds.) CIS 2005. LNCS, vol. 3801, pp. 957–964. Springer, Heidelberg (2005)
12. Liu, Y., Zheng, Q., Shi, Z., Chen, J.: Training Radial Basis Function Networks with Particle Swarms. In: Yin, F.-L., Wang, J., Guo, C. (eds.) ISNN 2004. LNCS, vol. 3173, pp. 317–322. Springer, Heidelberg (2004)
13. Chen, J., Qin, Z.: Training RBF neural networks with PSO and improved subtractive clustering algorithms. In: King, I., Wang, J., Chan, L.-W., Wang, D. (eds.) ICONIP 2006. LNCS, vol. 4233, pp. 1148–1155. Springer, Heidelberg (2006)
14. Vakil-Baghmishah, M.T., Pavesic, N.: Training RBF networks with selective back-propagation. Nerocomputing ScienceDirect. 62, 39–64 (2004)
15. Gao, H., Feng, B., Hou, Y., Zhu, L.: Training RBF neural network with hybrid particle swarm optimization. In: Wang, J., Yi, Z., urada, J.M., Lu, B.-L., Yin, H. (eds.) ISNN 2006. LNCS, vol. 3971, pp. 577–583. Springer, Heidelberg (2006)
16. Sun, J., Xu, W., Liu, J.: Training RBF neural network via quantum-behaved particle swarm optimization. In: King, I., Wang, J., Chan, L.-W., Wang, D. (eds.) ICONIP 2006. LNCS, vol. 4233, pp. 1156–1163. Springer, Heidelberg (2006)
17. Wang, X.G., Tang, Z., Tamura, H., Ishii, M.: A modified error function for the back-propagation algorithm. Neurocomputing 57, 477–488 (2004)

Author Index